MICHAEL TAYLOR

Impossible Monsters

How the Discovery of Dinosaurs Changed the World

VINTAGE

3 5 7 9 10 8 6 4 2

Vintage is part of the Penguin Random House group of companies

Vintage, Penguin Random House UK, One Embassy Gardens,
8 Viaduct Gardens, London SW11 7BW

penguin.co.uk/vintage
global.penguinrandomhouse.com

Penguin Random House UK

First published in Vintage in 2024
First published in hardback by The Bodley Head in 2025

Typeset in 9.35/12.46pt Sabon LT Std by Jouve (UK), Milton Keynes
Printed and bound in Great Britain by Clays Ltd, Elcograf S.p.A.

The authorised representative in the EEA is Penguin Random House Ireland, Morrison Chambers, 32 Nassau Street, Dublin D02 YH68

A CIP catalogue record for this book is available from the British Library

ISBN 9781529931341

Penguin Random House is committed to a sustainable future for our business, our readers and our planet. This book is made from Forest Stewardship Council® certified paper.

FSC www.fsc.org MIX Paper | Supporting responsible forestry FSC® C018179

BRITAIN'S IMPOSSIBLE MONSTERS (AND THEIR DISCOVERERS)

- Ichthyosaurus (Anning/Conybeare/De la Beche, 1819)
- Kirkdale Hyena (Buckland, 1823)
- Red Lady of Paviland (Davies/Buckland, 1823)
- Megalosaurus (Plot/Buckland, 1824)
- Plesiosaurus (Anning/Conybeare, 1824)
- Iguanodon (Mantell, 1825)
- Dimorphodon (Anning/Buckland, 1829)
- Hylaeosaurus (Mantell, 1833)
- Thecodontosaurus (Riley/Stutchbury, 1836)
- Cetiosaurus (Kingdon/Owen, 1841)
- Acanthopholis (Griffiths/Huxley, 1867)
- Gigantosaurus (Seeley, 1869)
- Hypsilophodon (Mantell/Huxley, 1869)
- Dacentrurus (Shopland/Owen, 1875)

N.B. Year refers to official date of announcement

Did not learned men, too, hold, till within the last twenty-five years, that a flying dragon was an impossible monster? And do we not now know that there are hundreds of them found fossil up and down the world? People call them Pterodactyles: but that is only because they are ashamed to call them flying dragons, after denying so long that flying dragons could exist.

Charles Kingsley, *The Water-Babies* (1863)

ALSO BY MICHAEL TAYLOR

The Interest: How the British Establishment Resisted the Abolition of Slavery

MICHAEL TAYLOR

Michael Taylor is the author of *The Interest: How the British Establishment Resisted the Abolition of Slavery*, which was shortlisted for the Orwell Prize 2021, chosen as a *Daily Telegraph* Book of the Year, and described as 'riveting' (*The Times*) and 'compulsively readable' (*Guardian*). He was born in 1988 and graduated with a double first in history from the University of Cambridge, where he earned his PhD. He has since been a lecturer in Modern British History at Balliol College, Oxford, and a visiting fellow at the British Library's Eccles Centre for American Studies. He now works at PwC LLP.

'Well-paced . . . fascinating . . . and it has a charming leitmotif, namely the periodic discovery of yet another dinosaur, each seemingly larger and more monstrous than the last . . . Taylor movingly tells us of the agony inflicted by scientific discovery on the "honest doubters"'
A. N. Wilson, *The Times*

'An extraordinary and important tale of a seismic moment in intellectual history. Epic in scale yet intimate in detail, Taylor's *Impossible Monsters* is a masterclass in combining peerless erudition with superb storytelling'
Matthew Parker, author of *One Fine Day*

'Amazing . . . Taylor paints the complex picture of the fundamental tension between religion and geology through the nineteenth century with verve and humour . . . An important story that still affects us today'
Michael Benton, author of *The Dinosaurs Rediscovered*

'Taylor's . . . lively prose make[s] for an extremely enjoyable excursion into a world where impossible monsters emerged from beneath the ground and forced us to rethink all our beliefs about the origins of our planet'
Times Literary Supplement

'Eminently readable and well researched . . . He writes well, knows his subject and has a fine eye for detail'
Spectator

'Such an attractive book . . . a sympathetic, charming, beautifully written guide through a pivotal part of history'
John van Whye, *BBC History Magazine*

'The tale of how evolution became a dominant paradigm is rousing . . . Taylor's book breathes new life into its dons and explorers. The writing is crisp, the handling of the scholarship graceful and precise'
The Critic

'Excellent . . . Everything that popular scholarly history should be . . . written with clarity, zest, and wit'
Piers Brendon, *Literary Review*

'Michael Taylor has produced a greatly informative, meticulously researched and exciting read, tracing the relationship between Christianity and the explosive effects of scientific theory'
Church Times

'Eloquent and authoritative, we're shown how the discoveries of ancient reptiles shook the very foundations of conservative nineteenth-century Britain'
Paul Barrett, Natural History Museum

'Brilliant, entertaining, noteworthy'
Ben Miller, actor and comedian

'Ambitious, readable, informative'
Atlantic

'Elegantly written, compellingly readable'
Wall Street Journal

An *Economist*, *Times Literary Supplement*, *Bloomberg* and Waterstones Book of the Year

'This book confirms what I've suspected for a while, that Michael Taylor is the most talented young historian around. This book dazzles in its originality and there is something you want to commit to memory on every page. A triumph'
Sathnam Sanghera, author of *Empireworld*

'An astonishing book about an extraordinary subject. Michael Taylor tells the story of the collision of science and religion in an age of change with authority, wit and verve. A delight'
Peter Frankopan, author of *The Earth Transformed*

'A truly marvellous book: superb research and a sparkling narrative dramatise an epic battle of ideas and an intellectual thriller. Michael Taylor succeeds in reanimating those famous dinosaur wars of the nineteenth century with real brilliance, and makes them as fresh and furious as ever. Exuberant, stylish and brilliantly sustained throughout'
Richard Holmes, author of *The Age of Wonder*

'An account of the discovery of deep time that is as thrilling as it is sweeping, populated by a brilliantly drawn cast of characters, and vivid with a Mesozoic bestiary'
Tom Holland, host of *The Rest is History*

'Marvellous . . . *Impossible Monsters* is a work of remarkable range. Taylor . . . belongs to that rare class of writers who can effortlessly encompass both scientific arcana and intellectual currents. It is also to his credit that he every so often takes us away from the high tables to show us what ordinary people made of these huge strides in thinking' Pratinav Anil, *Guardian*

'One of the most interesting stories in the world . . . Taylor has had the brilliant idea of crafting his history against the background of the "bone wars" . . . The story is told with brio and humour, but not without a sense of the pathos of Doubt . . . I relished every word'
A. N. Wilson, *Times Literary Supplement*, Books of the Year

'Skilfully blends an impressive array of sources into a highly readable, almost novelistic narrative. In particular, it features many women who played crucial roles but are too often invisible . . . Including gripping tales as well as serious commentary, *Impossible Monsters* chips out a fascinating slice through the strata of Victorian society'
History Today

'In writing *Impossible Monsters*, the task of Michael Taylor . . . was to tell a much-told tale better than it had been told before. He has succeeded splendidly . . . Mr Taylor also conveys a sense of just how risky it was to believe in and promulgate the new ideas tied to the rocks and tropical forests where people hunted for specimens'
The Economist

'The emotional impact on the Victorians . . . was profound . . . Taylor recounts not just the interventions of palaeontologists and geologists but also those stricken by events as their faith evaporated . . . he marshals his cast expertly and shows lucidly why it mattered so much'
New Statesman

'In this stunning work of popular history, historian Michael Taylor shows how the discovery of dinosaurs triggered a domino effect that shook the foundations of western culture. A most engrossing book of surprises and revelations'
Steve Brusatte, author of *The Rise and Fall of the Dinosaurs*

'Tremendously entertaining. Michael Taylor brings to splendid life the scandal and skulduggery that ensued as Victorian society came to terms with the existence of the shockingly unbiblical dinosaurs'
Catherine Fletcher, author of *The Roads to Rome*

'A sweeping account of the discovery of dinosaurs and the horrifying depths of time, and their impact on god-fearing Victorians. Taylor marches us with panache from Bishop Ussher's impossibly young world to today's incomprehensibly old planet. We feel the awe and fright across society as the vast reptilian empires are brought to light'
Adrian Desmond, author of *Darwin's Sacred Cause*

Contents

List of Illustrations xi
Map of Britain's 'Impossible Monsters' xiii
Preface xv
Prologue: The House of Ussher 1

PART ONE
Giants in the Earth

1 Shells by the Shore 13
2 Undergroundology 24
3 Of Caves and Paddles 37
4 A Bone of Prodigious Bigness 47
5 Free the Science from Moses 59
6 The Parliament of Science 71

PART TWO
The Lords of Creation

7 A Shropshire Lad 87
8 The Goodness of God 99
9 Terrible Lizards 111
10 Abominable Books 123
11 Nemesis 132

PART THREE
The March of Progress

12 The Palace 151
13 Death and the Dinosaurs 164

14 The Branching Tree 177
15 The Devil's Gospel 189

PART FOUR
The Gorilla War

16 Monkeyana 203
17 Going the Whole Orang 216
18 X 228
19 Feathers 240
20 Tides of Faith 255
21 The Angel and the Ape 269

PART FIVE
Entente

22 Learned Baboons 285
23 Badlands 296
24 The Citadels Fall 309

Epilogue: Rex 323

Acknowledgements 335
Bibliography 337
Credits 375
Notes 379
Index 437

List of Illustrations

INTEGRATED IMAGES

Skull of the '*Proteosaurus*' 17
The fossilised skeleton of the *Plesiosaurus* 46
Thigh-bone misnamed as the '*scrotum humanum*' 48
Jawbone of the *Megalosaurus* 50
Teeth used to identify the *Iguanodon* 58
The Temple of Serapis 66
Remains of the *Dimorphodon* 74
Fossilised remains of the *Hylaeosaurus* 95
Charles Darwin's first illustration of the 'tree of life' 109
The *Iguanodon* Dinner at the Crystal Palace 172
Darwin's barnacles 183
Owen's illustrations of the human and chimpanzee brains 211
Neanderthal skull from the Neander Valley 218
Frontispiece to *Evidence as to Man's Place in Nature* 225
Fossilised *Archaeopteryx* feather 242
Almost complete *Archaeopteryx* skeleton 243
Representations of the *Hadrosaurus* 248
Wallace's Bedford Level experiment 266
Remains of the *Hypsilophodon* 277
Hip-bones of birds, dinosaurs, and crocodiles 278
G.W. Foote's *Comic Sketches* 291
Edward Cope's Cretaceous landscape 301
The *Dimorphodon* and the *Pterodactyl* 308
Comic News's mockery of Owen's proposal for a new natural history museum 311
Charles Darwin's funeral 322
Gertie the Dinosaur 329

PLATE SECTION

William 'Strata' Smith's geological map of Britain
James Ussher
Mary Anning
Georges Cuvier
Gideon Mantell
William Buckland
Henry De la Beche's *Awful Changes* and *Duria Antiquior*
John Martin's *The Deluge* and *Country of the Iguanodon*
Charles Lyell
Young Charles Darwin
The orangutans of the Regent's Park zoo
Richard Owen
Thomas Huxley
The Crystal Palace dinosaurs
Alfred Russel Wallace
Charles Kingsley
Vanity Fair's depiction of Samuel Wilberforce and Thomas Huxley
John Tyndall
Charles Bradlaugh being arrested in Parliament
Othniel Charles Marsh and the first Yale expedition to the American Badlands
Edward Drinker Cope
Charles Knight's *Leaping Laelaps*
The Carnegie *Diplodocus* at the Natural History Museum
Natural History Museum and Royal College of Science

Preface

Behold now behemoth, which I made with thee;
He eateth grass as an ox.
Lo now, his strength is in his loins,
And his force is in the navel of his belly.
He moveth his tail like a cedar:
The sinews of his stones are wrapped together.
His bones are as strong pieces of brass;
His bones are like bars of iron.
He is the chief of the ways of God.

The Book of Job (40:15–20)

This is a book about the discovery of dinosaurs in the nineteenth century and the consequent revolutions that took place in science, religion, and society. And if all history is to some extent autobiography, I have written this book as a child of the *Jurassic Park* generation. Steven Spielberg's blockbuster was the movie event of 1993 and while I might have been too young to see the film in the cinema, I was the perfect age for the merchandise which accompanied its release. There were lunchboxes, pencil-cases, and schoolbags, and everyone but everyone had to have them; my parents bought me an army of sturdy dinosaur figurines with which I restaged the great battles of the prehistoric world on the living-room floor; and I filled light-blue binders with magazines that described the heroic deeds of fossil-hunters in the American West. In the context of these obsessions, December 1996 was a momentous time for me. Not only did I turn a mighty eight years of age, but that month also witnessed the premiere of *Jurassic Park* on terrestrial television. Even now, the scene in which Sam Neill and Laura Dern come across a herd of brachiosaurs fills me with a sense of wonder; less happily, the appearance of the *Tyrannosaurus rex* and its attack on the stricken vehicles still intrude upon the occasional nightmare. Growing up – well, growing up a bit – has done nothing to diminish my fascination with dinosaurs. It remains a matter of puerile pleasure that my favourite guitarist, R.E.M.'s Peter Buck, played gigs with an arrangement of model dinosaurs on his amp. As I read books by Steve Brusatte and Michael Benton, or true-crime stories about the clandestine traffic in dinosaur fossils, I become jealous of the lives that palaeontologists have led, hunting for prehistoric treasure in caves and canyons.[1] And I still pay good money to watch the sequels to *Jurassic Park*, movies of such diminishing quality that I can only conclude, to quote the original, that the producers have been so preoccupied with whether they could, they never stopped to think if they should.[2]

Yet as much as I love dinosaurs for their beauty and brutality, they were the wellspring of another curiosity. At primary school in Northern Ireland, teachers taught me that God had made the world in six days, 6,000 years ago. In the Ulster Bible Belt, it was a matter of regional pride that James Ussher, the seventeenth-century Archbishop of Armagh, had pinned the date of Creation to a Saturday night in October 4004 BC. There was no sense here that the early chapters of Genesis were metaphorical, or that one 'day' of Creation could have represented more than twenty-four hours, and the prominence of creationist theory in the National Trust's displays at the Giant's Causeway was testament to the persistent power and prevalence of these beliefs.[3] The conflict between these two great fonts of authority led me, even as an eight-year-old, to ask questions. If the world was only 6,000 years old then what about the dinosaurs, who had lived more than 66 million years ago? If biblical history was accurate and if God had created humankind in his own image, did that mean that the theory of evolution was wrong? Were my ancestors really monkeys? Should I believe the Bible or the scientists?

This book does not attempt to answer those questions, nor is it a polemic against scriptural literalism. Instead, it explores and narrates the first period in human history where new knowledge of the Earth and its prehistoric inhabitants collided seriously and persistently with Christian belief in the accuracy of the Bible.[4] It is therefore a history of the geologists, palaeontologists, and biologists who during the nineteenth century discovered dinosaurs, expanded our knowledge of the earliest ages of the Earth, and transformed our understanding of humankind's descent. Equally, it examines the ways in which new scientific knowledge interacted with religious assumptions about Creation, God's role in directing the world, and humankind's relationship to other life on Earth. Accordingly, this book tries to reconstruct the lives and the mindsets of the men and women who made these discoveries, and how their patterns of thought and behaviour changed – or did not – in response to the knowledge they created. How did nineteenth-century Christians, whose belief in the Old Testament was often implacable, react to compelling evidence that the world was probably millions if not billions of years old? Did they reject this evidence, or did they attempt to reconcile it with their religious convictions? If they accepted that the Bible was fallible as history, did they think it could be morally fallible too?

Although the Prologue to this book takes place in the seventeenth century, when Archbishop Ussher pronounced that Creation had occurred in 4004 BC, it really begins in 1811 when Mary Anning, the impoverished daughter of a cabinetmaker, found the fossilised skeleton of a bizarre 'fish-lizard', an *Ichthyosaurus*, on the Dorset coastline. To be clear, this was not the first discovery of a dinosaur. For one thing, ichthyosaurs are not dinosaurs; rather, they belong to the category of extinct marine animals known as thunnosaurs. For another, the creatures that we call dinosaurs had long since figured in the classical imagination as gryphons, chimera, and dragons, reports of their fossilised remains having reached the Mediterranean from the Silk Roads of Central Asia.[5] Even so, the Anning discovery launched the systematic study of these giants in the earth, these 'impossible monsters'; the seventy subsequent years – of scientific breakthroughs and intellectual turmoil, and the attendant transformation of society – are the subject of this book.

The early chapters focus mostly and unapologetically on Britain because this was a field in which British naturalists took the lead. Although there had been fossils found in North America, Patagonia, and the Low Countries, these had occurred in relative isolation and without sustained analysis. And though Paris was the global centre of zoology, French rocks had not yielded many fossils. In Britain, however, there was not only a critical mass of scholarship but also an industrial interest – in quarries, canals, and mines – in churning the earth that released these ancient bones. As these forces combined to foster the new disciplines of geology and palaeontology, Anning found the *Plesiosaurus* in December 1823; within a few weeks of that discovery, the Oxford clergyman William Buckland had unveiled the first 'proper' dinosaur, the *Megalosaurus*; on the very same day, the surgeon Gideon Mantell announced that he had found another, which he later called the *Iguanodon*. At the ancient universities, in the gentlemanly societies of London, and in the capital's dens of freethinking dissent, Britons of the late Georgian and early Victorian periods contemplated the significance of these new discoveries.

And they were truly significant – quite literally ground-breaking – for even if dinosaurs are now an established feature of both science and popular culture, the earliest discoveries of dinosaur fossils took place in a radically different context. Perhaps most importantly, early nineteenth-century Britain was a deeply and sincerely Christian place: most Britons attended religious services on a weekly basis; the King James Bible occupied a spot on almost every bookshelf; and churchmen enjoyed the prestige and dignity of an exalted social station. This everyday religiosity was reflected in national politics and, for the best part of twenty years, Britain had regarded itself as a Christian bulwark against the infidelity of Revolutionary France. Indeed, there had been violent reaction to Jacobin impiety: 'Throne and Altar' rioters had threatened suspected atheists; conservative vigilantes had burned radical literature in public; and by printing heretical works such as Thomas Paine's *The Age of Reason*, publishers risked prosecution and imprisonment for the crime of 'seditious libel'. Political rights, moreover, depended on being the right kind of Christian: discriminatory legislation prevented Dissenting Protestants and Roman Catholics – let alone atheists – from sitting in Parliament, holding public office, and graduating from university.

As in form, so in belief. The English Enlightenment had been cautious and conservative, suspicious of the continental radicalism which sought to purge society of superstition, and the Bible remained the supreme authority on moral and intellectual questions. 'Science', such as it was, was still the preserve of leisured gentlemen and Oxbridge-educated clergymen who explored the natural world as a means of better understanding the physical works of the Lord. Despite the geologist James Hutton warning that there was 'no vestige' of the planet's beginning, few people questioned biblical history or that the Earth was only a few thousand years old. In the same way, there were no serious grounds for questioning the account of Creation in Genesis, where the Lord had created humankind specially and separately in His own image.

This was why the discoveries of the 1810s and 1820s – the eruption of such impossible monsters from the soils and rocks of southern England – posed such difficult questions. If the Lord made the land and the seas on the third day of the week, and animals on the fifth day, why were fossil-hunters finding the remains of these creatures so deep within the earth? If the Lord had seen fit to save pairs of every animal from the Noachian Flood, meaning that the very notion of extinction was sacrilegious, why had nobody seen an *Ichthyosaurus* or a *Plesiosaurus* in the wild? And if the world had transformed fundamentally since the earliest days of Creation, how had those changes occurred, and how long had they taken? For these reasons, it was the concern of palaeontology's early pioneers – Buckland, Mantell, and the clergyman William Conybeare – not to disturb or subvert, but to reconcile their new science with existing religious assumptions; they understood perfectly well that accusations of heresy and social disgrace could attend radical thinking. Nonetheless, their discoveries had set in train a process of inquiry that would ultimately revolutionise human understanding of the world and of humankind itself.

By the 1830s, the former barrister Charles Lyell had established himself as the world's leading geological authority, arguing for a history of the Earth in which change had taken place gradually and consistently and which required much longer than James Ussher's 6,000 years allowed; it was a potentially perilous theory that found respectability in the personal and political conservatism of its author.

Lyell's travelling companion Roderick Murchison meanwhile sought to harness the power of this blooming science in the service of the British Empire and, in turn, imperial supremacy encouraged scientific exploration.[6] It was no coincidence that, in these halcyon years of the Pax Britannica, three young men could safely take journeys to the ends of the Earth to satisfy their curiosity. Charles Darwin circled the globe on HMS *Beagle*, observing new species and their peculiar distribution across South America; Thomas Huxley sailed the seaboard of Australia, dissecting jellyfish in a makeshift laboratory on HMS *Rattlesnake*; and Alfred Russel Wallace spent years in the jungles of Brazil and the Dutch East Indies, describing and collecting thousands of varieties of birds, bees, and butterflies.[7] Building on the work of the previous generation, all three would develop startling ideas about natural life.

As Darwin built up the courage to publicise his dangerous ideas on the variation of species, and as the Great Exhibition of 1851 demonstrated British ascendancy in science and industry, liberal thinkers had begun to question religious orthodoxies. By the mid-century, inspired in part by the 'higher criticism' that was emanating from German universities – and that George Eliot was translating into English – they were challenging the consistency and even the morality of the Bible. George Holyoake would call his preferred approach to ethical matters 'secularism'; years later, Huxley would coin the word 'agnosticism' to describe his own position. In the meantime, more prehistoric creatures had arisen from the earth: Mantell had discovered the armoured *Hylaeosaurus*, Anning the winged *Dimorphodon*, and the anatomist Richard Owen the enormous *Cetiosaurus*. Owen had also coined the term 'dinosauria' to describe the land-dwelling animals among them, but he did not deploy the dinosaurs in the scientific vanguard; instead, he argued that if such massive and complex creatures had been alive at the beginning of time, theories of continuous and progressive development among species simply had to be wrong.

That, of course, was precisely the argument that Darwin sought to refute with his theory of natural selection in *On the Origin of Species* in 1859. Thereafter, Huxley took up the cudgel, making war on the old order, and within a few short years British society would experience a moment of doubt that historians have canonised as the

Victorian crisis of faith. All the while, more fossils were being found, and not just among British rocks. There were strange, human-like skulls from the Rhineland and Gibraltar, the bones of an enormous two-legged dinosaur from New Jersey, and fossilised feathers from Bavaria. These discoveries of the late 1850s and 1860s were the keys to the kingdom. Now, the daring soul could ask: were the feathered *Archaeopteryx* and avian-hipped dinosaurs related to the common bird of the present day? In parallel, was the Neanderthal the ancestor of the human, and did the descent of man involve fraternity with the ape? On a more prosaic level, these fossils were engaging an ever-greater audience. Huxley founded new colleges and encouraged new journals that were independent of the aged Anglican gatekeepers; the Royal Institution sold out scientific lectures for working men; the model dinosaurs at the Crystal Palace were an enduringly popular attraction; and reports of the Bone Wars, where fossil-hunters fought over the bones of the American Badlands, commanded countless column inches. Dinosaurs were no longer the impossible monsters of a fevered imagination, but a simple fact of life.

As intimated in the section above, this book seeks to address and explain changing approaches to three connected issues, and to demonstrate that the discovery of dinosaurs and the developing science of palaeontology played a fundamental role in bringing about those changes.

First, the age of the Earth. In 1811, few people questioned Ussher's calculation that Earth was only a few thousand years old, yet by the 1880s all but the most reactionary had accepted that the age of the planet should be counted not in thousands of years, but tens and maybe hundreds of millions. (The real age of the Earth, around 4.4 billion years, is discussed in the Epilogue.) It was a vital correction because Lyell's uniformitarian geology and Darwin's theory of natural selection, each of which worked through incremental and barely noticeable changes, both required ages to take effect. So how did the geologists and palaeontologists of the nineteenth century persuade their contemporaries to dispense with biblical chronology and to embrace this longer timeline?

Second, the means of Creation. In the early nineteenth century, the prevailing theory on humankind's origins was known as 'monogenesis', according to which there had been a single creation as described in Genesis. However, Lyell's geology allowed for a much longer history of the world, Darwin's theory suggested that species had changed over time, and the fossilised remains of the Neanderthal and *Archaeopteryx* appeared to be proof of that theory. And if new sciences could articulate natural mechanisms for the development of species beyond the supervision of Providence, what were the implications for 'Creation' itself? Was humankind really the superior manufacture of the Lord? Had He been involved in Creation at all?

Third, and directly contingent on the second issue, was humankind's relationship to other species. Were humans truly and innately superior or were they merely the most advanced among myriad forms of life? Did it matter that other species – orangutans, cats, and dogs among them – shared in human emotions such as fear, anger, happiness, and affection? During a period when radicals, abolitionists, feminists, and Chartists were striving to eradicate the formal boundaries that society had erected *within* humanity, this was a biological question with profound political implications: if nothing but evolution separated humankind from brute animals, what was there naturally to separate the ruling classes from the masses?

All three issues feed into one of the defining questions of social, cultural, and intellectual history: secularisation. Although some scholars have argued that British society was not 'secular' before the inter-war period or even before the cultural revolution of the 1960s, many others have identified the commencement of that process in the nineteenth-century conflict between science and religion.[8] So, did science liberate itself from the strictures of religious conformity? Did science evolve from the leisurely pursuit of moneyed amateurs into a regimented profession? Did scientific evidence at last surpass the scripture as the ultimate font of knowledge? And if so, when? This book suggests that several key moments in 1881–82 may symbolise religion's concession of supremacy: the near-simultaneous foundations of Richard Owen's Natural History Museum and of Thomas Huxley's Normal School of Science; the re-election as MP for Northampton of the atheist Charles Bradlaugh in defiance of the religious requirements of the House of Commons; and the interment in

Westminster Abbey of Charles Darwin, the author of the devil's gospel. This is not to say that most Britons had become atheist or even agnostic, nor does it suggest that science had 'defeated' religion. Rather, it was now clear that, regardless of private feeling, Christianity no longer commanded the public sphere; conversely, by offering secular explanations of natural processes that eliminated divine agency as a consequential force, science had claimed an intellectual authority that it would never relinquish. As one historian has put it, Britons of the late nineteenth century now lived in 'a new[ly] secularized civil polity'.[9]

The aim of this book is to explain how and why these changes had occurred, and how the discovery of dinosaurs and the development of geology and palaeontology were essential to these developments. It is not, therefore, a straightforward history of 'science' or 'religion', if either can be described in such monolithic terms; instead, I have tried to write a history of the interaction between science and religion within the wider context of society. And while the scholarly and popular literatures on these subjects are vast, this is I think the first history of the whole century to bring together the biographies of the leading figures of Victorian science, the social and intellectual contexts in which they lived, and the essential role that the discovery of dinosaurs played in transforming the way that people thought about the world. It is also primarily – despite detours to France, Germany, and especially the United States in later chapters – a history of Britain. This is not just because the pioneers of palaeontology were British or because the most influential scientific theorists of the day were British too; it is also a reflection of the simple fact that, at the Victorian height of British power, British opinion often mattered more.

The term 'war', which appears in the subtitle of this book's hardback edition, may require explanation. Although the idea of a conflict between science and religion has been popular since the 1870s, tinged with an anti-Catholicism that harked back to the persecution of Galileo, historians have long sought to present a more nuanced assessment of the subject, with one recent volume regretting this 'conflict thesis' as an idea that will not 'die'.[10] Perhaps this is fair comment: just as the eighteenth-century divine William Paley had presented 'natural theology' as a means of unveiling God's wisdom through the study of nature, it was possible even after the florescence of Darwinian theory for science and religion to cohabit.[11] Even so, the sincere Christianity of some scientists and the scientific inquiries of some clergymen did not preclude tension between religious belief and scientific evidence; indeed, the very process of reconciling one to the other presupposes an inherent conflict. Moreover, as these pages will show, latent tension flared repeatedly into open hostility. Scriptural geologists denounced the heresies of colleagues who doubted the biblical age of the Earth; radical secularists condemned all priestcraft as delusion and bigotry; Huxley prosecuted his 'gorilla war' against the conservative Richard Owen before becoming 'Darwin's bulldog'; and by campaigning for scientific freedom from religious interference, the nine men of the X Club were self-appointed generals in a battle against the old order.[12] Science and religion were not implacable enemies, but the nineteenth century still played host to a culture war between the guardians of orthodoxy and the agents of change; there was conflict enough.

Prologue: The House of Ussher

> The World, which his hand made, is aged: but of what age, who can justly tell? The Sacred Writ is the best Register: Therein its Age possibly may be found.[1]
>
> William Greenhill, 9 June 1658

> A twentieth-century generation of Mankind ... would have revolted ... against any seriously intended suggestion that a six-thousand-years' plan was being imposed on them by a dictatorial Deity. The grotesque precision with which the term of this alleged sentence of penal servitude on Mankind had been dated by the pedantry of an archbishop, who had constituted himself the self-appointed clerk of God's court, would have been the last straw on a twentieth-century camel's back if this human beast of burden had any longer taken Ussher's calculations seriously.[2]
>
> Arnold J. Toynbee, *A Study of History* (1954)

Frail and cold, sitting on his horse in the winds of the Welsh autumn, James Ussher, the elderly Archbishop of Armagh, rode slowly along the rough, wet paths that led west into the Vale of Glamorgan. He had been safe in Cardiff, under the roof that he had shared with 'many persons of good Quality' and even, briefly, with the king himself.[3] It was September 1645 and civil war had engulfed the British Isles.

Earlier that summer, Parliament's New Model Army had inflicted a shattering defeat on Charles I at Naseby, and the royalist campaign was in disarray: 'All the foot [officers and soldiers]', recorded one royal aide, 'remained prisoners'.[4] Since that fateful day, the king had ridden more than 1,200 miles across Britain, trying desperately to rally troops to his banner, but this errand had failed in south Wales. The so-called Peaceable Army, representing Welshmen who were 'much insulted and harassed by the rapacious insolence' of Charles's officers, would not bow to royal authority.[5] By August, local support for the king had collapsed. So, when Charles decided to leave Cardiff for the north, taking the castle's garrison with him, Archbishop Ussher had been forced to seek sanctuary elsewhere.[6]

At first he considered going abroad but, with Parliament's navy patrolling the coastline, preventing escape by sea, Ussher looked instead to St Donat's Castle, the ancestral home of the royalist Stradling family. Gathering his possessions, including reams of scholarly material, he set out in the company of his daughter and 'other Ladies' who required shelter. The way was far from clear. As many as 10,000 rebels were thought to be roaming the Welsh countryside and, in seeking to avoid their attention, Ussher and his companions would move carefully 'by some private ways near the Mountains'.[7] It would not keep them safe. In a place where 'it was Crime enough that they were English', or simply not Welsh, they came upon a 'Party that were scouting thereabouts'. Seized and delivered up to the main body of bandits, Ussher and his daughter were 'barbarously used' and 'treated with great cruelty'.[8] Thrown from their horses, they watched as their captors plundered their baggage, ransacking Ussher's chests of books. In an instant, they dispersed his papers 'into too many hands to be retrieved'.[9]

There is no record of how Ussher and his daughter recovered from this distress, but in time they reached St Donat's, where the archbishop began to grieve: a friend remarked that he never saw him 'so much troubled in my life'. In conversation with his daughter, Ussher saw 'God's hand' in events and regarded himself as the victim of divine displeasure: 'I am touched in a very tender place', he lamented, for 'He has thought fit to take from me at once all that I have been gathering together, above these twenty years, and which I intended to publish for the advancement of Learning, and the good of the

Church'.[10] This was the cause of Ussher's pain: not the crisis of royal power, nor being thrown from his horse, but the loss of his books and papers. Yet when local grandees rallied around to restore much of Ussher's library, donating precious volumes from their own collections, they allowed him to resume the work of a lifetime: a chronology of the Earth from the moment of Creation.

Study had consumed James Ussher ever since his childhood in Dublin in the 1580s. Taught to read and write by two blind aunts who had learned 'much of the Bible by heart', he was among the first students to enrol at the city's new Trinity College, which had been founded as a finishing school for aspiring Protestant churchmen.[11] After graduating as the star of his class aged only sixteen, Ussher rose quickly in both academic and clerical circles, even forsaking his inheritance to avoid the distractions of running the family estate. Appointed professor of divinity at Dublin in 1607, he entered the senior ranks of the Church of Ireland when he became Bishop of Meath in 1620. Five years later, having curried the favour of King James I by preaching zealously against Catholicism – he had long thought that Rome was 'the principall seat of Antichrist' – Ussher secured the highest office in the Irish Church, the archbishopric of Armagh.[12]

All the while, he pursued his interests in chronology, in early Celtic Christianity, and the founding fathers of the Church. This was often at the expense of Ussher's episcopal duties: as one contemporary put it, he had 'too gentle a soul for the rough work of reforming' and was 'not made for the governing part of his Function'.[13] Nor was Ussher much interested in society. 'It is a sad thing', he once lamented, 'to be forced to put one's foot under another's Table, and . . . to be obliged to bear their follies'.[14] A man 'indifferent tall and well shaped', whose once-brown hair escaped in grey wisps from his clerical skullcap, Ussher preferred to eat 'wholesome Meat without Sauce', to avoid sitting 'long at Table', and to relax only by walking alone on riverbanks.[15] He therefore lived in a large measure 'the life of a recluse, his vision . . . bounded by the walls of his splendid library'.[16]

As Ussher developed his 'rich magazine of solid learning', he became a prominent figure in the network of theologians, antiquarians, and philosophers who populated the European republic of letters.[17]

Friendly relations with bibliophiles in France, the Low Countries, and the patchwork quilts of the German and Italian states meant that, if Ussher ever needed a copy of a rare book, he usually got it. There was no doubt of Ussher's renown, and Cardinal Richelieu, the *éminence grise* at the court of Louis XIII, was a noted admirer; the Frenchman sent a gold medal and his compliments in 'a very respectful letter' to Ussher, who in reply presented Richelieu with a gift of several Irish greyhounds 'and other rarities of the country'.[18]

In 1641, having spent forty years flitting between his archdiocese and English libraries, Ussher was in London when rebellion convulsed Ireland. His properties did not escape the fury of the insurgents: 'In a very few days', reported one of his colleagues, they 'plundered his houses in the country, seized on his rents, quite ruined or destroyed his tenements, killed or drove away his numerous flocks and herds of cattle . . . and left him [nothing] but his library and some furniture'.[19] Even the survival of the library was a close-run thing, for Ussher's steward reported that the rebels 'made no question of devouring us' and 'talked much of . . . burning [the library] and me by the flame of books'.[20] Thus was the archbishop stranded in London as tensions rose between Charles I and Parliament, and though he shared many Calvinist beliefs with parliamentary leaders, his sympathies lay with the king: 'he made it his business,' recalled one colleague, 'as well by preaching as writing, to exhort [MPs] to Loyalty, and Obedience to their Prince'. When war broke out in 1642, Ussher left London for the royalist stronghold of Oxford, where in the aftermath of the Battle of Edgehill he preached against the 'Rebellious Proceedings' in the presence of the king himself.[21] He stayed there for the best part of three years, burying himself in his work, but when rumours spread of a parliamentary assault on the city, he found himself moving further west, first to Cardiff and then – as we have seen – to St Donat's Castle.

As Ussher convalesced under the Stradlings' roof, the royalist plight grew 'every day more desperate'. If Wales fell, where else could Ussher go? Once again, he toyed with going 'beyond the Seas'. He even procured the early-modern equivalent of a passport, but this document held no sway with Richard Moulton, the roundhead admiral who commanded the Bristol Channel. Asked by Ussher's allies whether the archbishop could depart, Moulton 'returned a rude and threatening answer, absolutely refusing'.[22] It was only the charity of the Countess of Peterborough that rescued Ussher and, late in 1646, he travelled to London to lodge with her near Charing Cross. Safe in the capital, Ussher resumed work on his chronology, breaking only to preach at Lincoln's Inn and occasionally to the king, who by now was a prisoner of Parliament.

The one great event which punctuated this twilight life of study and prayer was the execution of Charles I in January 1649. Standing on the leaden roof of Lady Peterborough's house, Ussher could see down Whitehall to the scaffold outside Banqueting House. As Charles spoke his last words, the archbishop 'stood still and said nothing but sighed, lifting up his hands and eyes (full of tears) towards Heaven'. When the time came for the king to remove his cloak and doublet, and as 'the villains in vizards' pulled up his hair to clear a path for the executioner's blade, Ussher could not bear to watch. Possessed by grief and horror, he grew pale and fainted. Carried down to his bedroom, he spent the rest of the day in tearful contemplation; he would fast on the anniversary of Charles's death 'so long as he lived'.[23]

By 1650 James Ussher was sixty-nine and in failing health. Having lost many of his teeth and no longer able to speak in public, he had retired from preaching. Increasingly short of sight as well, he now spent his days chasing the sunlight from east to west, from window to window, that he still might read. Despite these infirmities, Ussher at last realised more than five decades of ambition when, at several presses around London, the publisher James Flesher printed the first volume of his chronology, *Annales Veteris Testamenti*. By 1654, when the second volume arrived, Ussher had produced almost 2,000 pages of densely written, exhaustively researched history, all in Latin.[24] The central conclusion of this chronology, which had used the biblical record to reach back to the moment of Creation, would have profound influence on the human understanding of the Earth – and life on Earth – for more than 200 years.

Readers of the twenty-first century might scoff at 'chronology', at the very idea that analysing the Bible could reveal hidden, precise truths about the history of the Earth. Yet in the early modern period, in a world where God's hand was thought to be the key force in driving human events, chronology represented a sincere and sophisticated attempt to understand the history of mankind and the Lord's own works. As one historian has explained, 'If world history was to be related to a divinely ordered plan, it was essential to begin by establishing the order of events, even in the remotest ages.'[25] Numerous scholars had already tried to do this. As early as the second century AD, Theophilus of Antioch had argued that the world had existed for 5,698 years; the Spanish orientalist Benito Arias Montano had suggested that Genesis began in 4084 BC; and the Cambridgeshire parson John Swan believed that 3,997 years had elapsed between Creation and Christ. By far the most revered chronologist was Joseph Justus Scaliger, the French Calvinist who had posited 3950 BC as the date of Creation within a framework of time that he called the Julian Period. But in terms of influence and longevity, Ussher would outdo them all.

Although he had begun his investigations at a young age, getting as far as the Books of Kings by the age of fifteen, it was only when Ussher ascended to the episcopacy (and its revenues) that he commenced his great work in earnest, and he did so by collecting materials from across Europe and the Near East. He 'looked particularly for assistance in his Biblical researches' to the chaplain of the Levant Company in Aleppo, who obliged by procuring precious copies of the Syriac Old Testament and – better still – a copy of the recently discovered Samaritan Pentateuch.[26] One linguist in Berlin was paid an annual retainer of £24 (some £7,000 in today's money) to source more than 300 manuscripts; from John Bainbridge, the Savilian professor of astronomy at Oxford, he acquired evidence on ancient astronomical cycles; and in his pursuit of biblical knowledge Ussher even deigned to use Catholic sources, seeking out a copy of a chronology issued at Rome.[27] Inevitably, 'his expences were much in Books'.[28]

Understanding Ussher's chronology demands working backwards in time. And though the most recent event that he describes is the destruction of the Temple at Jerusalem in AD 70, the first truly significant date is 4 BC, the birth of Christ. Of course, identifying any year 'BC' as the date of the Nativity appears paradoxical, but this was the obvious choice for two reasons. First, the 'Anno Domini' dating system assumed that Christ was born during the twenty-eighth year of the reign of Caesar Augustus, but even if Augustus was technically proclaimed emperor in 27 BC (which would have given a Nativity of AD 1), he had really reigned since the Battle of Actium in 31 BC, meaning that Christ had been born in 4 BC. The second reason was Matthew 2:1, which states that 'Jesus was born in Bethlehem in Judea, during the time of King Herod'. Yet if Christ was born in AD 1, Matthew 2:1 could not have been correct, since Herod was known to have died no later than 4 BC. By deciding on the earlier date, Ussher sought to settle the matter.[29]

Attaching the Nativity to a precise date, however, was a simple task when compared to reconstructing the period between 4 BC and – moving backwards – the end of the Old Testament, which had concluded in historical terms with the books of Ezra, Nehemiah, and Esther in the fifth century BC. Where the Bible was silent, Ussher called upon ancient and classical history, and more than three-quarters of his chronology concerns the intertwined histories of the Greeks, Romans, Persians, and Babylonians. It was only by pegging certain events to certain dates, such as Alexander the Great's victory at Gaugamela in 331 BC, that Ussher inched his way backwards into the 'historical' Old Testament. He established the death of Nebuchadnezzar the Great, who had carried into exile all Jerusalem, at 563 BC; after this, he fixed the accession of Nabonassar in Babylon to 747 BC. Making further progress towards Creation required sifting through the morass of regencies, interregna, and overlapping reigns which characterised the history of the kingdom of Judah. Ussher also had to bear in mind several remarkable events such as Joshua commanding the sun to stand still over Gibeon, and Hezekiah turning back the sundial of King Ahaz, both of which affected the progress of time and therefore his calculations.[30]

Ussher emerged at length into the genealogies of the Old Testament, into those dreary verses where the patriarchs 'begat' each other. Although he departed from Scaliger on exactly when – and to which mother – Abraham was born, and though he ignored some discrepancies between versions of the Torah, these genealogies constituted the final stretch of his chronology. With the Bible giving the age of each patriarch at the birth of his first son, it required only simple addition to reach back, past the Flood and the Exodus in Egypt, to the birth of Adam and the week of Creation.[31] And in which year did Ussher find himself? 4004 BC. Rather conveniently, this allowed him to reconcile his chronology with an established Christian belief that there had been 2,000 years of 'nature' and 2,000 years of 'law' under the Torah before the birth of Christ. Even more precisely, Ussher held that 'the beginning of time . . . fell upon the entrance of the night preceding the twenty third day of October in the year [4004 BC]'.[32]

But why the night of 22–23 October? First, Ussher assumed that Creation had occurred on a Sunday, given that Saturday was the Jewish Sabbath on which the Lord would have rested. Why the autumn? Well, the Jewish New Year is determined by the autumnal equinox, and this led Ussher – guided by astronomical calendars – to the night of 21–22 September. And the jump to 22–23 October? This was because of a discrepancy between the Julian and Gregorian calendars. In the former, every year divisible by 100 had been a leap year; in the latter, it is only where such years are also divisible by 400 (that is, 400, 800, 1,200, and so on) that an extra day is added to the year, which we still do on 29 February. This meant that between the beginning of time and the adoption of the Gregorian calendar in 1582, there had been roughly thirty 'too many' days, and so the archbishop needed to move the date of Creation by a month. And this was why, according to Ussher, the world began on the night of 22–23 October in 4004 BC.[33]

When Ussher died in 1656, succumbing to pleurisy, London mourned. The clergymen of the city escorted his hearse from Somerset House to Westminster Abbey, where they interred the archbishop in the chapel of St Erasmus. Despite political differences with the deceased, Oliver Cromwell contributed £200 to Ussher's burial. As Cardinal Mazarin and the King of Denmark fought over the remains of his Irish library, which at one stage had consisted of '10,000 Volumes, Prints, and Manuscripts', Ussher and his chronology only grew in stature.[34] When writers of the 1690s dared to dissent from its conclusions, one clergyman roared back in defence of 'that great ornament of his Age, See and Country, the incomparable Arch-Bishop Usher' [*sic*].[35] By the 1770s, Samuel Johnson would tell James Boswell that Ussher was 'the great luminary of the Irish church; and a greater . . . no church could boast of, at least in modern times'.[36] His chronology influenced great historians too: for Edward Gibbon, 'the Annals of Ussher distinguished the connection of events, and engraved the multitude of names and dates, in a clear and indelible series'.[37]

But why did Ussher's chronology – and not the work of, say, the French genius Scaliger – dictate the assumptions of so many people? It was not down to language, since both men wrote in Latin. Nor was there any sectarian motive, since Scaliger too was Protestant. Rather, it seems that Ussher's predominance is owed to John Fell, who as Bishop of Oxford in the 1670s proposed 'an Edition of the H. Scriptures . . . to which will be added . . . Chronological observations'.[38] The first such Bible appeared in 1679 and the fashion for biblical dates took root. When Ussher's dates were added to a new version of the King James Bible in 1701, 'the idea that the world was created roughly 4,000 years before Christ became fixed in popular consciousness in the English-speaking world'.[39] Generation after generation of Protestants would now open their Bibles and see the date of 4004 BC beside the first verse of Genesis, and 'for more than a century thereafter it was considered heretical to assume more than 6,000 years for the formation of the earth'.[40]

Of course, this is not to say that Ussher's theory prevailed universally, or that other scholars did not contemplate a longer history of the Earth. As long ago as the eleventh century, one Chinese philosopher had suggested 'that geological and astronomical evidence pointed to an immensely long history of the world'.[41] Moreover, as Enlightenment naturalists began to examine the world around them, and especially the ground beneath their feet, the notion developed – albeit within limited intellectual circles – that the six 'days' of Creation could have been metaphorical. Upon examining the Alpine landscape in the 1770s, the Genevan scholar Horace Bénédict de Saussure assumed that major changes in the Earth's structure had occupied 'a long succession of ages'; and by assessing how long it took for molten balls of lead to cool and harden, the French naturalist the Comte de Buffon estimated that the Earth had formed almost 75,000 years ago.[42] The Scottish traveller Patrick Brydone meanwhile related the embarrassment of a Sicilian bishop who realised that Etna's lava was truly ancient: 'Moses hangs like a dead weight upon him,' he wrote, 'and blunts his zeal for inquiry . . . he has not the conscience to make his mountain so young as that prophet makes the world'.[43]

Most importantly, at least for what became the science of geology, James Hutton would formulate a theory that the earliest ages of the Earth could be explained only by reference to geological processes that were *ongoing*. A Scotsman who had abandoned careers in both law and medicine, Hutton had studied in Edinburgh, Paris, and Leiden before making a fortune through the manufacture of ammonium chloride. Using this wealth to purchase extensive farmland, he took a close interest in the earth underlying his crops. Living off the rentals yielded by his estates, Hutton spent decades – often in the company of great Enlightenment figures such as Adam Smith and James Watt – discussing and exploring the Earth and its composition.[44] He rode through England, Wales, and much of Scotland, lamenting of his enthusiasms: 'Lord pity the arse that's clagged to a head that will hunt stones.' One critical excursion took place in 1785 in the Cairngorms, where he found 'branching veins of granite' piercing deep into layers of another metamorphic rock known as schist; it was a phenomenon that he could explain only if the granite were much younger than the schist. Three years later, at Siccar Point north of Berwick, he found sandstone lying horizontally across vertical strata of schist, an 'unconformity' that, in his view, suggested that the churning force of the ocean had been forming, folding, raising up, and setting down these layers of rock over an immeasurably long period of time.[45]

In a paper before the Royal Society of Edinburgh, Hutton proposed that 'the land on which we rest is not simple and original, but . . . a composition, and had been formed by the operation of second[ary] causes' other than the Creation and the Deluge.[46] When his theory appeared in print, he claimed that the natural history of the Earth even evinced 'a succession of worlds'. More distressing still for scriptural literalists, Hutton stated that examining 'the system of nature' led directly to this conclusion about the age of the Earth: 'We find no vestige of a beginning, no prospect of an end.'[47] It was a daring rebuke of biblical chronology and of the belief that only divine agency had shaped the physical world. Yet if this concept of what has been called 'deep time' shocked the readers of the 1790s, and even if it contradicted James Ussher's calculations, it would pale in comparison to a series of astounding discoveries which began, in 1811, with two teenagers finding something strange in a beachside cliff on the south coast of England.[48]

PART ONE

Giants in the Earth

DRAMATIS PERSONAE

Mary Anning, fossil-hunter, Lyme Regis
Henry De la Beche, geologist, Lyme Regis and Jamaica
William Buckland, clergyman and geologist, Oxford
William Conybeare, clergyman and geologist, Bristol
Georges Cuvier, naturalist, Paris
Charles Lyell, lawyer and geologist, London
Gideon Mantell, surgeon and geologist, Lewes
Roderick Murchison, soldier and geologist, London

I

Shells by the Shore

> The remarkable situation of the town, the principal street almost hurrying into the water, the walk to the Cobb, skirting around the pleasant little bay, which, in the Season, is animated with bathing machines and company, [and] the Cobb itself, its old wonders and improvements, with the very beautiful line of cliffs stretching out to the east of the town, are what the stranger's eye will seek; and a very strange stranger it must be, who does not see charms in the immediate environs of Lyme, to make him wish to know it better.[1]
>
> Jane Austen, *Persuasion* (1817)

There were many good reasons to visit Lyme Regis. The small Dorset town, which lay some twenty miles to the west of Weymouth, had a storied history: patronised by Edward I in the 1280s, it had sent ships to defeat the Spanish Armada and, almost a hundred years later, it was where the Duke of Monmouth landed his troops during the ill-starred rebellion against James II. There was the seaside too, and when the doctor Richard Russell popularised saltwater as a cure 'for the affections of the glands' in the 1750s, 'a residence at Lyme [was] rendered highly favourable ... to invalids who require to bathe'.[2] There was also the Cobb, the medieval harbour wall of boulders and pebbles which curled an arm around the town: 'There is not any one like it in the world,' declared one observer, 'for no stone that lies there was ever touched with a tool'.[3] But Lyme's real treasure lay in the grey-blue cliffs that rose up around it. Deep within that stone, and shaken loose onto the beach by storms and landslides, was one of the richest collections of fossils in the world. Every summer – then, as now – Lyme played host to 'a race of perambulators that, swallow-like, set out [to] ramble along the shore', to sift through the sand and the rocks of the coastline, and to collect 'specimens of sea-weed and shells'.[4] In hosting these seasonal 'fossilists', as fossil-hunters were called, Lyme stole a march on its neighbour Charmouth, which, as 'a very genteel village, inhabited by persons of small fortunes . . . would not condescend to . . . take in boarders'.[5]

For the people who lived the year round in Lyme, many of them working in mills on the River Lym, finding fossils and selling them to wealthy visitors was a vital means of boosting meagre wages. They hawked the ancient shells of ammonite molluscs, which they promised would cure blindness and impotence; the remains of the long, pointed squids known as belemnites were thought to confer similar benefits on horses. Besides these precious fossils, they sold fragments of ancient 'crocodile' backbones as 'verteberries', and romanticised other shells as 'John Dory's fingers' after the sailor of folklore.[6] As this cottage industry bloomed, visiting collectors learned which locals to trust. The Irish doctor James Johnson advised one friend to call upon 'a confounded rogue' by the name of Lock and to buy him 'a Grog or Pint', a gesture of friendship that Lock would reward with first choice of 'all the crocodiles he may meet with'.[7] It was good advice: Lock's ability to divine curiosities had earned him the nickname 'Captain Cury'.

Johnson also recommended 'a person who collects for sale by the name of Anning, a cabinet maker', who enjoyed an equally solid reputation.[8] This was Richard Anning, a bearded giant of a man who had moved to Lyme from the nearby village of Colyton, and who lived with his wife Molly and their children in a terraced house on Bridge Street beside Lyme's jailhouse. Theirs was not an easy life. Besides the want, disease, and deprivation which blighted the lives of most British workers, Richard, as a self-employed artisan, was vulnerable to the caprices of his clients. In 1805, he lost the business of an especially notable customer when Jane Austen, who was seeking to repair a broken box in her holiday home, scorned his quote of five shillings as 'beyond the value of all the furniture in the room together'.[9] Life was precarious in other ways. Living on the river, close to the seafront, the

Annings woke one morning to find that the ground floor of their home had been 'washed away ... [by] an exceptionally rough sea'.[10] Tragedy touched the family too: in December 1798, the four-year-old Mary had been 'left by the mother for about five minutes ... in a room where there were some [wood] shavings' and an open fire. In a scene of horror, 'the girl's clothes caught fire and she was so dreadfully burnt as to cause her death'.[11]

Always hopeful that fresh finds among the rocks might keep his family in bread, Richard Anning continued to hunt for fossils along Lyme's beachfront, often taking his son Joseph and another daughter, also called Mary, on his missions. It was a matter of luck that *this* Mary Anning had lived long enough to join her father in the hunt. In the summer of 1800, when only fourteen months of age, she had been taken by her nurse to a field outside Lyme to see a performance by a travelling band of horsemen. Having given 'an extraordinary display of vaulting' on the previous evening, the equestrians had attracted a sizeable crowd. But as the people of Lyme gathered in the late afternoon the weather became 'intensely hot and sultry' and, just before five o'clock, a storm rolled in.[12] According to one eyewitness, 'it thundred and lightninged [*sic*] at such a degree as not remembered by the oldest person then in the town, also accompanioned [*sic*] with heavy rain'. While some spectators fled home, others kept dry under the linhays where the cattle slept; others still 'sought shelter under cover of some lofty elms', and Mary's nurse was among them. It was a fatal mistake. One historian records that after 'a vivid discharge of electric fluid' a cry went up, and three people lay motionless under the elms. All three were dead, including the nurse: her husband recalled that 'on the right side my wifes [*sic*] hair cap and handkerchief [were] much burnt and the flesh wounded'.[13] Mary was thought to have perished as well, but when taken home and dunked in a warm bath, she revived. The local doctor proclaimed the miracle of her resurrection to an anxious crowd and, from this moment, a legend took root: supposedly 'a very dull girl' before the lightning strike, Mary 'now grew up lively and intelligent'.[14]

Although Richard Anning's wife was 'wont to ridicule' his beachfront excursions, not least when he risked social censure by setting out on Sundays, there was a steady turnover of fossils: 'If [he] found anything worth purchasing, it was usually exposed in front of his shop . . . on a little table.'[15] Molly Anning might also have worried for her family's safety, for hunting fossils was a dangerous business. The seafront at Lyme was prey to the cold, quick tide of the Channel; moreover, because the best material was wrested from the rocks by stormy weather, fossiling meant standing beneath an unstable cliff-face. The Swiss naturalist Jean-André Deluc, who visited Lyme frequently in the 1800s, narrated one of Anning's close escapes: 'One day,' he recorded, 'as he was walking along [a] path [on a cliff-top], he felt the ground give way under him, and at the same time he saw a section of the part which still remained firm, rising, as it were, above his level . . . He threw himself with a spring on the upper ground, and ran immediately back towards the town.'[16] But Richard Anning's luck did not hold.

On a winter's night in 1809, having arranged to sell his recent discoveries to the patrons of the Pilot Boat Inn at Charmouth, Anning was walking over the Black Ven cliff when he 'diverged from the path in a field on the summit of the hill'.[17] The sources do not record whether Anning was the worse for drink or if he simply could not see under dark skies, but he lost his footing and plunged onto the rocks below. It was a measure of Anning's strength that, despite being knocked out and seriously hurting his back, he picked himself up and returned to Lyme on foot, taking to bed with only a hot water bottle. But this fall from Black Ven had broken him. Over the next twelve months, Anning fell prone to violent coughing fits; when he started spitting blood, it was clear that he was too weak to fight the consumptive disease that was endemic at the time. He died in November 1810, leaving debts of £120 (some £10,000 today) and a family who soon relied upon parish relief for survival. While Joseph found an apprenticeship in upholstery, eleven-year-old Mary was left to her own devices because 'her mother, being in great distress, did not attend to her'. Lost and alone, Mary took solace in Lyme's beaches, 'wander[ing] along the seashore' until she 'picked up a fossil' for which a passing lady gave her half a crown.[18] This was not much, about £10 in today's money, but for a family living close to destitution it was enough. Mary and her brother would keep hunting for fossils.

*

In 1811, Joseph Anning found the most magnificent 'cury' on the beach between Lyme and Charmouth. Buried two feet deep in the sand, it had clearly fallen from the cliff-face and it was, he thought, the fossilised skull of a crocodile. This was a reasonable assumption. After all, bits and pieces of crocodiles had long been found on the Dorset coast, but this specimen was different. It was much longer than usual and it was terrifying: from a deep, powerful skull with cavernous eye-sockets, a long, curved snout extended some four feet in length, hosting close to a hundred sharp, conical teeth. It was extraordinary, but where was the rest of it? Before returning to work, Joseph took his young sister, the twelve-year-old Mary, to the spot where he made the discovery and charged her with tracking down the rest of the skeleton. It would take almost a year, but Mary Anning would find the larger part of the creature. Disinterring the fossil, however, was another matter, for while the skull had fallen to the beach, the body was lodged in the rocks above the strand; its removal would require the labour of three strong men and a great deal of care.

Today, when palaeontologists find something in the field, they use a meticulous procedure that has been refined ever since its development in the late nineteenth century. First, they strengthen exposed bones with a varnish to prevent them from shattering; next, they use hammers and chisels to dig a trench both around and then underneath the fossil, which sits like a mushroom on a pedestal of rock. After this, they apply a jacket of plaster and hessian cloth to the top of the

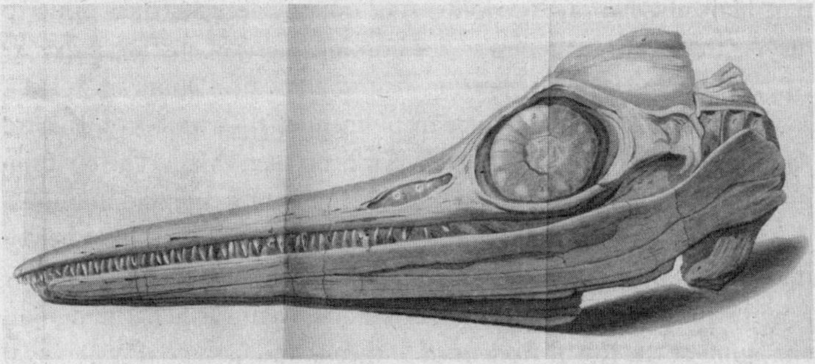

The skull of the marine reptile, discovered by the Annings in 1811–12, that Everard Home would christen the '*Proteosaurus*'

mushroom, which is flipped over so they can repeat the process on the underside. The palaeontologists then move this 'fossil jacket' to their laboratory, where they whittle away the rock that surrounds the bones using fine drills and brushes, at last releasing the fossils.[19] The Annings, of course, had none of these methods, nor modern equipment. Instead, they carved out the bones as best they could: where the rock was soft, they could lift them out individually; where it was hard, they chiselled out larger slabs and lugged them home along the beach. Either way, reassembling these ancient skeletons was a laborious, complicated affair.

News of the beast which emerged from the Dorset limestone spread quickly. On Thursday, 12 November 1812, the *Bath Chronicle* reported that 'the complete petrifaction of a crocodile, 17 feet in length' had been wrested from the cliffs.[20] Over the following weekend, the *Statesman*, the *General Evening Post*, and *Bell's Weekly Messenger* reprinted the same report in London.[21] By the next year, news of the Annings' discovery had reached even the *Madras Courier* in India.[22] It also caught the attention of Henry Hoste Henley, the proprietor of Colway Manor near Lyme – and of Sandringham House in Norfolk – who paid £23, almost £2,000 today, for the fossilised skeleton; it was a finders' fee that would let the Annings eat well for months.

Buying the 'Lyme crocodile' did not mean that Hoste Henley would take possession of it. Rather, the privilege of presenting the specimen to the world belonged to William Bullock, the proprietor of the Egyptian Hall museum at Piccadilly. As London's learned men came to inspect the creature, there was one burning question: just what *was* it? At the December 1814 meeting of the zoological Linnean Society, James Johnson – who had once recommended the services of Richard Anning – argued that the 'crocodile' was no such thing: 'The bones in question', he declared, were in fact 'the bones of a new and unknown species of amphibious animal'.[23] Over the course of four papers before the Royal Society, the most prestigious scientific institution in the English-speaking world, Sir Everard Home went further. The dominant figure at the Royal College of Surgeons and George III's personal physician, Home agreed that the Lyme creature was no crocodile: the spine and the teeth were too distinct. Nor was 'this very extraordinary animal' a fish or whale, since neither the ribs nor the shoulder blades conformed to type.[24] Instead, Home came to the conclusion that 'among the animals destroyed by the catastrophes of some remote antiquity [that is, the Flood], there had been some at least that differ so intirely [*sic*] in their structure from any which now exist, as to make it impossible to arrange their fossil remains with any known class of animals'.[25] Deciding that the creature most closely resembled the small, cave-dwelling salamander known as the proteus, he proposed the name of *Proteosaurus*.[26] Joseph and Mary Anning, two destitute teenagers from Lyme Regis, had not just discovered an animal that was unknown to British science; they had discovered an animal which no longer existed.

Home's conclusion was alarming, for the very idea of extinction was thought to verge on blasphemy. According to the book of Genesis, Noah had taken seven pairs of every 'clean' animal, two pairs of every 'unclean' animal, and seven pairs of every bird onto the Ark, thereby ensuring that every land species had survived the Flood. It was also widely assumed that when 'the fountains of the great deep burst forth', marine creatures would not have been unduly affected. Most British naturalists therefore shared the view of one Irish doctor that 'no real Species of Living Creatures is so utterly extinct, as to be lost entirely out of the World, since it was first Created'.[27] But if the Lyme specimen belonged to an extant species that had survived the Flood, where were the rest of them? Some observers took comfort in Robert Hooke's suggestion that there could be 'divers parts of the World' or of the deep sea, 'such as have been [. . .] not yet discovered by the *Europeans*', where these creatures would be 'frequent and plentiful enough'.[28]

Even so, the idea that proteosaurs could have survived in some remote part of the world could not explain the *depth* at which the Annings made their discovery, since they had found the Lyme creature 'about 100 feet below the surface of the earth'.[29] The problem here was the order of Creation laid down in Genesis: the land and the sea had come into being on the third day of Creation, yet God had created marine creatures – and Home believed that the *Proteosaurus*'s 'constant residence [was] in the water' – on the fifth day of the week.[30] So, if the Lyme creature had been created *after* the land, how could the

Annings have found it *beneath* one hundred feet of rock? Moreover, even if the biblical Flood had swept this particular animal over land and marooned it to die in south Dorset when the waters receded, why had it not been found on the surface of the earth? The logical inconsistency between biblical history and the new fossil record was painfully obvious, even to children. When the six-year-old Frank Rawlins was drawn to 'a fine display of fossils' at Lyme, he would ask his father 'why all these creatures should have burrowed their way down so deep into the rocks until they were smothered by them'.[31] From the perspective of the twenty-first century, the obvious solution is to disregard Genesis as an accurate account of the world's origins. Yet for many Britons of the 1810s, this was not an option.

It may be tempting to think that the Enlightenment, that awesome intellectual movement of reason and rationalism, had discredited biblical literalism during the long eighteenth century, thus opening worlds of inquiry that were unbenighted by religious dogma. Yet except among the most radical philosophers, this did not happen. The English Enlightenment had been immersed in Christianity, with Isaac Newton declaring that the very purpose of scientific investigation was 'to discourse of God'.[32] It followed that William van Mildert, who became Bishop of Durham in the early nineteenth century, took inexhaustible pride in the fact that Newton, Francis Bacon, and Robert Boyle were '*Christians*: these were Believers in *Revelation*, and were not led by Philosophy only'.[33] There had also been a British counter-Enlightenment, where the forces of conservatism sought to repress and expunge the most dangerous elements of new philosophy. Infamously, the British courts had sentenced the elderly schoolmaster Peter Annet to the pillory and a fatal year of hard labour for translating Voltaire into English;[34] and for fear that he was a member of the Illuminati, the Bavarian society which had promoted secular egalitarianism, British conservatives had condemned the Prussian philosopher Immanuel Kant.[35] The world in which the Annings discovered the Lyme creature was therefore deeply religious. One historian of the period has written that 'religious feeling and biblical terminology so permeated *all* aspects of thought . . . that it is hard to dismiss them',[36] and another that 'Protestant literalism still prevailed and Englishmen read both Old and New Testaments as a miraculously unified whole'.[37]

Early British forays into geology encountered this fervour directly. The Dublin chemist Richard Kirwan, a noted eccentric who removed his door knocker at seven o'clock each evening to prevent the intrusion of callers, decried James Hutton's geological theory of 'deep time' as contrary to 'reason and the tenor of the Mosaic history, thus leading to an abyss'.[38] James Parkinson, the doctor who described the eponymous palsy, was also a celebrated geologist and his *Organic Remains of a Former World* (1804–11) was an avowedly Christian attempt 'to find out the ways of God in forming, destroying, and reforming the Earth'. For Parkinson, the Bible provided irrefutable proof that 'after the complete formation and peopling of this globe, it was subjected to the destructive action of an immense deluge of water'.[39] As for Mary Anning, one of the few volumes that she is known to have possessed as a teenager was the 1801 edition of *The Theological Magazine*, which insisted on a literal, six-day interpretation of Creation.[40]

This enduring intellectual commitment to Christian values and biblical teaching pertained in British society more widely. Although the Church of England suffered badly from inept administration, meaning there were nowhere near enough ministers or churches to serve its parishioners, it claimed the nominal affiliation of almost ninety per cent of the English population, with one clergyman in Hampshire estimating that eighty-one per cent of his flock attended church at least 'occasionally'.[41] Beyond the Anglican supremacy, Dissenting Protestants – such as the Annings – constituted the most vibrant religious community in Britain: despite laws which prevented Baptists, Methodists, and other sects from graduating from the ancient universities and holding public office, Dissenting congregations more than doubled in size between 1800 and 1820.[42]

British reading habits were deeply religious too, thanks in part to the Bristolian writer Hannah More. Living alone but in comfort thanks to annual payments from a former fiancé, More had become concerned with the publication of irreligious works in the late eighteenth century: 'Vulgar and indecent penny books were always common,' she explained to an ally, 'but speculative infidelity brought down to the pockets and capacity of the poor forms a new era in our history'. It was an ungodly sensation that required 'strong counteraction'.[43] From 1795 to 1798, therefore, More had churned out 114 *Cheap Repository Tracts* which urged Britons to embrace piety and deference: she advised readers to be more like her character Mr Johnson of Wiltshire, who was content to go about life placidly 'that he might have leisure to admire God in the works of His creation'.[44] In the first year of publication alone, the *Tracts* sold two million copies.

The French Revolution only galvanised hostility towards unbelief. Although many Britons were at first sympathetic to the enemies of the Bourbon monarchy – Edmund Burke, who railed against 'this new conquering empire of light and reason', was an isolated voice of warning – opinion hardened quickly.[45] In 1791, the 'Throne and Altar' rioters of Birmingham marked the second anniversary of the storming of the Bastille by razing several Dissenting chapels and destroying the laboratory of the radical chemist Joseph Priestley. Following France's declaration of ideological warfare upon Europe in 1792, and especially after the National Convention's execution of Louis XVI in January 1793, William Pitt's conservative administration embarked upon an extensive programme of political and intellectual repression. When, for example, the radicals Thomas Hardy, John Horne Tooke, and John Thelwall denounced the government's suspension of habeas corpus, they were tried for sedition. And when the London Corresponding Society sought to promote fraternity with its French comrades, the government put its weight behind the conservative Reeves Associations, which restricted membership to 'persons well affected to the government of these realms'. In seeking to disarm 'the wicked and senseless reformers of the present time', these Associations attained a fearsome influence. At Leeds, one loyalist claimed that 'sedition is crushed to death'; at Richmond outside London, they forced a liberal vicar into the public recantation of reformist views; in Somerset, one bookseller so feared retribution that he gave up his entire stock of seditious literature for burning by the local Association.[46]

There was an explicitly religious dimension to British reaction too, for the political revolution in France had unfolded alongside the 'de-Christianisation' of the country and especially Paris.[47] As early as

October 1789, the National Constituent Assembly had seized and sold the properties of the Catholic Church, whose clergy were now answerable to the French government, not the Pope. By 1793, many French churches, whether Catholic or Protestant, had become temples of the 'Cult of Reason', a civic religion which made no room for God or Christ. The Cult of the Supreme Being, which Robespierre decreed as the state religion, might then have proclaimed belief in the divinity of nature and 'the immortality of the soul', but for many in Britain it looked like the French had lapsed into infidelity.[48] Even when Napoleon restored the Catholic Church to its privileges, British suspicion endured: the third edition of the *Encyclopaedia Britannica* fumed against French philosophy for having 'disseminated, far and wide, the seeds of Anarchy and Atheism'.[49] It followed that if France was the enemy, and if French revolutionaries were deists or atheists, 'true Britons' were bound to defend the Church and the scriptures. For the British men and women who were exploring the history of the Earth, the religious implications of their work carried serious social and political risk. How could they communicate their theories and findings without provoking the wrath of either the Church or the government? In the 1810s, nobody would feel these pressures more keenly than the Anglican minister and reader in mineralogy at the University of Oxford, William Buckland.

2
Undergroundology

He, who with the pocket hammer smites the edge
Of every luckless rock or stone that stands
Before his sight, by weather-stains disguised,
Or crusted o'er with vegetation thin,
Nature's first growth, detaching by the stroke
A chip, or splinter, – to resolve his doubts;
And, with that ready answer satisfied,
Doth to the substance give some barbarous name.[1]

William Wordsworth, *The Excursion* (1814)

William Buckland, like Mary Anning, was a child of south-west England. The son of an Anglican clergyman, he was born in 1784 in Axminster, only five miles north-west of Lyme, and he spent his childhood exploring a landscape of pits and fossils, hunting ammonites with his father. He 'could not take a stroll in the neighbouring fields', a friend related, 'without stumbling . . . on lias quarries'. In this way, in learning where the treasures of the earth abounded, 'all the circumstances of his early life were calculated to impress on him that character of mind which so peculiarly qualified him' to study geology. Wherever he went, the countryside remained his laboratory. At school at Winchester, he collected fossilised sponges from the chalk of the South Downs; later, he spoke fondly of Bristol as another 'geological school' where the rocks appeared to speak to him: 'They wooed me, and caressed me, saying at every turn, "Pray, pray be a geologist!"'[2]

Buckland would not study geology formally, for nobody did.

Instead, as the senior scholar for Devonshire at Corpus Christi College, Oxford – an honour secured by his uncle, who was a fellow there – Buckland studied Classics and theology, graduating in 1805 before joining the fellowship at Corpus Christi and entering the Anglican clergy. It was an easy and well-trodden path, for in the eighteenth and early nineteenth centuries Oxford, as Gibbon had famously lamented, was an incurious place, saturated in port: the master of Balliol College, John Parsons, was not alone in succumbing to 'suppressed gout'.[3] The lack of intellectual vigour was so manifest that in 1785 the university authorities had felt the need to enact a statute 'for putting down [the] unacademical expenses' which had served to distract 'the younger members' of the colleges. Henceforth, the dons forbade their students from keeping horses or dogs, and from indulging their 'unbridled and ruinous fondness for games wherein there is a monied stake'.[4] It was not until 1800 that Oxford introduced a common system of examination.

A few enterprising scholars were kicking against the pricks, and one of them was the London-born doctor John Kidd who, in 1803, at the age of only twenty-seven, became Oxford's first Aldrichian professor of chemistry. Later described as a 'sensible, homely man' who conducted his medical rounds in unfashionable, 'obsolete' coats, Kidd was commissioned to deliver lectures in the 'Medicinal and Philosophic' aspects of chemistry; at the same time, it was implied that he should steer clear of the radicalism which had forced the Bristolian physician Thomas Beddoes from a previous chair in the subject.[5] Kidd therefore did not seek to overturn the traditional dominance of theology, mathematics, and Classics; rather, he presented chemistry 'in the same light as the supernumerary war-horses of Homer's chariots, which were destined to assist, but not to regulate, the progress of their nobler fellow-coursers'.[6] But as the author of an extensive syllabus of lectures on chemical attraction, compounds, and analysis, Kidd bridled at suggestions that Oxford science was moribund: 'The Physical and Experimental Sciences', he protested, 'are not neglected in this place'.[7] Besides chemistry, Kidd also gave voluntary lectures in mineralogy, often in the dark rooms beneath the Ashmolean Museum, and in his audience were two firm friends: the Classics student William Conybeare, and William Buckland.

Though his fellowship was tied to theology and the classics, Buckland used whatever time he could find for geological research. Confined to the British Isles by the war in Europe, he rode alone through Berkshire and Wiltshire in 1808; over the next couple of years, he inspected the granite of Dartmoor and the gravel of the Midlands. It was all good training for his horse, an old black mare who became 'so accustomed to the work that she invariably came to a full stop at a stone quarry'. Buckland also took in Lyme Regis, where the fossil-hunters attended him at breakfast. On his table were 'beefsteaks and belemnites, tea and terebratula, muffins and madrepores, toast and trilobites'; sprayed across the room were 'fossils whole and fragmentary, large and small, with rocks, earths, clays, and heaps of books and papers'.[8]

By 1814, during Napoleon's exile to Elba, Buckland and Conybeare began to consider going 'to Paris, to see Kings and Emperors, and . . . Crocodiles'. This grand tour took place in 1816, when the pair undertook 'five months of intense labour . . . in seeing every collection and professor that could be heard of'. They met Goethe at Weimar and geological pioneers at Freiburg; they descended into Hungarian goldmines, walked through Viennese museums, and scaled extinct volcanoes in northern Italy. They revelled in this expedition as if heroes of a scientific epic: Conybeare thought of himself as Don Quixote; Buckland was dubbed 'Sir Ammon Knight'. By the time that Buckland had shipped home the fossils that took his fancy, his rooms at Corpus Christi resembled a prehistoric warehouse. An academic colleague recalled climbing a narrow staircase and entering a long, corridor-like room 'filled with rocks, skulls, and bones in dire confusion'. Buckland himself reposed 'in a sort of sanctum at the end . . . in his black gown, looking like a necromancer, sitting on one rickety chair . . . and cleaning out a fossil bone from [its] matrix'.[9]

Now in his late twenties, and often seen ranging the fields of southern England with a bright blue bag in which he stored his specimens, Buckland was the coming man. By 1813 he had succeeded John Kidd as Oxford's reader in mineralogy, and he soon ascended to membership of the Geological Society of London. Established over dinner in Covent Garden in 1807, the Society was less a vehicle for disruptive science than a club for like-minded gentlemen. It followed a 'routine of breakfasts and dinners, punch and beef-steaks', and it kept the quality of its membership under review.[10] In 1809, its secretary expressed grave concern that 'we are thought to be going on too rapidly in the admission of members, and are not sufficiently discriminate'.[11] Membership of the 'JollSoc', as it was known, bestowed valuable social cachet.

Yet Buckland was not content. He was also poor, at least by the standards of gentlemanly science. 'I am Lecturer in Mineralogy', he told his father, and 'for that I receive £100 per annum', which is less than £7,000 in today's money. Besides this, Buckland received a stipend from Corpus Christi and earned a little more through informal tutoring, but it was not enough. Indeed, his relative penury was a major obstacle to social advancement and, more painfully, his prospects of marriage. Buckland knew what he needed, or at least what he wanted. 'I crave to be Lecturer in Geology', he wrote, and 'for that I ask £100 for my lectures'. If ever appointed to such a lectureship, Buckland figured that he would need another £100 to curate Oxford's geological collections and for 'exchanging specimens with foreigners of all countrys' [*sic*]. However, at a time when fiscal retrenchment and paying down the post-war debt were the essential government policies, funding for new scientific ventures was limited. This meant that Buckland needed to apply to the Prince Regent, the future George IV, for the money.[12]

This petition entailed the kind of torturous application that has become an eternal burden of academic life. First, Buckland required the support of Oxford's proctors, who met weekly 'to deliberate upon all matters relating to the preservation and liberties of the University'; if they approved, they would forward the petition to the university's chancellor, who in 1818 was the former prime minister Lord Grenville; it would then be Grenville who would plead Buckland's case to the Treasury. Before the proctors, Buckland submitted that, by establishing lectures at Oxford, and so by bringing geology under the supervision of the ancient university, it would prevent improper inquiries from undermining biblical authority: 'Geology is a Branch of Knowledge ... of so much National importance', he argued, 'and so liable to be perverted to Purposes of a tendency dangerous to the Interests of Revealed Religion, that it is ... a proper and desirable subject for a Course of Public Lectures in Oxford'. It was Buckland's good fortune that Grenville agreed, writing to the current prime minister Lord Liverpool that 'the advantages of this particular science, as forming one essential link in the great chain of Natural History', were evident. As for Buckland himself, Grenville vouched that he was 'a Person who in pursuit of that Study has already done much credit to the University and to his Country', concluding that there could not be a more deserving recipient of 'His Royal Highness's bounty'. Though Buckland soon complained that £200 was nowhere near enough, he had his money. Yet with his new position came one immediate obligation: the delivery of an inaugural lecture.[13]

William Buckland would deliver this lecture on geology – Oxford's first on what some were calling 'undergroundology' – in May 1819 at a time of social, political, and intellectual turmoil. Luddites had been ranging across the Midlands, destroying the machinery that they blamed for job losses and lower wages; the Prince Regent had survived an attempt on his life; and the Cato Street conspirators were soon plotting the assassination of the cabinet. Worst of all, the Manchester and Salford Yeomanry would run their sabres through the massive, peaceful rally for parliamentary reform at St Peter's Field, thereby committing the massacre of Peterloo. Lord Liverpool's ministry responded to this distress by pursuing a legislative agenda that was no less repressive than the government's reaction to Jacobinism in the 1790s. The Six Acts, as they were known, were among the most draconian ever placed on the statute book. Under the Seditious Meetings Act, any public meeting of more than fifty people which intended to discuss religious or political matters required the sanction of a sheriff or magistrate; the Blasphemous and Seditious Libels Act allowed the courts to sentence writers to transportation for fourteen years; and the Newspaper and Stamp Duties Act imposed heavier taxes on publications that contained either news or opinion.[14]

One of the *causes célèbres* of this fractious period was the middle-aged London bookseller William Hone, whom the authorities charged in 1817 with seditious blasphemy over three parodies of Church and state. In *John Wilkes's Catechism*, Hone wrote ten new commandments for the Tory regime and praised 'Our Lord who art in Treasury'.[15] In *The Sinecurist's Creed*, he sang 'Glory be to Old Bags [Lord Liverpool], and to Derry Down Triangle [Castlereagh]'.[16] And in *The*

Political Litany he prayed on behalf of 'all Bishops, Priests, and Deacons, that their fleshly appetites [might be] reduced'.[17] Hone was no threat to the constitution: he was described as having 'a half-sad, half-merry twinkle in his eye', always seen with 'a clutch of books under his arm', and he was desperate to provide for twelve children at home.[18]

Britain's Home Secretary had nonetheless singled out Hone for persecution, mentioning him by name when moving the parliamentary motion to suspend habeas corpus yet again. Although Hone withdrew the offending pamphlets from sale, government agents bribed his employees to 'sell' them spare copies; the same agents then planted illicit material on his premises. After his eventual arrest, the Lord Chief Justice ordered that Hone would have to pay for the privilege of seeing the charges brought against him: 'If a copy of the information were given to you,' he reasoned, 'by the same rule every person charged with a crime might claim a copy of the indictment'. On three successive days in December 1817, Hone stood trial on each of the charges. Representing himself before the court, he argued that the target of his ridicule was the government, not Christianity, and observed that many other scurrilous publications, some of them penned by government ministers, had not provoked a similar wrath. When the jury acquitted Hone of the first charge, the Lord Chief Justice took over proceedings in order to secure a conviction, but to no avail; defending both himself and the liberty of the press, Hone was acquitted of the remaining charges. When news of the verdicts reached the poet John Keats, he celebrated Hone as 'William the Conqueror, the Game Cock of Guildhall'.[19]

Despite these humiliations, the British authorities did not relent in their persecution of supposedly blasphemous thinking. Thomas Paine's *The Age of Reason* (1794–1807) was a totemic work of secular thought in which the author had sounded forth against the supposed idiocy of the holy writ. Though he scrutinised each book of the Bible, Paine had reserved some of his sharpest criticism for the scriptural accounts of Creation and the Flood. 'Take away from Genesis the belief that Moses was the author,' he wrote, 'on which only the strange belief that it is the word of God has stood, and there remains nothing of Genesis but an anonymous book of stories, fables, and traditionary or invented absurdities, or of downright lies'. He had concluded that 'the story of Eve and the serpent, and of Noah and his ark, drops to a level with the Arabian Tales, without [even] the merit of being entertaining'. For Paine, the early chapters of the Bible were a tissue of nonsense, so what faith could a reader place in their history of the Earth?[20]

The initial reaction to *The Age of Reason* on both sides of the Atlantic had been ferocious. When the London publisher Thomas Williams printed 2,000 copies in 1797, he was indicted for 'blasphemously, impiously, and profanely devising and intending to asperse, vilify, and ridicule . . . the Holy Bible'.[21] But at the peak of Britain's social unrest in the post-war years, the reprinting of *The Age of Reason* by the free-speech radical Richard Carlile provoked even greater aggression. When Carlile was tried in 1819 for blasphemy, the judge warned the jury that anarchy and bloodshed would attend the toleration of such impiety. 'In their own times,' he advised them from the bench, 'in an adjoining country [France] . . . the worship of Christ was neglected . . . They knew what took place; the bands of society were torn asunder, and a dreadful scene of anarchy, of blood, and confusion followed.'[22] Carlile did not endear himself to the court by reading out excerpts from *The Age of Reason* as part of his defence, and the jury followed its instructions: the court fined him £1,500 and sentenced him to six years in Dorchester gaol.

The perils attending anyone who dared to question the Bible were obvious, but some guardians of the faith had focused more precisely on the dangers of geology. In his *Geological Essay* of 1815, Buckland's old lecturer John Kidd had warned that scientific inquiry could lead humanity too far from the scriptures: 'All may possibly be lost', he wrote, 'by a renunciation of Revealed Religion, while no equivalent is offered by the establishment of some physical proposition'. Kidd was calm enough to concede that 'there is not necessarily and always a natural variance between Philosophy and Religion', but he urged caution upon any geologist who might contemplate 'such propositions, as may possibly admit of an interpretation subversive of the authority of the Scripture'.[23] That same year, in a furious letter to the *Monthly Magazine*, the art writer George Cumberland derided those geologists who sought 'to attain the age of the earth as a planet'. Advising them to put 'their intellects, and their time of existence here [to] better uses', Cumberland, a friend of William Blake's, protested against geological speculation: 'We want no better guide than Moses.'[24] All this provided a very serious challenge to William Buckland. How could he present geology to the Oxford hierarchy, many of whom were senior members of the Anglican clergy, at a university whose major function was to educate aspiring members of the clergy, without undermining biblical authority? How could he reconcile the growing fossil record with the account of Creation in Genesis? Here, for help, he looked to Europe, and to the flame-haired colossus of Parisian science, Georges Cuvier.

Born in 1769 in what is now eastern Burgundy, Cuvier had been regarded as a genius since childhood: 'blessed with a memory that retained every thing he saw and read,' wrote one biographer, 'when twelve years old he was as familiar with quadrupeds and birds as a first-rate naturalist'.[25] After sheltering from Robespierre's Terror in Normandy and navigating the political maze of Revolutionary Paris, Cuvier took up residence at the grand new institutions of French science: the Collège de France, the Jardin des Plantes, and the National Museum of Natural History. The last of these was Cuvier's fiefdom and its resources inspired awe in British visitors. 'I got into Cuvier's sanctum sanctorum yesterday', reported one geologist: 'In the museum is a library disposed in a suite of rooms, each containing works on one subject. There is one where there are all the works on ornithology, in another room all on ichthyology, in another osteology . . . [His] ordinary studio . . . is a longish room, comfortably furnished . . . with eleven desks to stand to . . . like a public office for so many clerks. But all is for the one man, who multiplies himself as author . . . and moves . . . as fancy inclines him, from one occupation to another.'[26] From these eleven desks, Cuvier had produced the most remarkable body of work: *Le Règne Animal* (1816) is a seminal moment in the history of zoology, while every anatomist operated in the shadow of *Recherches sur les Ossemens Fossiles* (1812–23).[27] Less edifyingly, he had concluded from an autopsy of Sarah Baartman, a woman from southern

Africa whom merchants had paraded across Europe as part of a freak show, that she, 'the Hottentot Venus', was closer anatomically to monkeys than to humans.[28]

Critically, for Buckland's purposes, Cuvier had applied his powers to species of animal which appeared to be extinct. In 1796, he had concluded that the enormous bones in his museum's collection did not belong to elephants, but to a species that he called 'mammoths'; in the same year, following the excavation of a bizarre skeleton in South America, he announced the discovery of the *Megatherium*, a massive, ground-dwelling sloth. Cuvier was also the first to describe the *Mosasaurus*, a vast marine lizard whose remains had been seized by Revolutionary soldiers during the invasion of the Netherlands: the French general at Maastricht had offered '600 bottles of best wine' for the fossil.[29] But how did the strictly Protestant Cuvier reconcile his belief in the divine creation and ordering of the animal world with the disappearance of species, and the burial of their fossils in the earth? His answer was 'catastrophism', a theory which held that incongruities in the fossil record, the apparent extinction of prehistoric creatures, and the very surface of the Earth were explicable by the occurrence of sudden, repeated cataclysms. This was not entirely new thinking: 'Neptunist' geologists had long believed that a series of floods had reshaped the globe, while the 'Vulcanists' (or 'Plutonists') contended conversely that volcanic explosions had moulded the Earth, and neither camp paid much heed to James Hutton's theory that gradual, uniform forces had worked changes over the ages. Yet when Cuvier published his *Theory of the Earth*, and when the Scottish naturalist Robert Jameson translated it into English, the reconciliation of biblical history with the fossil record appeared possible, even plausible.

First, the Frenchman stated that 'the crust of our globe has been subjected to a great and sudden revolution, the epoch of which cannot be dated much farther back than five or six thousand years ago', and readers understood this to be Noah's Flood. But second, and more importantly, Cuvier argued that the dry land which had emerged after the Flood was 'formerly inhabited *at a more remote era*' and that 'at least one *previous* revolution had submerged [the land] under the waters'. Here, Cuvier advanced the idea that the 'days' of Creation were metaphorical, and so that the Lord could have made and remade several worlds which now lay beneath the one in which he and his readers lived. In the preface to his translation, Jameson exulted that 'the six days of the Mosaic description are not [now] inconsistent with our theories of the earth'. In Jameson's view, the 'motions of the earth' at the beginning of time could have been much slower than they were in the early nineteenth century, meaning that the length of a day could have been 'indefinitely longer than it is at present'.[30] The *Proteosaurus* that the Annings had found at Lyme Regis could therefore have been the vestige of a *previous* world, but a world that the Lord had created nonetheless. Historians agree that Jameson took liberties while translating Cuvier from the French, with all his discrepancies tending towards a biblically consonant interpretation of the Frenchman's theory, but British critics received his translation in rapture. The *Perthshire Courier*, for one, declared that 'the Christian may furnish himself from this production of a Parisian philosopher, with armour to defend his faith against those writers who have endeavoured to overturn it by objections against the Mosaic account of the deluge'.[31] If the France of the Revolution had threatened Britain with its atheism, the France of the Bourbon Restoration now offered intellectual succour.

It was in this context, on 15 May 1819, that Buckland delivered his inaugural lecture on geology, and he had designed every part of it to reconcile Oxford's clerical establishment to the new science. The Latin title *Vindiciae Geologicae*, meaning 'in defence of geology', was a conciliatory nod to the university's classical traditions. The subtitle, 'the Connexion of Geology with Religion Explained', reassured his audience that nothing here would undermine scriptural authority. A deferential dedication to Lord Grenville meanwhile praised the former prime minister's 'sincere regard for the inseparable interests of science and religion' and thanked him for endorsing Buckland's efforts 'to shew that the study of geology has a tendency to confirm the evidences of natural religion'. Within the lecture itself, Buckland presented geology as a field of study that, by tracing 'the finger of an Omnipotent Architect' through natural history, could 'unite the highest attainments of abstract science and literature with the much more important purposes of Religious Truth'. Far from being a subversive and dangerous discipline, geology in fact contributed 'proofs to Natural Theology' and demonstrated that humankind had not appeared on Earth 'before that time which is assigned to it in the Mosaic writings'.[32]

Crucially, and borrowing heavily from Cuvier, Buckland observed that Moses had confined himself to 'the preparation of this globe for the reception of the human race, [and did] not deny the prior existence of another system of things'. Quoting one churchman's observation that 'we are not called upon to deny the possible existence of previous worlds', Buckland suggested that 'the word "beginning" as applied by Moses in the first verse of the Book of Genesis' could well have referred to 'an undefined period of time which was antecedent . . . to the creation of its present animal and vegetable inhabitants'. Buckland therefore proffered the idea that 'the days of the Mosaic creation are not to be strictly construed as implying the same length of time which is at present occupied by a single revolution of our globe, but PERIODS of a much longer extent'. In other words, geological conclusions would not be inconsistent with Genesis, so long as readers of Genesis allowed for a much longer 'beginning' or a much longer 'day'.[33]

In persuading his audience that geology had come not to bury religion but to praise it, Buckland had triumphed. He had secured for this new science a place of safety not only at Oxford, but also in the pages of the review journals where polite opinion formed. Yet in the very act of pleading for the reconciliation of geological theory with biblical history, Buckland had acknowledged an implicit tension between those two great fonts of authority. So how long could he and his allies sustain such liberal readings of Genesis? How long would the evidence of the earth suffer this alliance to the holy writ?

At the same time, the Anning family was welcoming ever greater numbers of fossilists and geologists to Lyme Regis. One of them was Henry De la Beche (pronounced 'Beach'), a handsome, dashing young man who had been expelled from the Royal Military College for allegedly fomenting a 'dangerous spirit of Jacobinism' among the cadets.[34] Coming to Lyme Regis with his mother, whom the scandalised locals dubbed 'Madame Trois-Maris' on account of her conjugal profligacy, the young De la Beche was a talented artist and became a provincial celebrity.[35] Lampooned in the anonymous *Lymiad* of 1818 as 'Sir Fopling Fossil', a 'most accomplished youth' who sailed the coast in the company of 'a damsel young and passing fair', he had both the time and the money – as the heir to a sugar plantation in Jamaica – to explore the Dorset landscape.[36] By 1816 he had taken membership of the Geological Society in London but often returned to Lyme, where he and Mary Anning went fossil-hunting under the dark, dangerous limestone of the Black Ven cliff.

Other events of the 1810s revealed the 'wild romance' of Anning's character.[37] In March 1815, the East Indian merchant vessel *Alexander* arrived in the English Channel, having set sail from Bombay the previous October. Carrying cotton, coffee, and sugar, in addition to 150 crew and passengers, the *Alexander* was only two miles from its destination of Portland when a strong south-westerly wind forced the ship onto a 'ridge of pebbles'. This happened at two o'clock in the morning. Within a couple of hours 'she was a complete wreck, and every soul on board . . . were consigned to a watery grave' except, as the *Salisbury and Winchester Journal* put it, 'four Malays and one Persian' who had been serving as crew. The residents of the coast woke to death and carnage: 'the whole line of coast', it was reported, 'has been strewed with the vestiges of the hull, dead bodies, &c'.[38] Among the flotsam was the body of Lady Jackson, the wife of an army major who had been stationed in India since 1806. It is unclear whether Mary was the first to find the corpse, but she vowed to attend it. Crouching on the beach, Anning took care to remove the seaweed from Jackson's hair; when the body was taken to lie in a local church, awaiting burial, she cut fresh flowers each day to place beside it.

As for many in Britain, where unemployment and depressed wages awaited the soldiers returning from the continent, the 1810s were a decade of hardship for the Annings. The money from the sale of the 'crocodile' had not lasted long and, by the spring of 1820, they were in considerable difficulty. Indeed, they were 'on the act of selling their furniture to pay the rent – in consequence of not having found one good fossil for near a twelvemonth' – when Colonel Thomas Birch came to their rescue. A retired army officer, Birch had been spending his post-war pension in pursuit of fossils in the West Country. First meeting the Annings in 1818, he had struck up a friendship with the family and with Mary especially, an amity that was not lost on the geological community: 'Col. Birch is generally at Charmouth', carped one cynic, and 'they say *Miss Anning attends him*'. Upon finding the Annings in poverty in 1820, Birch resolved to help. 'The fact is', he told a friend, 'that I am going to sell my collection for the benefit of the poor woman and her son and daughter at Lyme who have in truth found almost *all* the fine things, which have been submitted to scientific investigation'.[39]

This was no small gesture. Over the course of the past few years Birch had amassed a considerable collection of fossils and geological curios, and the thought of parting with them caused him considerable pain. 'I may never again possess what I am about to part with', he realised. But there was consolation in the cause: 'In doing it', he wrote, 'I shall have the satisfaction of knowing how the money will be applied'.[40] The auction took place at the Egyptian Museum in London and, for three days from 15 May, bidders from across the British Isles, Germany, and the Austrian Empire competed for 102 lots, among which were a 'fragment of a fish, the reverse side beautifully seen', 'a most interesting illustration of the . . . Proteo-Saurus', and 'part of the foot of a prodigiously large animal'.[41] Even the great Cuvier sent an emissary, who took a skull and a thigh-bone back to Paris. The last lot was the Annings' most recent bonanza, a complete fossil skeleton of the *Proteosaurus*. This alone went for more than £100, and the auction raised more than £400 in total, almost £30,000 in today's money. But this windfall did not make the Annings secure. Having found another crocodile in the limestone in 1821 and sold it to the British Museum, Molly Anning – the mother of the family – would plead with Charles Konig, the keeper of the Museum's fossil collection, for prompt payment. 'As I am a widow woman and my chief dependence for supporting my family being the sale of fossils,' she wrote to London, 'I hope you will not be offended by my wishing to receive the money for the last [one], as I assure you, Sir, I stand in much need of it'. Despite this sale, Birch's charity, and increasing renown – in July 1821, Henry De la Beche paid handsome tribute to 'the Annings who search for fossils here' – life remained difficult for Molly and Mary. It was not until the winter of 1823 that they made another staggering and truly profitable discovery.[42]

3
Of Caves and Paddles

Have ye heard of the woman so long underground?
Have ye heard of the woman that Buckland has found,
    With her bones of empyreal hue?
O fair ones of modern days, hang down your heads,
The antediluvians rougèd when dead,
    Only granted in lifetime to you.[1]

Philip Duncan, 'On the Woman in Paviland Cavern'

Even today, entering the Goat's Hole Cave at Paviland is a potentially deadly mission. It sits within a limestone cliff, facing the sea, high above the rugged shoreline of the Gower Peninsula in south Wales.[2] Though local farmers had long known of its existence, nobody had ventured into the cave – at least in modern times – until 1822, when a band of local antiquaries braved the jagged rocks to intrude upon an ancient burial scene. They found a cave floor covered in a 'diluvial loam of a reddish yellow', decorated not just with sea-shells but the teeth and bones of several animals. Six inches beneath the soil, they found something else: 'the entire left side of a human . . . skeleton'. Curiously, the bones were dark red, 'enveloped by a coating of a kind of ruddle'; sitting around the ribcage were forty or fifty fragments of small ivory rods.[3] The land at Paviland and therefore the cave belonged to Lady Mary Cole, who was the widowed owner of Penrice Castle, the mother of six fossilist daughters, and a frequent correspondent of William Buckland. When Cole reported the discovery to the Oxford don, she received unequivocal instructions: barricade the entrance and let nobody disturb the site. Buckland would conduct the rest of the excavation himself.

Now entering his late thirties, Buckland had established himself as one of Britain's leading geological authorities. Having taken another trip to Paris, where he dined with Cuvier and the Prussian naturalist Alexander von Humboldt, he had settled into a syllabus of Oxford lectures that was attracting a devoted audience among students and other academics. His lecturing style was a curious blend of farce and obfuscation. The physician Henry Acland recalled one lecture where Buckland 'paced like a Franciscan Preacher up and down behind a long show-case' with a fossilised skull in his hands before charging at a terrified undergraduate on the front bench, barking the question, 'What rules the world?' The answer that he demanded was 'the stomach'.[4] Explaining Buckland's eccentricity was far from straightforward. The socialite Caroline Fox thought that it stemmed from a deep-lying shyness, observing that he was 'very nervous in addressing large assemblies till he has once made them laugh'.[5] Other listeners were less charitable. Some scoffed at Buckland's 'undignified buffoonery' and 'coarse joking manner', others that he used comedy simply to prevent his audience from taking offence.[6]

In any event, when Buckland came to Paviland he confronted a mystery: whose was the skeleton? Staying with Lady Cole as he explored the cave, Buckland and his host toyed with the idea that these were the remains of an exciseman who had been murdered for policing smugglers too vigorously, the ivory rods a primitive form of currency. As for the skeleton's age, Buckland knew that British tribal armies had camped in the area, so there was 'reason to conclude that . . . these human bones [were] coeval with . . . the Roman invasion of this country'.[7] Yet by the time he got back to Oxford, Buckland had changed his mind. He wrote to Cole that the Paviland body was in fact 'a Woman, whose History wd afford ample Matter for a Romance to be entitled the Red Woman or the Red Witch of Paviland'. Abandoning the theory of the customs officer, Buckland now proposed that the burial of the woman with a 'Blade Bone of Mutton' was a sign of sorcery. A colleague reinforced this idea by referring to the 'small British encampment immediately above the cave' and by supposing that the 'enchantress may have stimulated the former inhabitants of Gower to warlike deeds'.[8] We now know that Buckland was wrong about almost everything. Since he believed that no person could have lived before 4004 BC, he could not have fathomed that the bones were palaeolithic, almost 29,000 years old. Nor was he right to talk of a 'Red Lady': the remains in fact belonged to a man. Nonetheless, Buckland and his colleagues had found the oldest human remains in Britain, a record which stands today.

Paviland, however, was far from the most important cave that Buckland explored in the early 1820s, for rumours had spread south from Yorkshire of a peculiar discovery near Kirkdale, some twenty miles west of Scarborough. There, a handful of quarrymen had come across 'the mouth of a long hole or cavern, closed externally with rubbish, and overgrown with grass and bushes'. Deep within the earth they had found the bones of dozens and dozens of animals and, thinking them only the remains of cattle which had somehow fallen into a crevasse, they had thrown them onto the roads with the common limestone. Yet when the local surgeon Thomas Harrison passed by, he recognised the bones as belonging to more exotic animals. Among the twenty-three identifiable species were the tiger, hyena, rhinoceros, elephant, and hippopotamus.[9] But just what were these bones doing in Yorkshire? For geologists whose conceptions of time and the Earth were still informed by the Bible, the Kirkdale discovery was of potentially seismic significance; for Buckland, it was 'not easy to conceive that anything short of the common calamity of a simultaneous destruction could have brought together in so small a compass so heterogeneous an assemblage of animals'. In other words, Kirkdale was potential proof that the biblical Flood had swept these animals from their natural habitats into this one small place in the north of England. Even Cuvier was impatient for intelligence, his secretary begging Buckland 'in all haste . . . to procure for him some of the bones found in such quantity in Yorkshire'.[10]

Buckland arrived in Yorkshire in December 1821 and found, in the face of the quarry, a small hole measuring some three feet by five. Getting down on his hands and knees, he crawled into the darkness. With only a candle lighting his way, Buckland emerged some hundred yards later into the cave's main chamber. In a long, narrow space, where stalagmites and stalactites studded the walls and roof, there was a floor of 'soft mud or loam, covering entirely the whole bottom [of the cave] to an average depth of about a foot'.[11] Within this mud, miraculously preserved, were more of the bones that rumour had promised. There were shin-bones, thigh-bones, shoulder blades, ribcages, skulls, and vertebrae, and for Buckland their presence was explicable by one of two causes. The first was that they had 'drifted into a fissure by the [force of the] diluvian waters', but from everything that Buckland saw, the cave was 'covered all over at the top with continuous beds of limestone', meaning there was no way that the floodwater could have carried the bones to their resting place.[12] Buckland could therefore 'only suppose the bones to be the wreck of animals that were dragged in for food by the hyenas'.[13] It was a beguiling theory that he had based on the evidence before him, in defiance of any desire to ascribe the discovery to Noah's Flood. But how could Buckland prove the hyena hypothesis? Noticing a series of consistent, repetitive toothmarks on the Kirkdale bones, he devised a plan.

First, he enlisted the celebrated explorer and naturalist William John Burchell, whose travels across and beyond the Cape Colony, not to mention his collection of more than 50,000 specimens, had rendered him the British expert on African fauna. What Buckland wanted, and what Burchell delivered to Oxford, was a Cape hyena. Soon enough, the animal 'Billy' was feeding royally on prime cuts – on the bone, of course – from Oxford's butchers. Within days, Buckland observed that Billy had 'performed admirably on shins of beef, leaving precisely those parts which are left at Kirkdale . . . and leaving splinters and scanty marks of his teeth' just like those on the bones from the cave.[14] If this were not enough, the fossilised dung that Buckland had recovered from Yorkshire matched the new 'evidence' that Billy delivered to his master, although one of Buckland's colleagues warned him that 'too frequently and too minutely examining faecal products' would do nothing for his reputation.[15] Despite this potential embarrassment, Billy had proved Buckland's theory. It was not Noah's Flood that had washed the bones to their resting place; instead, hyenas had killed their prey in the open before dragging them into the cave, where they tore at the flesh and gnawed at the bones. Buckland put his conclusions into a paper that he sent 'directly to the Royal Society', where the chemist Humphry Davy could not 'recollect a paper . . . which has created so much interest'.[16] The next year, the Royal Society would award Buckland its highest honour, the Copley Medal, for work that was ground-breaking in every sense.

Buckland wasted no time in preparing an account of his subterranean adventures – both at Kirkdale and Paviland – for wider circulation with the publisher John Murray, who in 1823 released *Reliquiae Diluvianae*, meaning 'the relics of the flood'. Given that Buckland had in fact repudiated the diluvial theory for Kirkdale, his title was misleading, but the public did not care: the book was immensely successful. Writing to the Archbishop of York, Buckland trumpeted 'the rapid sale my book has had. Not a copy has been left for some time. Mr Murray', he boasted, 'is very busy in bringing out a second edition of one thousand copies more'. Within a month, Murray had sold 400 copies of the new edition and expected the whole thing to be 'out of print'.[17] There was critical praise too. The *Quarterly Review*, arguably the most influential journal in the English language, recognised Buckland as 'brilliant beyond example' and 'the first' to establish 'a distinct and detailed view of a *state of animal life* previous to the deluge'.[18] Buckland's fame soon crossed Europe: Cuvier worked the hyena theory into the next volume of his monumental *Recherches*, while Alexander I of Russia rewarded the Oxford man with 'a gold snuff-box set with mosaic'. Embracing his unofficial role as geology's new champion, Buckland could not help but think himself 'very successful'.[19]

Rare among this early generation of British fossil-hunters, William Conybeare was a child of the city. The son of a wealthy Anglican clergyman, he spent the most part of his youth 'in the old rectorial house of Bishopsgate . . . in a most ghoulish atmosphere'. It was only at Bexley, at the time a small village to the south-east of London, that he could enjoy 'three summer months of emancipation' in the fresh air. There, at the turn of the nineteenth century, Conybeare wandered the fields and heaths, discovering 'shafts sunk about sixty or seventy feet' through the sandy soil, and networks of caves in the chalk that lay beneath. Though he later learned they were in fact fossilised shells, he also delighted in collecting 'miniature unicorn's horns'.[20] Study of the classics at Westminster and Oxford would not prevent scientific pursuits from occupying Conybeare's thoughts and, by the 1810s, thanks to the £500 that his grandmother allowed him each year, he could devote much of his time to geology. Admission to the Geological Society came in 1811 and, five years later, he joined Buckland in touring Europe. Regarded by his peers as the most gifted theorist among them, Conybeare was Buckland's intellectual touchstone, especially during the preparation of the Oxford lecture; with Buckland reproducing large parts of Conybeare's advice verbatim, it is thought that less than one-third of *Vindiciae Geologicae* was original.

Upon taking up a clerical lectureship in Bristol – he was yet another ordained geologist – Conybeare helped to establish that city's Literary and Philosophical Institution, but he did not make a major contribution to science until 1821, when he and Henry De la Beche presented a paper on the 'Lyme crocodile' to the Geological Society. As we have seen, Sir Everard Home had proposed the name *Proteosaurus* for the creature found by the Annings in 1811–12, but Home's star was waning. For one thing, Georges Cuvier thought him an idiot: in correspondence with British colleagues, the French titan's secretary castigated Home's 'last ridiculous paper' on the Lyme creature; among the other 'insignificant memoirs' that Home had foisted on the scientific community was 'a most stupid one of the [sea cow known as the] Dugong'.[21] More damningly, there was growing suspicion about Home's integrity. In 1800, the British government had purchased the unpublished manuscripts of the distinguished anatomist John Hunter and donated them to the Royal College of Surgeons, where Home – who was also Hunter's brother-in-law – was president. Over the next two decades, Home delivered more than a hundred anatomical papers to the Royal Society and, for those who listened carefully, it was increasingly clear that he had not just taken inspiration from the Hunterian manuscripts, but was passing them off as his own work. In time, Home would be damned by the cover-up: he set his house on fire while trying to burn the evidence of plagiarism, then protested that he had destroyed Hunter's work only because it was 'unfit for the public eye'.[22] Later hearings before a parliamentary select committee exposed 'irrefutable evidences' of 'the piracy' that Home had committed in pursuit of 'imperishable fame'.[23]

All this meant that, in the spring of 1821, Conybeare and De la Beche were quite ready to contradict Home and to mount their own case about the nature of – and the appropriate name for – the Lyme crocodile. They recognised that the creature had commanded 'a considerable share of attention among the scientific public', much of which was down to the striking illustrations which had accompanied Home's papers before the Royal Society. Yet they could not accept the name of *Proteosaurus*: 'On a full and careful review of its whole structure', they reported, 'it will not be found to possess analogies sufficiently numerous or strong with the peculiar organisation of the Proteus'. Instead, they described in detail 'a marine quadruped, nearly resembling the crocodile in the osteology of its head and its mode of dentition', but 'much more analogous to . . . fishes' in respect of its backbone. Consequently, they resurrected a name coined by Charles Konig of the British Museum; henceforth, the Lyme creature would be known as the *Ichthyosaurus*, the 'fish-lizard'.[24] It was not, however, the only ancient creature that Conybeare would christen.

In its Christmas Eve edition of 1823, the *Taunton Courier and Western Advertiser* carried an extraordinary variety of headlines. There were reports from the Greek War of Independence, of French diplomatic mischief in Latin America, and columns on the mysteries of sleepwalking; the glad tidings that a leopardess in an Exeter menagerie had 'whelped three remarkably fine cubs' meanwhile compensated for the sad news from London that a violent 'storm of wind, rain, and hail' had caused a Mrs Cooke of Pentonville to expire 'from excessive fright'. There had also been a geological bonanza at Lyme Regis. On the Tuesday night of the previous week, the retired naval captain Henry Waring had 'removed from a slaty part of the blue lias ledges . . . a fine portion of organic remains'. The *Courier* affirmed that 'the part which has been cleared is particularly beautiful; and there is no doubt of this specimen proving to be that of some rare antediluvian animal'. On the next night, 17 December, Mary Anning made a much more significant discovery. 'East of the town, and immediately under the celebrated Black Ven', Anning worked through the night and into the morning to excavate a specimen which appeared 'to differ widely from any which have been before discovered'. According to the

Courier, the new creature 'approache[d] nearly to the structure of the *Turtle*'. It was massive, too: from tail to snout it was more than nine feet long; across its front, from paddle to paddle, it measured another five.[25]

Intelligence of the new creature went first to Buckland, who wrote in turn to Conybeare at Bristol. The latter had long suspected that another ancient animal, besides the *Ichthyosaurus*, had lurked in the western limestone. In 1821, he had even alluded to 'a new Fossil Animal forming a link between the Ichthyosaurus and Crocodile ... in the vicinity of Bristol', coining the name 'plesio-saur', meaning 'near-lizard', for the elusive beast.[26] He refrained at the time from any lengthy exposition because there endured 'reasonable ground for suspicion that ... [he] had been led to constitute a fictitious animal from the juxtaposition of incongruous members', but when Buckland sent him 'a very fair drawing by Miss Anning of the most magnificent specimen', all doubt evaporated.[27] The illustration so excited Conybeare that he could not settle to compose his sermon for the coming Sunday; as he related to De la Beche, who had quit Lyme to attend to his family's plantation in Jamaica, he left it half-written, to be finished by one of his sisters-in-law, and rushed to Bristol's Philosophical Institution to break the news from Lyme. 'Such a communication could not fail to excite great interest', he noted: 'I did not get home till midnight.'[28]

The Annings had no trouble in finding a buyer for their new fossil. Richard Temple, the Marquess of Chandos and later Duke of Buckingham, was a notorious spendthrift and paid £200 for it. Yet as with the *Ichthyosaurus*, the *Plesiosaurus* was not delivered to its owner – which in this case would have added to the riches of Stowe House – but to the Geological Society, which had laid claim to its examination. Getting the fossil from Lyme to London would prove an operation of considerable complexity. Because the Annings had encased the skeleton in plaster before setting it within a wooden frame, ten feet tall by six feet wide, it was too heavy and too fragile to move by road. Workmen therefore hauled the *Plesiosaurus* onto the Cobb, then onto a cargo ship bound for the capital. Waiting on the London dockside was Conybeare, who had been charged with conveying the fossil securely to the Society's premises lest it fall into the hands of Everard

Home. As the new wonder came by water, Conybeare grew impatient. 'When I came to town, I found the specimen delayed in the Channel', he complained; 'Nor did it arrive for ten days afterwards.' Further problems arose when the fossil finally made it to the Society's premises: it was simply too large to go upstairs. Conybeare regretted that 'after wasting a day in vainly attempting to move it . . . we were constrained to unpack it in the entrance passage'.[29]

Comically marooned in the hallway, the *Plesiosaurus* attracted gaggles of geologists who crowded round it in the gas-lit gloom of the London winter. George Cumberland, who had once declared that the science needed 'no better guide than Moses', dispatched an agent to inspect the new fossil with the instruction: 'Tell me your opinion of its being one Fish or not.' Charles Konig, the Brunswick-born naturalist, described the creature in awe: 'This really is an extremely curious object', he wrote to a colleague. 'It is no doubt perfectly genuine, in spite of the apparent disproportion of the parts, especially the neck and head.' This was true: the *Plesiosaurus* had 'the smallest brain-box known in proportion to the bulk of [other] animals' and, as a result, Konig thought the animal likely 'to have been none of the wisest'. At the same time, Conybeare rejoiced that his suspicions of a second, different creature had been correct: 'The magnificent specimen at Lyme has confirmed the justice of my former conclusions.'[30] But not everyone was convinced, not least Georges Cuvier. Part of this was due to the extraordinary nature of the new fossil: it had thirty-five vertebrae in its neck, which was far more than in any living reptile. Cuvier had also grown suspicious of British science generally, due in part to Home's mangling of the *Ichthyosaurus* and in part to the fabrications of the fossilist Thomas Hawkins, who was 'such an enthusiast that he [made] things as he imagines they ought to be, and not as they really are found'.[31]

This was the scepticism that Conybeare had to overcome when, on 20 February 1824, Britain's geologists met in their rooms off Regent Street to hear him 'lecture on [his] M[on]st[e]r'.[32] Listening to Conybeare was no great pleasure: his credentials were unimpeachable, and his *Outlines of the Geology of England and Wales* (1822) was a celebrated manual for the science, but his speaking voice was stilted, ungraceful, and oddly high-pitched.[33] Still, the Geological Society

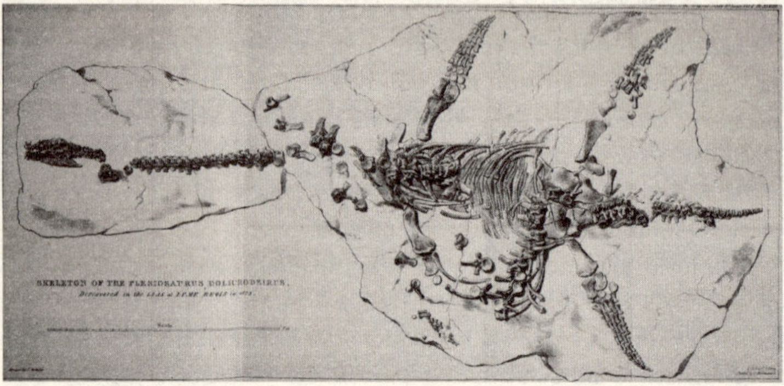

The fossilised skeleton of the *Plesiosaurus*, discovered by Mary Anning and announced by William Conybeare

listened in rapture that night as he described 'one of the most curious and important additions which geology has yet made to comparative anatomy'. There was the head, remarkably small, which rendered the *Plesiosaurus* 'a very unequal combatant' against its neighbours; there was the neck, longer than 'that of the longest-necked birds, even the swan, [which] deviates from the laws which heretofore were regarded as universal in quadrupedal animals'; and at the extremities of the creature's limbs were great paddles which propelled it through the waters.[34] Memorably described as 'a serpent threaded through the shell of a turtle', the *Plesiosaurus* later leapt from the rocks of Lyme Regis into the fiction of Jules Verne and Arthur Conan Doyle, and into the popular imagination as the Loch Ness monster.[35] With good reason would Conybeare reflect on that evening as being 'one of the pleasantest public meetings I have ever attended'.[36]

Mary Anning had now discovered two great prehistoric beasts, the *Ichthyosaurus* and the *Plesiosaurus*, and it does not stretch credulity to describe Lyme Regis as the geological frontier of the early nineteenth century. Yet on that auspicious night in February 1824, immediately after Conybeare had finished speaking, William Buckland rose to announce the discovery of a third ancient animal; in doing so, he would solve a mystery that had been puzzling Oxford scholars for more than 150 years.

4
A Bone of Prodigious Bigness

So Cuvier says;–and then shall come again
  Unto the new creation, rising out
From our old crash, some mystic, ancient strain
  Of things destroy'd and left in airy doubt:
Like to the notions we now entertain
  Of Titans, giants, fellows of about
Some hundred feet in height, not to say miles,
And mammoths, and your winged crocodiles.[1]

Byron, *Don Juan*, Canto IX (1823)

In the late 1670s, when the alchemist, antiquarian, and Oxford alumnus Elias Ashmole donated a cabinet of curiosities to his old university, it took twelve full wagons to transport his collection of 'beasts, fowle, fishes, serpents, wormes ... pretious stones and other ... Coines, shells, [and] fethers ... of sundrey Nations' from London to its new home on Broad Street.[2] The Ashmolean was one of the first public museums in the world, and the Oxford authorities knew exactly who should serve as its keeper: Robert Plot was a celebrated naturalist, a sound 'tory and establishment man', and besides commencing a series of lectures on 'philosophical natural history' he had just published *The Natural History of Oxfordshire* (1677).[3] In this work, the 'learned Dr Plot' had explored the landscape, rocks, and minerals of his home county, and its most intriguing passage addressed a huge bone from a quarry in the parish of Cornwell in the north-west of the county. This bone, which Plot illustrated in fine detail, had 'the figure of the lowermost part of the *thigh-bone* of a Man, or at least of some other *Animal*'. But there was a problem: it was simply too large to be credible. 'It will be hard to find an *Animal* proportionable to it', Plot wrote, with 'both *Horses* and *Oxen* falling much short of it'. He even considered whether it belonged to an elephant that the Romans might have brought to Britain, but he found no evidence for this in the Latin scrolls of antiquity. Plot was at such a loss to explain the presence of such a specimen in Oxfordshire that it appeared in his index merely as a 'Thigh-bone of a prodigious bigness'.[4]

It was not for another eighty years that the Cornwell bone was reconsidered, when the physician Richard Brookes published the fifth volume of his *System of Natural History*. This book focused on the 'waters, earths, stones, fossils, and minerals' that Brookes, of whom little else is known, had encountered on his travels in Britain, America, and Africa. In a plate section which also featured 'a stone ... which represents the leg and foot of a man cut off above the ankle' and 'other stones which exactly resemble the heart of a man', Brookes faithfully reproduced Plot's engraving of the Cornwell bone, to which he appended the label '*scrotum humanum*'. If there was any doubt about what Brookes meant, he had observed on the previous page

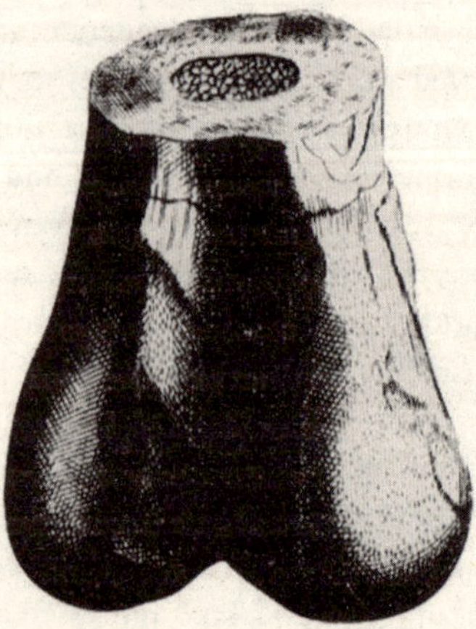

The thigh-bone that Richard Brookes mistook and misnamed as the '*scrotum humanum*'

that 'stones have been found exactly representing the private parts of a man'.[5]

This could have been the inglorious height of the Cornwell bone's role in natural history, were it not for a spate of initially unconnected discoveries being made nearby. Only ten miles or so from Cornwell, the parish of Stonesfield was 'hilly & verdant' but with 'soil [that was] entirely rocky', and its quarries had long provided slate for local roofs. There, in the 1690s, Edward Lhuyd, who had succeeded Robert Plot at the Ashmolean, found the fossilised tooth of a carnivore; in the 1750s, a customs official excavated 'a fossile thigh-bone of a large animal'; and in 1797 an Oxford don acquired a 'Large jaw bone with two serrated teeth in calc. schistus' for the Anatomy School at Christ Church.[6] It was not until the 1810s, however, when 'Fossil Vetebra [*sic*] and [a] Rib from the Stonesfield slate' made their way into the collection of the Geological Society that the wise men of Oxford began to wonder if the Cornwell bone and the Stonesfield relics could belong to the same species.[7] But if the bones did not obviously belong to any existing animal, just what had they found?

The answer, or at least the spur to it, came from Georges Cuvier, who in 1818 paid British geology the great honour of visiting London and Oxford, where he declared that the Stonesfield assemblage was reptilian: 'I saw a very great specimen', he informed a colleague in Munich, 'a thigh-bone, remarkably similar to that of a crocodile, but more than three times larger' than expected. It was a fact which 'heralded a truly gigantic animal'.[8] Emboldened by Cuvier's confidence, British geologists now spoke more openly of the 'Stonesfield lizard'. In their 1821 paper on the *Ichthyosaurus*, for instance, De la Beche and Conybeare made passing reference to 'an immense Saurian animal, approaching to the character of the Monitor [lizards], but which . . . cannot have been less than forty feet long'.[9] The next year, James Parkinson expressed his anxious hope that 'a description may shortly be given to the public' from the drawings that lay 'in the Museum of Oxford', and even suggested a name for the new beast: *Megalosaurus*, meaning 'great lizard'.[10] Notwithstanding these glimpses of the new creature, the honour of announcing Oxford's own reptile – and of solving one of the university's oldest riddles – fell to William Buckland. It would take him a while, given the distraction of the caves at

Paviland and Kirkdale, but with a typical sense of showmanship Buckland resolved to describe the *Megalosaurus* at the February 1824 meeting of the Geological Society. Of course, this was the same night that Conybeare planned to describe the *Plesiosaurus*, and Buckland well knew that adding the *Megalosaurus* to the agenda would render the occasion a seminal moment in the history of the science; not coincidentally, it was also Buckland's first meeting as president of the Society.

When Conybeare had finished his paper, Buckland laid before his colleagues the 'representations of parts of the skeleton of an enormous fossil animal, found at Stonesfield near Woodstock, about twelve miles to the N.W. of Oxford'. He explained that the bones had been found in a quarry but not, 'as many supposed', in a fissure or cavity or conveniently 'in a superficial and merely local deposit'. Rather, Buckland described how the fossils came to light through the painful, dangerous labour of the Oxfordshire quarrymen who every day 'descend[ed] by vertical shafts through a solid rock of cornbrash and stratified clay, more than 40 feet thick, to the slaty stratum' which had yielded this prize. He took his colleagues through a variety of bones, including vertebrae and jaws adorned with sharp teeth, but the key to his conclusions was a thigh-bone that was 'two feet nine inches long'. Deploying Cuvier's principle of correlation, which held that the

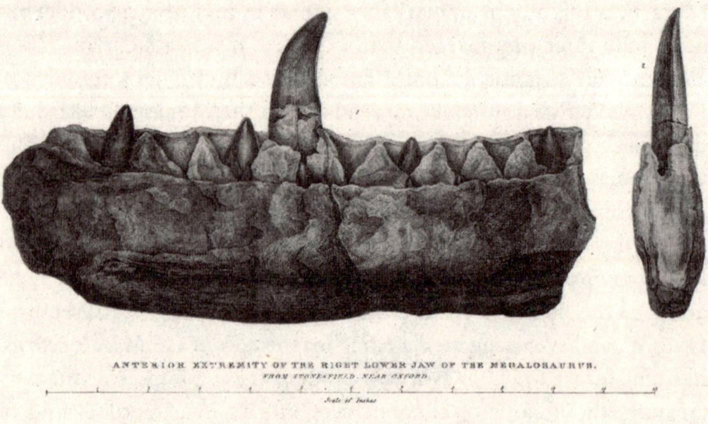

The jawbone of the *Megalosaurus* that Buckland announced in February 1824

size of each bone was related proportionally to every other, Buckland speculated that an adult *Megalosaurus* could be sixty feet from tip to tail. It would have equalled in height 'our largest elephants, and in length fallen but little short of the largest whales'.[11] In later lectures, Buckland allowed himself even greater licence: 'During this period of monsters,' he told his students, 'there [were] in the neighbourhood of what is now the lake of Blenheim huge lizards, their jaws like crocodiles, their bodies as big as elephants and their tails as long and as large as the steeple of [the church at] Kidlington'.[12]

Not everything in Buckland's 1824 paper would hold good. For one thing, research has shown that the *Megalosaurus* was probably thirty feet long, only half of what Buckland supposed. For another, he was wrong to suggest that it was 'probably an amphibious animal' and that further fossils would be found among 'the scales and teeth of crocodiles and . . . tortoises';[13] in fact, the *Megalosaurus* was a land animal, a bipedal predator which inhabited Europe and northern Africa some 160 million years ago. Nonetheless, on that late winter's evening in February 1824, Buckland achieved a first in human history: he had described a dinosaur, which modern science defines broadly as a land-dwelling reptile whose hind legs descend from the hips. One can well imagine Buckland beaming as he reported on the meeting to his friend William Vernon Harcourt, a canon of York cathedral: 'We had a great meeting in Bedford Street on Friday last,' he wrote, 'the largest I ever remember . . . With two monsters of such a kind, and so crowded an audience, my first evening of taking the chair as President was one of great éclat.'[14] But what Buckland did not mention to Harcourt, and what only added to the spectacle of that evening, was that a young, ambitious geologist from Sussex had stood up to say that he, too, had found another ancient animal.

Gideon Mantell was born in the county town of Lewes in 1790. Like Mary Anning, he was the child of a Dissenting artisan but, unlike Richard Anning, Mantell's father was successful, employing more than twenty men in his shoemaking business. This money meant that Mantell enjoyed a varied and thorough education at a 'Classical and Commercial Academy'. He learned arithmetic, geography, and history, and his public recitations – for Mantell's teacher also schooled his students in rhetoric – included an account of Hector parting from Andromache before the duel with Achilles. After a short spell in Swindon, where he embraced a teenage romance with a 'Miss Strange', Mantell returned to Lewes to take up a medical apprenticeship. With his father paying two instalments of £50 to the local surgeon James Moore, who had published a tract on the medicinal applications of steel, Mantell began five years of intensive training. Besides accompanying Moore on his daily rounds, he cleaned and swept the surgery, delivered medicines to patients, collected empty vials, and answered the bell at night while his master slept.[15]

By his early twenties, while the Annings were uncovering the *Ichthyosaurus*, Mantell had taken an interest in geology: he was corresponding with the antiquarian cleric James Douglas about 'the last great revolution of our planet', and he once took all three volumes of James Parkinson's *Organic Remains* as payment from a patient.[16] The young Mantell then had the honour of meeting Parkinson, who 'warmly encouraged [his] attempts to elucidate the nature of the strata ... of [his] county, Sussex'.[17] It was a watershed moment: 'Having at an early period imbibed a predilection for the study of natural history,' Mantell recalled, 'I resolved to devote my leisure moments to the investigation of the Organic Remains of a former World'.[18] The Sussex countryside was ideal for these explorations, for beneath the fields of wheat, barley, and cattle that surrounded towns such as Lewes lay the Weald, a swathe of sandstone, clay, and limestone which sat between the chalk of the North and South Downs. These were the sedimentary formations of rock that, in layer upon layer, had trapped and preserved the remains of prehistoric life.

All the while, Mantell's medical career was flourishing. Upon James Moore's retirement in 1818, he took over part of the High Street practice and, by the next summer, he was prosperous enough to buy the surgery outright for £725 (more than £70,000 today) and pay another £600 for the adjoining house. The result was a grand residence of four storeys with Ionic columns, a false balcony, and space for several domestic servants; he even decorated the house's façade with an ammonite motif. Now in the prime of life, Mantell was – in the description of a fellow surgeon – 'tall, graciously graceful, and flexible, a naturalist [who] realis[ed] his own lordship of creation'.[19] He had worked hard for these rewards. Waking most mornings at three o'clock, Mantell had a daily round which could take in as many as sixty patients along a twelve-mile route through the Sussex countryside. Those who could pay to see him suffered from the typical infirmities of a rural community. In 1819, he amputated the shattered fingers of a blacksmith's apprentice, strove to contain an outbreak of typhus, and struggled to save 'a man . . . runover by a Waggon'. It was difficult and exhausting: 'I am fatigued almost to death', he noted in his journal.[20]

Although Mantell made sure to examine the land he covered on his medical rounds, it was only in spare moments at night or at weekends that he could focus on geology. To prevent other fossilists from stealing a march, and to lay claim to whichever specimens emerged from the earth, he 'made some . . . arrangements' with Mr Leney, the foreman of the limestone quarry at nearby Cuckfield. Gradually, he built up an impressive collection of artefacts and geological curios. There were silver coins minted by Anglo-Saxon kings, green slabs of serpentine crystal, and brass that dated back to the Roman Empire. The deal with Leney then paid off handsomely when Mantell received 'a packet of fossils from Cuckfield, among them . . . a fine fragment of an enormous bone, several vertebrae, and some teeth of the Proteosaurus'. He exhibited this collection in the drawing room of his home in Lewes, and it soon attracted visitors from across the south of England: in July 1819 'an unknown gentleman from Brighton' spent several hours perusing Mantell's museum, while the young lawyer and geologist Charles Lyell became a frequent visitor and firm friend. Mantell even received the former Home Secretary, the Earl of Chichester.[21]

Despite this local reputation, Mantell did not fulfil his ambition of publishing a geological account of his neighbourhood until 1822. With the author spending heavily on 'maps from the lithographer' and 'copper-plate engraving', *The Fossils of the South Downs* promised much.[22] The list of subscribers, whose upfront payment allowed Mantell to finance the printing, indicated that the great and the good had seen merit in the work: there was Buckland, Conybeare, Mantell's old hero James Parkinson, and even George IV, whose patronage had been secured by the Cornish geologist Davies Gilbert. In the introduction, Mantell wrote evocatively about fossils that were 'no longer regarded merely as subjects of natural history, but as memorials of revolutions which have swept over the face of the earth'.[23] The book itself, however, was a disappointment. Comprising a succession of often-dull descriptions of the ancient ephemera of Sussex, it was replete with lists, notes, and question marks, all reading like marginalia. And though Mantell remarked that the work 'appears to have been well received', and that reviewers 'speak of it in terms of commendation', the commercial reception of *The Fossils* was muted: neither profit nor fame attended the author. On New Year's Eve 1822, Mantell lamented that 'the publication of my work on the Geology of Sussex . . . has not yet procured me that introduction to the first circles in this neighbourhood which I had been led to expect it would have done'. Bitterly, he blamed his failure on the 'humble situation' of his father, something that he felt would always count against him 'in the minds of the great'. The next day, his Christian faith unshaken either by his geology or his failure, Mantell made his new year's resolution: 'to make every possible effort to obtain that rank in society, to which I feel I am entitled both by my education, and by my profession'.[24]

It might be tempting to deride Mantell's concerns as those of a social climber, of a man who regarded science – and especially the fashionable science of geology – merely as a means of advancing himself. And perhaps there is some merit in this interpretation: his journal is pocked with entries about 'appear[ing] before the world as an author', boasts on the 'respectable' nature of his subscribers, and the 'éclat' that *The Fossils* should have given him.[25] Yet Mantell could not help but consider the connotations of class. Most geologists were gentlemen who could afford the time and money required to investigate the earth: Buckland had his college fellowship and university lectureship; Conybeare came from a wealthy family; even James Hutton, back in the 1790s, had enjoyed the rents from his farms as he explored the Scottish landscape.

Conversely, there was the example of William Smith.[26] A brawny, self-educated civil engineer who spoke with a strong West Country burr, Smith had spent decades developing canals and mines in Somerset and the bordering counties. In doing so, he had observed – often literally at the coalface – that the earth comprised successive, distinct layers or 'strata' of rock, that strata always appeared in the same order, and that younger strata were on top of the older. Smith also recognised that the same minerals and fossils appeared in the same strata, and by articulating these principles he effectively founded the geological sub-field of stratigraphy; indeed, he became known as 'Strata Smith'. In these peak years of the Industrial Revolution, it was a potentially lucrative science: knowing that certain minerals appeared in certain layers of rock meant that landowners could work out whether to establish quarries or collieries on their estates.

Yet Smith, despite publishing several astonishing, innovative, and truly beautiful maps of the rocks which underlay the British Isles, remained poor and uncelebrated: 'Mr Smith, at no time in his life', reflected his nephew, 'abounded in money'.[27] He was even jailed for his debts. The Geological Society, that pillar of polite civility, had played no small part in this misfortune: its members recognised the brilliance of Smith's work but, instead of welcoming him into their fold, they shunned him, marginalised his allies, and plagiarised his ideas, publishing their own maps in rivalry. Though Smith was honoured in time, he was for most of his life an outsider in a world of 'gentlemanly' science.[28] This was the fate that Gideon Mantell, the son of a shoemaker, feared above all else; but it would take him only a few years more to lay many such fears to rest.

Curiously, for someone who kept extensive diaries, there is no first-hand description of the discovery that secured Mantell's fame. Tradition holds that in the spring of 1822 Mantell was visiting a patient near Tilgate Forest in the company of his wife, Mary Ann, who was left to wait outside. They had been married for six years, blessed with three children, but it was not a happy marriage. On New Year's Eve 1820, Mantell had complained that the 'domestic events' of the past twelve months had presented 'but little interest or variety' despite the birth of a son; exactly one year later, he rued that he would not 'in all probability ... realize ... moments of leisure and happiness'.[29] Even so, Mary Mantell, a talented artist, was indispensable to her husband's ambitions: he might have derided his wife as 'a lady but little skilled in the art', but her illustrations of bones, rocks, and shells had improved *The Fossils* considerably.[30] Mrs Mantell also attended her husband in the field. As he examined the hills and swales surrounding Lewes and stopped their carriage to inspect the land beneath them, she waited. In the spring of 1822, as her husband tended to a patient, Mary Ann Mantell was waiting once again when she spied several walnut-sized objects 'in the coarse conglomerate of the [Tilgate] Forest'.[31] Historians now debate which of the Mantells saw them first but, either way, they had found what appeared to be fossilised teeth 'wholly unlike any that had previously come under [Mantell's] observation'.[32] Even the local quarrymen, who spent their working lives among the rocks, had seen nothing of the sort.

So where did they come from? Mantell did not think they belonged to a crocodile, an *Ichthyosaurus*, or a *Plesiosaurus* because they lacked the sharp, pointed crown that distinguished reptilian teeth; rather, they had broad, flat surfaces which suggested grinding and chewing, and it was assumed knowledge that reptiles simply were not 'capable of masticating their food'. But if the teeth, as they seemed to be, were those of a large mammal, why had they been found in 'such ancient strata' of sandstone which had only ever yielded reptilian remains? In a quandary, Mantell consulted the experts. He showed the teeth to Buckland, Conybeare, and William Clift of the Royal College of Surgeons, but they were nonplussed, declaring the teeth to be of 'no particular interest'. Buckland was notably disdainful, being of 'little doubt [that] the teeth belonged either to some large fish' or 'were mammalian teeth obtained from a diluvian deposit',[33] and therefore sure that nothing found by Mantell could trump his own discoveries at Kirkdale. Nor could Mantell persuade the great Cuvier of the teeth's importance. Having sent a specimen to France with Charles Lyell, who was attending one of Cuvier's soirées, Mantell received the disappointing news that the Parisian genius had looked at it once, 'promptly dismissed it as the upper incisor of a fossil rhinoceros, and went on with his party'.[34]

But Mantell did not give up. By 'stimulating the diligent search of [local] workmen [with] suitable rewards', he acquired more and better specimens, some with 'serrated edges [and] longitudinal ridges'.[35] Soon he had 'a graded series of deeply worn to nearly unworn teeth', and he was sufficiently confident to speak up at the February 1824 meeting of the Geological Society, an intervention which prompted Buckland to reconsider things and travel to Lewes 'to inspect [the] Tilgate fossils'. Buckland was impressed with Mantell's wider collection – so much so that the Geological Society had to warn him against plagiarising the Sussex surgeon's work – but his doubts about the Tilgate teeth endured. It was not until that summer, when Lyell took another pack of teeth to Paris, that Cuvier and the wider community agreed upon the novelty of Mantell's fossils. 'These teeth are certainly unknown to me', wrote Cuvier: 'They are not from a carnivore, and yet I believe they belong . . . to the order of reptiles . . . Have we not, here, a new animal, a herbivorous reptile?'[36] Mantell was ecstatic. Having received the blessing of the great man from Paris, he could raise a defence against Buckland's scepticism; vindicated, he spent the next six months preparing an exposition of the Tilgate reptile for the Royal Society.

Because he was not a fellow of the Society, Mantell could not read the paper himself, but he was in attendance – invited as a 'stranger' – for its delivery by Davies Gilbert. The teeth in question, he wrote, jibing at Buckland, 'possessed characters so remarkable, that the most superficial observer would have been struck with their appearance, as indicating something novel and interesting'.[37] Upon recounting his exchanges with Cuvier, Mantell then revealed his debt to Samuel Stutchbury, a conservator of the Hunterian Museum at the Royal College of Surgeons. Mantell had been rummaging through the Museum's collection in pursuit of a comparator for the Tilgate teeth when Stutchbury noted their similarity to a specimen that had just arrived from Barbados. That animal was the iguana and, when William Conybeare persuaded Mantell to cite the creature's teeth in its name, he settled on '*Iguanodon*'.

Within months of the paper's delivery, Mantell had received an honorary membership of a French learned society, secured election to the governing council of the Geological Society, and even been cited approvingly in the next volume of Cuvier's *Recherches*. Most importantly, for a man so concerned with social cachet, he became a fellow of the Royal Society that December. On Mantell's return to London for the formal ceremony, the mathematician Charles Babbage and the astronomer William Herschel introduced him as a 'Gentleman well skilled in science and particularly in geology, and well known for his remarkable paper on the iguanodon'.[38] Mantell gloried in placing his name 'in the Charter Book which contained that of Sir Isaac Newton and so many eminent characters' and, on New Year's Eve 1825, he reflected on 'the close of a year that has been so fruitful'. He had 'advanced [his] literary reputation', joined prestigious societies, and expanded his medical practice 'considerably'; he was more than just a shoemaker's son.[39]

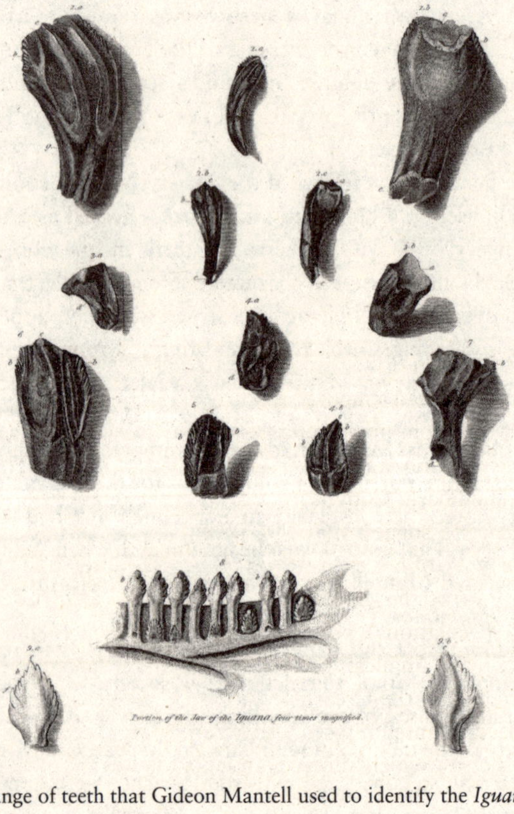

The range of teeth that Gideon Mantell used to identify the *Iguanodon*

5
Free the Science from Moses

> We cannot wonder that geologists have been suspected of a love for the marvellous . . . The Pterodactyls . . . or flying lizards . . . recall still more forcibly to our recollection the winged dragons of fabulous legends . . . The size of some of them, their long jaws armed with sharp teeth, and the hooked nails of their claws, would render them truly terrific were they to revisit Christendom, now no longer under the shield of the Seven Champions.[1]
>
> *Quarterly Review* (1826)

The first generation of British geologists had been devout Christians, all of them. None had sought to disturb Christian belief in the biblical account of Creation, nor to undermine the 'truth' of the scriptures.[2] Reconciling religion and geology was almost William Buckland's intellectual hallmark; Conybeare was an ordained minister; and there is no evidence to suggest that Mary Anning ever suffered a crisis of faith in the wake of her excavations. Gideon Mantell had meanwhile declaimed against the 'idle clamours that have been raised against geological speculations', prefacing *The Fossils of the South Downs* with a tract by an Anglican priest on 'the correspondence between the Mosaic account of the Creation, and the geological structure of the earth'. Indeed, Mantell refused to 'render any apology for [the] insertion' of this essay, which allowed 'an indefinite period of time' for the first three days of Creation; later, he honoured the 'wise provisions of the Supreme Cause'.[3] But for some readers, anxious about the health of the Church and religion itself, always on guard against unbelief, these professions of piety were not enough.

George Bugg was an Anglican clergyman who had 'lived nearly forty years under the full and firm belief that the Scriptures are strictly and literally true'.[4] Despite this zeal, or perhaps because of it, he had been sacked from three curacies in the space of fifteen years and so, living in Leicestershire in the 1820s, had plenty of time to read the works of Cuvier and Buckland with consternation.[5] Constructing his riposte 'under the full conviction that the *Mosaic narrative* was LITERALLY CORRECT', Bugg damned geological theory as not only 'alarmingly mischievous' but also as 'contradictory . . . to every known operation of nature, and every dictate of the rational understanding'. Dedicating the two volumes of his *Scriptural Geology* (1826–27) to the 'divinely appointed . . . patrons and guardians of revealed truth, and of legitimate philosophy', Bugg explained over 700 pages that the Lord had given the Bible to mankind 'to teach us what nothing else can teach us – the *time* and *manner* of the world's creation'. Moreover, Bugg believed that the same holy texts had offered more than 'enough respecting the *cause* and *operations* of the *Deluge* to prove the real *ground* and *principle* upon which we account for the actually existing state of [geological] strata'.[6] As one historian has noted, there was more than a little of the anti-Jacobin conspiracy theorist to Bugg, who represented 'Geologists', often in excited italics, as 'a secret society dedicated to the overthrow of the established order'.[7]

Another prominent scriptural geologist was Granville Penn. A classical scholar, civil servant, and grandson of the founder of Pennsylvania, he used the second edition of his *Comparative Estimate of the Mineral and Mosaical Geologies* to rage against the materialism of geological theory and to declare that the separation of the waters in the beginning and the Deluge were provable historical facts. Everything that geologists might observe or fossilists might find, Penn thought, belonged to 'the mysterious scheme of the great Dispensation which is now in actual and irresistible progress, and, by which, the *Truth* of that which God has been pleased to reveal to Man will be triumphantly upheld and continually increased in lustre, until the END'.[8] Joining Bugg and Penn in their crusade against geological infidelity was the Glaswegian chemist Andrew Ure, who from an early age had exhibited a 'combative and rancorous disposition'.[9] Having served as an army surgeon, then lecturing in natural philosophy and working for the Irish linen board, Ure caused a sensation in 1818 when, during a public demonstration, he used voltaic material to reactivate the muscles of a corpse. (This was the year, after all, that Mary Shelley published *Frankenstein.*) Ure's attention had thereafter turned to geology and, in *A New System of Geology* (1829), he sought to prove that 'the Great Revolutions of the Earth and Animated Nature are reconciled at once to modern science and sacred history'.[10]

The guardians of that sacred history were equally ferocious in their defence of other aspects of the biblical past. Henry Hart Milman was a clergyman and the professor of poetry at the University of Oxford, and his *History of the Jews* (1829) was one of the first English-language works to incorporate the 'higher' criticism of the Bible that was seeping out of German universities. Here, Milman had written a history of the Old Testament where miracles were absent, where Abraham was described as a nomad sheikh, and where the Israelites – God's chosen people – figured 'more or less [as] barbarians'.[11] This would have been dangerous in a minor academic tract, but Milman's crime was 'to have put forth [these ideas] in a *popular* book' within the publisher John Murray's inexpensive Family Library series. By writing a history from contemporary evidence in defiance of the obligations of faith, and by thus putting heretical ideas before an impressionable public, Milman provoked the fury of High-Church orthodoxy and jeopardised his entire career: 'His own parishioners have taken it up,' noted one sympathetic reader, 'and the once popular preacher is in danger of having his gown torn off'.[12]

The major source of these conflicts was not personal animus or academic disagreement, for the scriptural geologists regarded themselves as warriors in a much wider battle against godlessness. The potential collapse of the Anglican supremacy was the major case in point. In the years after the Civil War of the mid-seventeenth century, the restored government of Charles II had sought to prevent radical Protestants from wielding power in England and Ireland by legislating for Anglican monopolies over public office, army commissions, and even university education. It amounted to widespread religious discrimination, archly described by a government minister as a form of 'civil death', but by the 1820s there had developed serious and sustained opposition to this '*ancien régime*'. Even though one MP warned that repealing these laws would mean that true Britons would 'have to fight for the citadel', Protestant Nonconformist relief passed into law in 1828 without much controversy.[13] The status of Roman Catholics was a different matter. For one thing, there was the historical association of Catholicism with Spanish fleets, French absolutism, and the Gunpowder Plot; for another, there was the question of how Catholics, who owed fidelity to the Pope, could also be faithful subjects of the British Crown. This meant that when the Catholic reformer Daniel O'Connell won a parliamentary by-election in the west of Ireland in 1828 – even if Catholics could not enter the House of Commons, they could still run for it – the Duke of Wellington's government had a miserable choice: upholding discrimination and risking civil war in Ireland, or granting Catholic relief and enraging Protestants. In choosing the latter, the government provoked violent cries of 'No Popery' and the formation of militant 'Brunswick Clubs' which organised furious resistance to the end of Protestant days.[14]

In these years of religious fervour and millenarian zeal, when the age of the Earth was an article of faith for many ordinary Britons, it was no coincidence that crowds flocked to the exhibitions of John Martin, whose vast and awesome paintings depicted biblical scenes of wonder, doom, and earthly destruction.[15] The son of a north-eastern journeyman, Martin had been apprenticed to a coachmaker and worked with porcelain and glass before finding his niche in watercolour landscapes. His breakthrough came with the depiction of a tortured man falling into the waters of oblivion, and he was soon combining Romantic grandeur with the most dramatic events of the Bible. In 1816, he showed Joshua commanding the sun to stand still so that the Lord's army might triumph at Gibeon; by 1820, the writing had flared on the wall at Belshazzar's feast in Babylon; the seventh plague then brought fire and hailstone to Pharaonic Egypt; and in 1829 Nineveh fell, its palaces burning and its priests betrayed. With Martin also producing celebrated engravings of Milton's *Paradise Lost* – even now, his image of Satan presiding over the Infernal Council remains a trope of fantastical fiction – his art was the incarnation of the age; moreover, as the great portraitist Thomas Lawrence reflected, he was 'the most popular painter of the day'.[16]

As the scriptural geologists roared in the defence of biblical history, and as the British public lapped up the religious imagery of convulsion and catastrophe, it was a brave man who denied the truth and the accuracy of the Old Testament: who was the mere geologist to doubt that Noah had crested the waters of the Flood or that Moses had parted the Red Sea? It was a braver man still who designed a system of geology explicitly to 'free the science from Moses'. But in the late 1820s, the young lawyer Charles Lyell – the friend of Gideon Mantell who had taken the teeth of the *Iguanodon* from Sussex to Cuvier – was just that man.

The son of a Scottish laird, Lyell had grown up on the Hampshire coast at the height of the Revolutionary Wars against France. A clever but reserved child, he at first abjured 'wholesome exercise, football, cricket, fives, &c' in favour of the chessboard, by which his mind 'was excited with the same kind of feelings which are aroused by gambling'. At Oxford, however, William Buckland's early lectures on geology kindled a fierce and enduring love of the physical world; and even if Lyell left Oxford for the bar, it was geology that now commanded his devotion. Where time and money allowed, Lyell made trips to Paris, attending Cuvier's soirées as if he were an ambassador for British geology. At the same time, he made friends with Buckland and Mantell, visiting the doctor at Lewes and encouraging his research on the *Iguanodon*. And when, at the age of thirty, Lyell's eyesight began to fail, preventing him from poring over briefs and judgements by candlelight, he leapt at the chance to quit the law: 'You know', he had complained to Mantell, 'that half my time is now spent at Sessions, Circuits, &c, and must not therefore be surprised when you receive no immediate answers to your [geological] correspondence'.[17]

Lyell's early papers had concerned Scottish limestone, but in the late 1820s he embarked on a continental expedition with Roderick Impey Murchison and his wife, Charlotte. A veteran of the French wars and of the retreat to Corunna that Charles Wolfe immortalised in verse, Murchison was another gentleman geologist who had spent small fortunes on touring Italy and purchasing antiquities. A tall, striking man whom Mary Anning described as 'the handsomest piece of flesh and blood [she] ever saw', Murchison first made his mark in geology by describing the strata of Scottish coalfields and the Weald, and he would later engage Henry De la Beche in the 'Great Devonian Controversy' over the age of petrified plants in the rocks of south-west England.[18] Charlotte, whom the Oxford scientist Mary Somerville described as 'an amiable and accomplished woman, with solid acquirements which few ladies at that time possessed', had learned to sketch, collect, and classify specimens in assistance of her husband's inquiries.[19]

They undertook the first leg of their journey in the spring and summer of 1828 in the Auvergne, where Lyell marvelled at 'rich wooded plains, picturesque towns, and the outline of the volcanic chain unlike any other I ever saw'. It was fresh terrain for a British geologist and he realised that 'a paper might be written if we stopped here'. With Murchison striking deals for rooms and board for 'half of what John Bull does en route', the trio settled on a gruelling regime of exploration and analysis. They began work at six o'clock each morning, setting off into the countryside in a *patache*, 'a one-horse-machine on springs', and 'neither heat nor fatigue' stopped them for even an hour until sundown. As a sportsman and hunter, Murchison relished the physical labour and resolved with Lyell 'to direct all our force to where there is real work, to concentrate, and not to run'; Lyell thought the former soldier 'a very strong man' who could 'get through what could knock up most men who never need a doctor'. The only problem was Murchison's dependence on 'the sellers of drugs', a habit which on one occasion found the party offering their custom to the mother superior of a local nunnery, who furnished them with a 'quantity [that] would kill six Frenchmen'. Charlotte was meanwhile 'very diligent, sketching, labelling specimens, and making out shells'; if the men departed for the field without her, she would receive 'the [local] gentry and professors, to whom we had letters . . . and [spend her days] collecting plants'.[20]

As for Lyell, he wrote to his mother in June that he had 'really gained strength so much, that I believe I and my eyes were never in such condition before'. Six hours in bed, 'exercise all day long', and the '*vin du pays*' were in combination 'the best thing that can be invented in this world for my health and happiness'. He was also learning something from his French hosts: 'Their system of rising early, and spending a moderate time at dinner', he reflected, 'agrees famously with those who wish to make use of their evenings and still see society'. Yet by the height of the summer, the touring party began to flag. 'I do not think the Murchisons will stand fire', wrote Lyell from Puy: 'Symptoms of flinching from the heat, which makes scarcely any impression on me, begin to betray themselves.' With every mile further to the south in the 'dog-days' of summer, Lyell learned something more. From Nice, he wrote of 'analogies between existing Nature and the effects of causes in remote eras which it will be the great object of my work to point out'; in the foothills of the Pyrenees there abounded 'evidences of modern action ... of *proving* the positive identity of the causes now operating with those of former times'.[21]

By the autumn, Lyell and the Murchisons had crossed the Alps into Italy, and when the husband and wife departed to rest in the Tyrol, Lyell headed south again to Campania. Writing to his sister from Naples, he marvelled at finding 'shells, among other things (marine), 2,000 feet high on the old volcano, which had not been dreamt of here'. He also despaired of Neapolitan customs and rued the loss of Murchison's negotiating prowess: 'On an expedition, the hours spent in bargaining is a nuisance ... In the shops there are no fixed prices. Whatever contract you make, your driver or boatman finds he has done something more than was specified, into which he has designedly led you.' Still, it was at Naples that Lyell's thesis came together at last: 'My work is in part written,' he reported to Murchison, 'and all planned'.[22]

Lyell would find the emblem of his whole theory, his system of geology, at the ruins of the Temple of Serapis that lay west of Naples. Here, three marble pillars, some forty feet in height, first erected roughly one hundred years before the birth of Christ, had lain hidden beneath soil, rocks, and bushes until the mid-eighteenth century, when their excavation revealed a curious pattern. The top thirds of the columns were weathered, and the bottom thirds 'smooth and uninjured', but their middle sections played host to dozens of cavities made by bivalve molluscs known as *lithophaga* (literally, 'stone eaters'). But what could explain this pattern? Why was there evidence of a shoreline so high above the ground? For Lyell, only one chain of events could explain it. First, sometime after the columns had been restored under the Severan dynasty in the third century AD, 'the lower part was covered up and protected by strata of tuff and the rubbish of buildings': this would explain why there was no evidence of weathering or erosion at the bottom. Second, there must have been 'a long, continued immersion of the pillars in sea-water', which alone could explain the action of the molluscs on the middle third. Finally, 'after remaining for many years submerged', the pillars 'must have been upraised to the height of about twenty-three feet above the level of the sea', allowing for the weathering and their discovery by modern Neapolitans. In other words, the level of the land on which the temple was built had changed repeatedly in relation to the level of the sea.[23]

This might seem mundane, if a little convoluted, to a modern reader who is all too familiar with the concept of changing sea levels. But in the 1820s it was a radical theory since it allowed for a history of the world in which geological forces – and the transformation of the natural world – were not confined to the Almighty's sporadic bouts of

The Temple of Serapis bearing the erosion and watermarks that Charles Lyell used to prove that the relative levels of the land and sea had changed

creation and destruction as described in the earliest chapters of the Bible. For a geologist such as Lyell, who had been searching for evidence that geological change was explicable only by causes that could be observed in the present, the fact that these forces had been so plainly in action *after* the third century AD – *long* after the Flood – was revelatory. For Lyell, it was proof that he could rescue geology from the restraints of Genesis; he could 'free the science from Moses'.[24]

Between Johannes Gutenberg's invention of the printing press in the 1440s and the early nineteenth century, the means of making books had barely changed. First, printers took a wooden case and arranged within it the small metallic letters that a page of text demanded. A workman then took two smooth leather-covered pads, each of them lathered in ink, and pressed down upon the type; it was a precise business because the ink had to cover the letters consistently. After this, the printers folded paper over the top of – but not touching – the type, before rolling the whole thing under the press itself. At this point, another workman pulled on a lever which dropped a flat wooden platform onto the paper, pushing it into the type and impressing ink onto the paper in the shape of the author's words. Finally, as the author approved the proofs of each page, all were bound between hard leather covers to make the book.[25] Two developments – rotary cylinders and the application of steam-power – might have accelerated things, but the mechanics remained the same; and this was the process that, in the winter of 1829–30, at street-corner premises in south London, William Clowes undertook with Charles Lyell's *Principles of Geology*.

Not unusually for the time, the printing began long before Lyell had finished the book. In December 1829, having returned to London from another excursion to Italy, he wrote to Gideon Mantell that, although the first volume would not 'quite be finished by the end of the month', the first few pages would be 'in the press on Monday next'. By the first week of February, Lyell had 'corrected the press to page 80' and, three weeks later, he told his sister that he expected 'p. 128 home from the printer' that evening. At least one of the prospective reviewers of the *Principles* was receiving the book in the same piecemeal fashion. In June 1830, Lyell told the liberal economist and soon-to-be Whig MP George Poulett Scrope, who had been charged with writing up the book for the *Quarterly Review*, that he 'ought to have received more sheets long ere this'. Yet in writing to Scrope three times that month, twice from London and once from Le Havre, Lyell did much more than apprise him of progress at the press. Intriguingly, Lyell's letters are also replete with advice and directions on how best to review the *Principles*: 'Don't meddle much with that', he tells Scrope at one point; 'Mark, too, my argument' at another.[26]

Part of Lyell's motivation was self-interest. The *Quarterly* was the most influential journal in Britain – owned by Lyell's publisher, John Murray, to boot – and he realised that a favourable review in its pages could lead 'to a second edition' and another print run.[27] More importantly, Scrope was a fellow geologist and another veteran of the Auvergne who, in his own work, had sought to undermine the 'Neptunist' theory that oceanic forces such as the Flood had shaped the Earth.[28] As such, he was Lyell's intellectual ally, because the core thesis of the *Principles* was that geological change, even from the beginning of time, was referable only to forces that were active and observable in the present day. It was not an original idea, for James Hutton had propounded the theory of gradual, uniform change as long ago as the 1790s, but the *Principles* would be its most complete, most persuasive articulation. Pointedly and controversially, it was also a system which accorded little agency to divine interventions such as the Flood.

Lyell was therefore sincere in seeking Scrope's assistance to advance a system of geology which owed nothing to the Bible. In these letters, he complains about 'the mischief and scandal brought on . . . by the Mosaic systems', deriding the 'saints' who still adhered to scriptural geology.[29] And though he acknowledges that he could have been 'tenderer' in the *Principles* about Muslims, whom he accuses of 'dreading the effects of the diffusion of . . . the physical sciences', there was no question that Lyell regarded his book as a deliberate strike against religious dogma.[30] 'Should you hit it off well for Q.R.', he pleads, 'we shall be able in a short time to work an entire change in public opinion'.[31] In the 23,000-word, fifty-nine-page review which appeared in October 1830, Scrope did not let down his friend: 'To bring forward the scriptures as the foundation of geology, or geological hypotheses as a support to the scriptural relations', he began, 'is to degrade the sacred writings, as well as to impede the progress of knowledge'. He thanked Lyell 'for the great addition he has made in the present to our knowledge of nature, and the beneficial influence it is likely to have in communicating a right direction . . . to geological inquiry'.[32] Other reviews were flattering too, with the Anglican *British Critic* celebrating the *Principles* as 'generally elegant' and possessed of a 'candid' spirit, and the *New Monthly Magazine* celebrating the book as doing more to establish geology as a formal science than anything previously written.[33]

The specialists were more guarded in their praise. Though the Cambridge professor Adam Sedgwick knew that the *Principles* would assume 'a distinguished place in the philosophic literature of this country', he regretted that Lyell had come forward 'as the defender of a theory': conservative opinion still deplored abstract theory, even decades after the dangers of the French Revolution had passed.[34] In a similar vein, William Conybeare, who believed truly in the transformative power of the Deluge, thought that Lyell had acted 'contrary to a sound spirit of inductive reasoning' by arguing that geological forces had been constant since Creation. He nonetheless conceded that the *Principles* had inaugurated 'a new era in the progress of our science'.[35] *The Times* would soon proclaim Lyell's ascent to the chair in geology at King's College London; given the implications of his work, it was no small irony that King's had been founded as a pious counterweight to the 'godless' University College at Bloomsbury.[36]

It would take a little longer for the *Principles* to reach a wider, popular audience, but this was a deliberate commercial strategy that Murray and Lyell had formulated in the wake of the controversy which had attended Henry Hart Milman's *History of the Jews*.[37] First, in procuring positive reviews from conservative journals such as the *Quarterly*, and in garnering support from the geological establishment, they had secured precious respectability for a book that could otherwise, because of its implicit criticism of Genesis, have been damned as profane. Second, and until the book's reputation was assured, they had priced the *Principles* out of the reach of the ordinary Briton: by the time that its third volume arrived, Murray had published 1,400 pages on fine paper that altogether cost two pounds and five shillings, which was equivalent to the monthly wage of a London office-worker. In this way, Lyell – unlike Milman – could not be accused of sowing unseemly ideas in the minds of Britain's middling classes. Yet when the *Principles* was deemed safe for popular consumption, Murray could reprint it in four cheap volumes that cost five shillings apiece. Now, Lyell's easy, lucid style found favour with a broader readership: 'There are very few authors or ever have been', Murray reflected, 'who could write science and make a book readable'.[38] Not only had Lyell made a persuasive, elegant case for geological changes that were innocent of divine agency, he had grown the audience of the science beyond the cloistered universities, gentlemanly societies, and learned journals. But how far, now, were naturalists prepared to go when considering the role of the Lord in the development of the world? And just who would take the lead in expanding the frontiers of accepted scientific knowledge?

6

The Parliament of Science

> *No causes whatever* have from the earliest time to which we can look back, to the present, ever acted, but those *now acting*; and . . . they never acted with different degrees of energy from that which they now exert.[1]
>
> Charles Lyell to Roderick Murchison, 1829

As Charles Lyell propounded his new system of geology, Mary Anning was discontented, not least with the widening chasm between her reputation and the little she had to show for it. There was certainly no question of her standing in scientific circles. In June 1824, a Scottish mineralogist had written that her 'knowledge of the subject is quite surprising', that she was 'perfectly acquainted with the anatomy of her subject', and that British geologists were 'entirely indebted to her for the preservation of some of the finest remains of a former world that are known in Europe'.[2] Later that year, Harriet Silvester, the widow of the most senior judge at the Old Bailey and a summertime visitor to Lyme, paid greater tribute: 'The moment [Anning] finds any bones she knows to what tribes they belonged . . . By reading and application', Lady Silvester remarked, 'she has arrived to that degree of knowledge as to be in the habit of writing and talking with professors and other clever men on the subject, and they all acknowledge that she understands more of the science than anyone else in the kingdom'.[3] Even the Americans came to Lyme to pay their respects, with George William Featherstonhaugh, the curator of the New York Lyceum of Natural History, placing 'several orders for the grand specimens of the Lizard tribe'.[4]

Among the other visitors to Dorset were the Murchisons, Roderick and Charlotte. The latter proved a formidable companion for Anning and, as the pair spent long hours together on the coast, hunting for fossils among the rocks, they forged a lasting friendship. When the Murchisons returned to London, the two women developed an intimate correspondence: Anning confided in Charlotte that she was obsessed with a certain type of marine sediment – she called it 'green Sand Mania' – and thanked her friend for sending the latest scientific papers from London so that she, the provincial autodidact, could challenge and critique the university dons and society men. 'I do so enjoy', she wrote, 'an opposition amongst the big wigs'. At the same time, these letters reveal the real, practical difficulties that Anning faced in her life as a fossilist. On one occasion, she was so intent on excavating part of a plesiosaur that she failed to notice the incoming tide and was 'like to have been drowned'. Having walked ten miles and looking like a 'drowned rat', Anning was so tired that she could 'scarcely hold the pen'; offering 'an appolige' [*sic*] for the poverty of her handwriting, she begged Charlotte to throw the letter 'into the fire'.[5]

Their friendship saw Charlotte invite Anning to stay with her near Regent's Park in London. For Anning, who had 'never been out of the smoke of Lyme', a holiday to the great metropolis was a daunting prospect and, in her poverty, she would depend on the Murchisons' credit for the journey. But it was exhilarating: 'I think should anything occur to prevent my accepting [the invitation]', she wrote, 'it will be the death of me'.[6] The trip took place at last in 1829, and Anning marvelled at the sights of the city. In her own breathless, unpunctuated account, she rejoiced in the sights of the British Museum: 'the King's Library in which there are sixty thousand Volumes saw a Book of Queen Elizebeth's writing also King Ethelbert's prayer Book . . . Saw the Elgin Marbles though them remarkably fine the Gigantice Egyptian heads . . . mummies coffins & hethan Gods without number'. She thought Regent Street and Somerset Place especially beautiful and, on the weekend, she took in the Baker Street Bazaar 'in wich was the Diorama of St Peters at Rome'. There was also time to visit the Geological Society's new home at Somerset House, where she saw 'a model of a jaw of a large Megliosaurus also of the plesiosaurus sent to paris so like that that I could hardly distinguish the difference'.[7]

Yet as much as Anning relished the sights of the imperial capital, they only emphasised the gulf that lay between her and the clever men who made their way in London. When Anna Maria Pinney came to Lyme Regis – her brother, William, would stand there as the Whig candidate for Parliament – she observed a pointed bitterness in Anning's feelings towards the world. The fossilist might have been 'good humoured' when they went walking along the beachfront, but Pinney noted that Anning 'abused almost everyone in Lyme, laughing extremely at the young dandies, saying they were numbskulls'. More poignantly, Anning felt that 'men of learning ha[d] sucked her brains and made a great deal by publishing works of which she furnished the contents, while she received none of the advantages'. What use was respect, if it could not pay for bread?[8]

Anning expressed similar unhappiness to Fanny Bell, a sickly London teenager who had come to Lyme for the sea air. In one letter, she complained: 'The world has used me so unkindly [that] I fear it has made me suspicious of all mankind.' It cannot have helped her mood that a ferocious storm in the English Channel had wreaked havoc in Lyme in the mid-1820s. 'A great part of the Cobb is destroyed', she regretted, and 'two of the revenue men drowned . . . Every bit of the walk, from the rooms to the Cobb, is gone; and all the back parts of the houses, from the fish-market to the gun-cliff, next to the baths.' Worst of all, the floodwaters had washed away Lyme's stores of coal. With the Cobb 'so shattered that no vessel will be safe', Anning resigned herself to the 'cold prospect' of sitting 'without fires this winter'.[9]

Still, Anning had kept looking for and finding fossils. In 1824, the *Salisbury and Winchester Journal* reported on the discovery of a slender-jawed *Ichthyosaurus*; the next year, the *Taunton Courier* announced that she had 'brought to light another antediluvian animal from the West Cliff', an ichthyosaur with a dorsal fin and 'remains of the skin . . . covered with small prickles'.[10] Also in 1825, the *Bath Chronicle* reported on Anning's more peculiar discovery, without further explanation, of 'a sub-marine forest at the north of the Char, Dorset'.[11] The rocks also yielded the fossilised, ray-like fish known as the *Squaloraja*.[12] With the money made from these sales, Anning at last moved herself and her mother – her brother, Joseph, had long since begun his own family – to a new house on Lyme's Broad Street. This place would double as the premises of the Anning Fossil Depot, which one visitor recalled as 'a plain, unpretending little shop with a small white oblong board, whose primitive and crude lettering informed the public that it was richly stored with precious specimens of the saurian . . . denizens of primeval waters'.[13]

Anning's most exciting find in this period was Britain's first flying reptile, though – once again – she would take limited credit; instead, it was William Buckland who would announce 'the discovery of a New Species of a Pterodactyle' to the London set. Borrowing heavily from Cuvier's descriptions of European pterosaurs, which had emerged from the rocks of the Rhineland since the 1780s, he waxed lyrical about a creature which 'resembled our modern bats and vampyres, but had its beak elongated like the bill of a woodcock, and armed with teeth like the snout of a crocodile'. Other parts of the skeleton 'resembled those of a lizard', its fingers 'terminated in long hooked claws', and 'over its body was a covering, neither composed of feathers as in the bird, nor of hair as in the bat, but of scaly armour

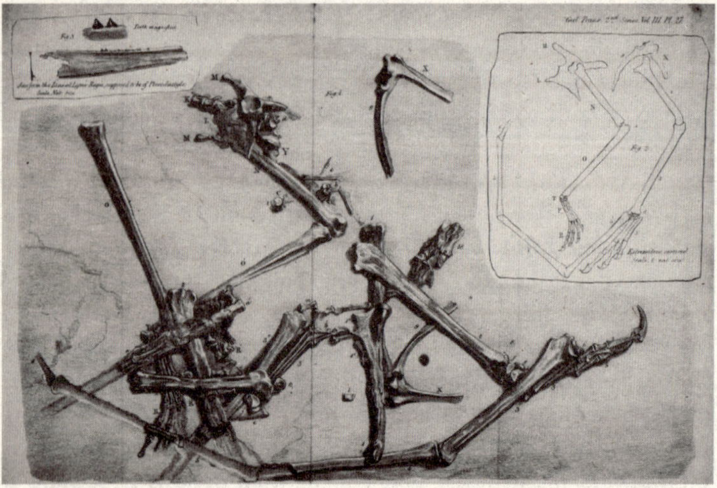

The jumbled remains of the *Dimorphodon*

like that of an Iguana'. The new creature, to which Buckland gave the binomial *Pterodactylus macronyx* – and which we now call the *Dimorphodon* – was in short 'a monster resembling nothing that has ever been seen or heard-of upon earth, excepting the dragons of romance and heraldry'. What a 'strangely tenanted' place was ancient Lyme, he marvelled, 'with flocks of such-like creatures flying in the air, and shoals of no less monstrous Ichythyosauri and Plesiosauri swarming in the oceans, and gigantic crocodiles and tortoises crawling on the shores of the primaeval lakes and rivers'.[14]

Despite the *Dimorphodon*, Anning and her mother were soon in financial trouble again, for the occasional payday, no matter how lucrative, did not bring long-term security. Once more, a benefactor intervened. Henry De la Beche had been busy during the 1820s. A year in Jamaica had produced two major publications, one of them a respected account of the island's geology, the other a defence of the treatment of enslaved people on his plantation.[15] When he returned to England he sought divorce from his wife, Laetitia, who had embarked upon a scandalous affair with the son of the Earl of Egremont and, by 1827, De la Beche was a wreck: 'Misfortune has followed me from my cradle,' he wrote, 'and it will follow me to my grave'.[16] A flurry of holidays to Europe revived him and, within three years, he had returned not just to London but to drawing. His cartoon *Awful Changes* saw 'Professor Ichthyosaurus' describing a human skull to his fellow reptiles and, on learning of Anning's distress, he produced the watercolour *Duria Antiquior: A More Ancient Dorset* for her benefit.[17] With an ichthyosaur biting down on the neck of a plesiosaur, with pterosaurs swooping in the sky above the water, and other ancient animals marauding through the scenery, it was the first realistic representation of prehistoric life. It was another world, almost magical in its difference from the 1820s, that a small band of fossil-hunters – Anning, Buckland, Conybeare, and Mantell – had reconstructed from the rocks of southern England. On commissioning a Bavarian artist to produce lithographic copies of his original illustration, De la Beche sold the image of this ancient Earth to his friends and colleagues for £2 10*s* each, and all proceeds went directly to Anning.[18]

*

Although, by the early 1830s, Charles Lyell was delivering a daring new system of geology into parlours and drawing rooms – reaching many more readers than, say, Gideon Mantell ever had – there was no sense yet that British science was a participatory community. The Royal Society was an elite establishment, its fellowship guarded closely; as we have seen in the case of William Smith, the Geological Society was comparably exclusive; and the universities remained sacred bastions of privilege where 'science' still deferred to the assumptions of Anglican orthodoxy. Mary Anning, moreover, remained almost at the mercy of her acquaintances and their charity. What was there, then, for the 'ordinary' Briton who had neither the money nor status to be a gentleman of science? Where was the scientific equivalent to the cause of Reform, which from 1830 was gaining rapid ground in the quest to broaden parliamentary politics?

Since the spring of 1829, the Edinburgh-based publisher and physicist David Brewster had been working on an answer to that question. Looking enviously at the continental academies, especially at 'the different meetings of German naturalists' which enjoyed the sponsorship of the Prussian monarchy, Brewster had in mind 'an association for the purpose of protecting and promoting the *secular* interests of science'. Writing to the mathematician Charles Babbage, the inventor of the difference engine, he suggested that 'the power of such a body to promote science' would be obvious. Babbage agreed to help: the Cambridge man was already writing a polemic on the perceived 'decline of science in England' and believed that 'scientific men [should] mix in more general society', not least among 'the wealthy manufacturers'.[19] At the turn of the 1830s, Brewster recruited other leading men to his cause, among them the physicist Michael Faraday. British geology put its weight behind the project too: Roderick Murchison signed up quickly, as did Buckland, Conybeare, and the Cambridge don Adam Sedgwick. There was also political aid from Henry Brougham, who had become Lord Chancellor under the new Whig ministry. There were *some* holdouts: the Cambridge scientist William Whewell, for instance, had taken great offence at the claims that university science was impotent, and so had 'no great wish to rally round Dr Brewster's standard'.[20] Even so, Brewster secured enough backing that, by the spring of 1831, he was arranging the first meeting of the British Association for the Advancement of Science.

Deliberately looking away from London and Oxbridge, Brewster chose York – 'the most centrical city' in the kingdom – as the venue. Working with the city's lord mayor and Buckland's friend William Vernon Harcourt, he put in motion a logistical masterplan for Britain's first academic conference. They secured accommodation for over one hundred delegates, acquiescing in specific requests: Murchison, for instance, wrote that he 'should not like to be so long in an inn', so Brewster found him 'a sitting room and bedroom in a private lodging'. They prepared a budget for 'subsisting the philosophers cheaply' in order that men of limited means – such as 'a little druggist of Preston (who has a splendid collection of . . . limestone fossils)' – could afford to attend. And they secured the favour of the Archbishop of York, who allowed them to hold a dinner in Bishopthorpe Palace. Of course, Brewster also took counsel to avoid the excessive 'eating and drinking and dining' that defined German societies, for that would not suit 'British phlegm'.[21]

When the meeting began on Tuesday, 27 September 1831, in the theatre of the Yorkshire Museum, the Association had sold 353 tickets for the lectures, debates, and dinners that were to follow. Over the next five days, delegates organised themselves into specialist divisions on chemistry, geology, the 'mechanical arts', and other fields besides. They discussed magnetism, optics, electricity, the relative weights of chemical elements, and the benefits of hourly meteorological reporting; and they feasted on the 'venison, game, and fruit' that the Earl of Carlisle had furnished as a gift.[22] That Saturday's *York Herald* acclaimed a 'Great Scientific Meeting', while newspapers as far away as Norwich reported that 'letters of approval were read from Professors of all the Universities and other eminent men'.[23] It was a glorious success, a meeting that was in Murchison's opinion 'very big with results'.[24] Perhaps the highest compliment came from Charles Dickens, who would soon parody the members of the BAAS as the well-fed and -watered patrons of 'the Mudfog Association for the Advancement of Everything'.[25]

The formation of the Association was not so dramatic as to mark the birth of 'public' science. After all, Humphry Davy's chemistry lectures at the Royal Institution in the 1800s had been so popular that, in seeking to manage the traffic, the London authorities made Albemarle Street the world's first one-way thoroughfare. Now, however, for an annual subscription of £1, any member of *any* philosophical institution could belong to the British Association, which undertook not only 'to give a stronger impulse and a more systematic direction to scientific inquiry', but also 'to promote the intercourse of those who cultivate science in different parts of the British Empire'; in time, the Association's annual meeting became known as the 'parliament of science', a great forum for debate and a rival to the genteel societies of London.[26] A further mark of its democratic mission was the question of whether it would be 'too academical', after visiting Oxford in 1832, to hold its 1833 meeting at Cambridge instead of Manchester. But as for the Oxford meeting, there was unanimous agreement on who should be its president.[27]

In 1825, only three years after receiving the Royal Society's highest honour for his account of Kirkdale Cave, William Buckland had considered quitting academic life. His university duties paid an annual stipend of £200 – roughly £20,000 today – but this was all he had to cover the expenses of geological fieldwork and acquiring fossils. He was also lonely: under the quasi-monastic lifestyle that colleges imposed upon their dons, he was forbidden from marrying. Buckland had therefore resolved to resign his fellowship at Corpus Christi and take up a living in Hampshire. It was a decision that would have boosted his income *and* allowed him to marry Mary Morland, 'a truly excellent and intellectual woman' who had shared in Buckland's geological inquiries ever since they had bonded over the works of Georges Cuvier in a Dorset stagecoach.[28] It was only the death of James Burton, 'an amiable, gouty, influential old gentleman' who had been canon of Christ Church cathedral since 1793, that changed his thinking.[29] Through the favour of his Tory friends in government and especially the Home Secretary Robert Peel, another Christ Church man, Buckland was preferred as Burton's replacement. It was a plum job, one of the best in the university, and it came with 'a good house, 1000*l.* per annum, and no residence or duty required'.[30] It meant that Buckland could stay in Oxford, deliver lectures as he pleased, and – because he would be a canon, not a fellow – he could marry. 'I'm speedily about to follow your example', he wrote to William Conybeare, 'in entering into the *holy* estate'.[31]

He wed Mary Morland at Abingdon that December and, naturally, their honeymoon was of a 'geological flavour'. During a year-long tour of Europe they met with Cuvier and Alexander von Humboldt at Paris, climbed the Swiss Alps, and almost caused a diplomatic incident in Sicily when Buckland informed local priests that their sacred relic, the bones of the patron saint of Palermo, in fact belonged to a goat. On returning to Oxford, the couple moved into an extensive set of rooms in the corner of Christ Church's great 'Tom Quad', where, after extensive renovations that saw 'the hunting of bricklayers and carpenters ... supersede that of crocodiles and hyenas', they would remain, with a growing family, for more than twenty years.[32]

Those rooms and much else of Christ Church played host to Buckland's extensive collection of fossils and even a zoo of live animals. One visitor recalled 'a wide and spacious staircase ... covered with ammonites, fossil trees and bones, and various other geological fragments'; in the several apartments, he found 'piles upon piles of books and papers ... spread upon tables, chairs, sofas, bookstands, and no small portion of the floor itself'. Buckland's son would reminisce that 'there were cages full of snakes, and of green frogs; in the dining-room, where the sideboard groaned under successive layers of fossils, the candles stood on ichthyosauri's bones'. As ponies galloped and guinea pigs scurried around the college, and as Buckland and his sons ducked lizards in the famous fountain of Mercury, porters were seen to carry Mrs Buckland from one side of the quad to the other in a sedan chair. Most curious of all was the menu that Buckland served to his guests, for during his time at Christ Church he set himself the mission of tasting every animal on Earth. Lyon Playfair, who later served in Gladstone's cabinet, recalled the 'successful experiment' of hedgehog but the 'utter failure' of 'a dish of crocodile', while another luncheon party ate 'joints of bear'.[33] When John Ruskin came up to Oxford, pursued by an overbearing mother who rented rooms on High Street to watch over him, he would regret that 'a day of unlucky engagement' caused him to miss out on a 'delicate toast of mice'.[34] A later legend told that Buckland even tried to eat the mummified heart of Louis XIV.[35]

By 1832, Buckland was in his late forties, his frame fuller and his hair thinner. Since taking up residence at Christ Church he had announced the Lyme Regis pterosaur and worked with Mary Anning on the mystery of the 'bezoar stones', hardened lumps of intestinal matter that adorned the beaches of Dorset. In 1829, he presented the somewhat delicate conclusion that these objects had 'passed undigested through the intestines of the Ichthyosauri' and coined the term 'coprolite' to describe this fossilised faecal matter.[36] Given the debt that Buckland and his colleagues owed to Anning, it was fitting – as the Oxford meeting of the British Association approached – that the most important question was whether women could attend. Buckland thought not. Writing to Murchison, he argued that 'if the meeting is to be of scientific utility, ladies ought not to attend the reading of the papers – especially in a place like Oxford – as it would at once turn the thing into a sort of Albemarle-dilettanti-meeting, instead of a serious philosophical union of working men'.[37] This meant that even Mary Somerville, the mathematician who had ascended into the 'first rank' of scientists by translating Pierre-Simon Laplace's astronomy into English, would not receive an invitation. Apparently, although some of her biographers think it out of character, Somerville agreed in this decision 'for fear that her presence should encourage less capable representatives of her sex to be present'.[38] In dissent, Charles Babbage opined that 'the male residents throughout the country will attend in greater number if their wives and daughters partake some share of the pleasure'.[39] As a compromise, the Association allowed women to attend its dinners and *conversazioni*, but not the 'business' of discussions and lectures.[40] There was no invitation at all to Mary Anning.

Over the course of one week in June 1832, with the Oxford fellows unburdened of their students by the long vacation, the Association gathered on Broad Street, dividing its time between the Sheldonian Theatre and the adjacent Clarendon Building. Following a strict routine, the men met daily at one o'clock to hear the divisional reports, and then at nine in the evening for specialist readings and lectures. On the Wednesday, before sessions on 'thermo-electricity' and 'the Philosophy of Sound', representatives of the Geological Society accorded overdue recognition to William 'Strata' Smith, whom they dubbed 'the father of English geology'. On the next day, the Association watched as Oxford conferred honorary doctorates on the Association's founder David Brewster, John Dalton, and Michael Faraday. On the Friday, in his report on the state of British geology, Conybeare paid tribute to the recent discoveries of the *Megalosaurus* and the *Iguanodon*: Buckland and Mantell, he claimed, had effected 'the complete restitution, and . . . almost . . . the resurrection, of the long-extinct and monstrous Saurians of the lias'. In describing these studies, Conybeare adopted a term that the French had coined ten years before: palaeontology.[41]

The week concluded with Buckland's presidential address, which he devoted to 'the Fossil Remains of the Megatherium, recently imported from South America'. With typical flair, he narrated how an Argentinian peasant, walking along a river during the dry season, had thrown his lasso at something 'half-covered with water' and dragged onto the shore an 'enormous pelvis'. Intrigued by this discovery, the locals then constructed a dam to 'turn aside the current' and reveal the remainder of the skeleton. The fossil presented 'an apparent monstrosity of external form'. Some eight feet high and twelve feet long, the huge, sloth-like creature possessed 'three gigantic claws' on the end of each arm, which the Almighty, clearly, had designed for 'scraping roots out of the ground'. On its side and back was an inch-thick coat of bony armour, designed 'to prevent the annoyance which this class of animals would feel . . . from the constant presence of sand and dirt'. It was proof, Buckland concluded, of 'the wisdom, and goodness, and care of the Creator over all his works', for this was a creature that the Lord had designed perfectly for the environment of the pampas.[42] The Association thrilled. 'Buckland was really powerful last night on the megatherium', Lyell reported to Mantell: 'A lecture of an hour before a crowded audience: only standing room for a third . . . It was the best thing I ever heard Buckland do!'[43] But Buckland's lecture was a eulogy too, for everyone in the Sheldonian that night was there to mourn the passing of 'the greatest naturalist and one of the greatest philosophers that have arisen in distant ages', a man whose name ranked alongside 'Aristotle, and Pliny', and the anatomist who had first revealed the *Megatherium* to Europe.[44] Only a month before, at the age of sixty-two, Georges Cuvier had died in Paris. But he had not departed without making one last bracing contribution to scientific debate.

Throughout his work on biology, anatomy, and zoology, Georges Cuvier had advanced two general principles concerning the organisation of animals. First was the principle of correlation, which held that every part of every animal had been designed in given proportions with a specific function in mind. In respect of carnivores, for instance, he wrote that 'if an animal's teeth are such as they must be, in order for it to nourish itself with flesh, we can be sure without further examination that the whole system of its digestive organs is appropriate for that kind of food'.[45] This principle of correlation applied to an animal's size, too: it was why William Buckland, when estimating the length and height of Oxfordshire's *Megalosaurus*, had been so confident in his conclusions, despite not having anything like a complete skeleton to examine. Cuvier's second great principle, following from the first, was the 'fixity' of species. This held that, if the Lord had designed each animal perfectly, with all parts of each body working in harmony to achieve a particular objective, there could be no reason for species to change in their constitution. It was a comforting, orthodox approach to zoology which meant that Cuvier, unlike many other continental philosophers, was adored and adopted by conservative British scientists.

Opposing Cuvier was Jean-Baptiste Lamarck. As a young man he had dropped out of Jesuit college and ridden across Europe to fight during the Seven Years War. As an adult, Lamarck devoted himself to botany and zoology, and his career peaked with *Philosophie Zoologique* (1809), where he put forward a theory known as the transmutation of species.[46] In short, Lamarck believed that species of animals varied in their nature and structure because they inherited certain characteristics that were determined by their environment, and that the strength of any such inheritance was determined by the use or disuse of that characteristic. In Lamarck's telling, giraffes had long necks because they needed to eat leaves from the highest branches of trees, while the subterranean mole rat had gone blind because, living underground, it did not need to see. The common human analogy was that a blacksmith, who might have spent decades performing arduous physical labour, would bequeath strong arms and broad shoulders to his children.[47] Although most naturalists celebrated Lamarck's work on invertebrate animals, his theory of transmutation ran into several obstacles. First, there was simply little evidence for it: the fossil record that was available at the time did *not* suggest that animals had changed in their physical structure over the generations. Second, it presupposed that the Lord had made mistakes during the Creation: if animals had 'transmuted' out of necessity then clearly God had got something wrong, and that was impossible. Third, and most offensively, was the ultimate implication of his theory: that mankind, in the beginning, might not have been made in the Lord's own image. As Charles Lyell had explained to Gideon Mantell as early as 1827, 'if pushed as far as it must go . . . [Lamarck's argument] would prove that men may have come from the Ourang-Outang'.[48]

There is no question that Cuvier had the better of Lamarck in the salons. In 1829, when the former's place in the pantheon of science was beyond question, Lamarck died in blindness, ridicule, and poverty. In death, he suffered further ignominy. With his family so poor that they could afford only five years of burial in a rented grave, Lamarck's remains were lost to the world when, at the expiry of the lease, the Parisian authorities exhumed his body and tossed it into a lime-pit. The acerbic eulogy that Cuvier gave to his departed rival was no less abusive. Addressing the French Academy of Sciences, Cuvier placed Lamarck in the rank of men who had been 'less severe in scrutinizing the evidence', who had mingled 'fanciful conceptions' with real discoveries, and who, 'believing themselves able to outstrip both experience and calculation', had 'constructed vast edifices on imaginary foundations, resembling the enchanted palaces of our old romances'.[49] As one historian puts it, the eulogy was 'one of the most deprecatory and chillingly partisan biographies' ever printed.[50]

Some 'Lamarckians' now rallied to defend the honour of their mentor, the most prominent of whom was Étienne Geoffroy Saint-Hilaire. Having served in the legion of scientists and artists that accompanied Napoleon's expedition to Egypt in 1798 and then pillaged the museums of Portugal at the emperor's direction, Geoffroy had enjoyed a conventionally successful career: he was a professor at the University of Paris; his *Philosophie Anatomique* (1818–22) was a standard synthesis of zoological theories; and he curated intellectual friendships across Europe, most notably with the Scottish academic Robert Grant, who in 1826 had dared to suggest that animals could 'have evolved from a primitive model'.[51]

In February 1830, when the Academy of Sciences asked Geoffroy to review a paper on molluscs, he spied a chance to redeem the honour of Lamarckian theory. In the epilogue to the review, he mocked Cuvier's ideas without naming him; incensed, Cuvier demanded the removal of the offending passages. The Cuvier–Geoffroy 'debate' ensued when, a week later, Cuvier came in arms to the next meeting of the Academy. With elaborate illustrations and his own paper on molluscs, he savaged Geoffroy's philosophy. The rebuttal came on 1 March, when Geoffroy declaimed against Cuvier's obsession with detail, calling on the Academy to adopt a properly 'philosophical' approach to the animal kingdom. It was to no avail. By early April, with Cuvier preparing an arsenal of evidence, Geoffroy 'retired from this direct contest'.[52] The great man appeared to have vanquished the Lamarckians, and the sound of his triumph echoed over Europe. In Weimar, when Goethe declared that the 'volcano has come to an eruption' and that Paris was 'in flames', he referred not to the revolution which had deposed Charles X from the French throne, but to the 'open rupture' in the academy.[53] In England, the effect of Cuvier's victory was so profound that Charles Lyell, who had once been sympathetic to Lamarckian ideas, made sure that the second volume of the *Principles* contained an explicit censure of transmutation. In the immediate aftermath of this debate, the idea that species could adapt to their environment appeared as a dead letter. But would Cuvier's victory endure? And how far would Lyell's *Principles*, which maintained its theory of gradual geological change over long ages of the Earth, shape the thinking of one young naturalist who had taken a copy with him onto HMS *Beagle*?

PART TWO

The Lords of Creation

DRAMATIS PERSONAE

Mary Anning, fossil-hunter, Lyme Regis
William Buckland, clergyman and geologist, Oxford
Charles Darwin, naturalist, Cambridge and at sea
J. A. Froude, academic, Oxford
George Holyoake, journalist, London
Charles Lyell, geologist, London
Gideon Mantell, surgeon and geologist, Brighton and London
Richard Owen, anatomist, London

7
A Shropshire Lad

> My theory would give zest to recent & Fossil Comparative Anatomy: it would lead to [the] study of instincts, heredity, & mind heredity, whole metaphysics, it would lead to [the] closest examination of hybridity & generation, causes of change in order to know what we have come from & to what we tend.[1]
>
> Charles Darwin's notebook, 1837

On the shoreline lay the wreckage of the earthquake and the waves that came after it. There were chairs, tables, and bookshelves, sacks of cotton and tea. Inland, on the banks of the Bío Bío River, the city of Concepción was in ruins. 'Great cracks' wider than a yard had opened in the ground, and the cathedral's walls were dust. The beds of a boarding school for girls lay under bricks, beams, and slate, and more than one hundred people were missing. It was March 1835 and one of the great cities of South America now compared to 'Ephesus or the drawings of Palmyra'. Yet for Charles Darwin, the twenty-six-year-old naturalist who had been sailing on HMS *Beagle*, the desolate, 'terrible' aftermath of the earthquake also revealed a profound, unsettling truth about the movement of the earth: 'The permanent level of the land & water' had changed.[2] Another visitor to the Chilean coast deduced from the 'visible evidence of dead shell-fish, water-marks, and soundings . . . that the land had been raised about eight feet'.[3] When Darwin left Britain in 1831, one of his Cambridge mentors had recommended that he take to sea *Principles of Geology*, in which Charles Lyell had articulated a theory of the Earth where change occurred gradually, over time. Though Darwin had been warned 'on no account' to accept Lyell's theories, the *Principles* had beguiled him; he even ordered the second volume for delivery to Montevideo.[4] Now, in Chile, Darwin was witnessing further evidence of geological change for himself because, even if the earthquake was terrible, it was not the sort of cataclysm that Genesis had described. Only days later, in the foothills of the Andes, he would find further compelling proof that the biblical history of the Earth was fallible: there were 'fossil shells on the highest ridge'.[5] Writing home that summer, he declared himself 'a zealous disciple of Mr Lyell's views'.[6]

The son of a wealthy Shropshire doctor, Darwin had grown up in comfort at The Mount, a Georgian mansion on the outskirts of Shrewsbury.[7] There was a greenhouse of luxuriant variety, a library of politics and science, and a coterie of elder sisters who mothered young Charles with love and attention. In his teenage years, Darwin shot game on the estate of the local squire and conducted chemical experiments in the grounds of the family home, but he struggled at school with rote learning of the classics: 'Nothing could have been worse', he reflected, 'for the development of my mind'.[8] Frustrated by this academic stagnation, the young Darwin left school at the age of sixteen and, in the autumn of 1825, joined his brother Eras at Edinburgh to study medicine. It would be a rigorous, thorough education, for this was the leading medical school in the country. Unlike at Oxford and Cambridge, where Anglican convention prescribed the limits of inquiry for professors such as Buckland, continental heresies had infiltrated Edinburgh's syllabus; it was where Robert Grant, a leading expert on marine invertebrates, was importing Lamarckian theories of transmutation into British zoology. Yet if Darwin relished Edinburgh's debating societies, where students and their masters pushed at the boundaries of acceptable discourse, the cold and chill of the northern winters bade little welcome to the Shropshire lad; even worse, the blood and viscera of dissection turned his stomach. In April 1827, he abandoned medicine and left Edinburgh without a degree.

What to do with an aimless son? With Darwin's father 'properly vehement against' his son becoming 'an idle sporting man', the new plan was for Charles to take holy orders in the Church of England, and this meant an ordinary degree at Cambridge. It was a happy compromise. The prospect of a tranquil country parish back in Shropshire, with plenty of time for shooting and walking, appealed immensely; more to the point, Darwin 'did not then in the least doubt the strict and literal truth of every word in the Bible'. He arrived at Cambridge in January 1828 and though, as at Edinburgh, he despaired of mundane studies, the company of professors such as the geologist Adam Sedgwick and the botanist John Stevens Henslow inspired him to take up natural history. Indeed, Darwin followed 'no pursuit at Cambridge ... with nearly so much eagerness ... as collecting beetles'. Ranging across the fenland, crawling through the grass of the meadows, he obsessed over entomology. These ventures were not always successful. During one expedition, upon finding a rare creature on the ground, Darwin freed his right hand by popping one of the two beetles he was already holding into his mouth; alas, it was a bombardier beetle, which protested at this course of action by emitting 'some intensely acrid fluid' down the young man's throat. It was a technique for collecting that he would not repeat.[9]

Darwin continued to resent the grind of a degree, preferring his rambles and practising his aim by firing empty guns at candles in his rooms, but he read deeply of at least two major authors. The first was William Paley, the English divine whose system of natural theology held that man's experience of nature supplied evidence for a divinely ordained Creation. Darwin later admitted he would have thought differently if he had considered Paley's 'premises' but, as a younger man, he delighted in 'the long line of argumentation' that Paley set out in support 'of the existence and attributes of the Deity'.[10] The second great influence was Alexander von Humboldt, the Prussian polymath whose *Personal Narrative* described five years of travel around and over South America. Here, Humboldt told of sailing through schools of electric eels on the Orinoco, scaling snow-capped heights in the Andes, and riding mules across the Mexican plateau.[11] Darwin's plans for a clerical life now dissipated; instead, he dreamed of an odyssey and especially of Tenerife, which Humboldt had described as 'a pretty specimen'. This 'Canary scheme' would fall through but, by the summer of 1831, Darwin had secured a berth on another exotic voyage.[12]

The second mission of HMS *Beagle* owed much to South American politics and to the aligning interests of British imperialism and British science.[13] In 1808, when Napoleon enthroned his brother at Madrid, Spanish colonists from Mexico to the River Plate had lost their loyalty to their motherland and, in the Bolivarian revolutions that followed, new republics emerged. British interest in these developments went far beyond the collapse of Spanish power. South America was rich in gold, silver, and minerals, and it yielded luxury crops such as sugar and coffee; by the mid-1820s, British enthusiasm for South American projects was so intense that several hundred Scottish investors paid for passage to Poyais, a supposedly prosperous country which did not in fact exist.[14] Yet if British merchants were to make the most of these opportunities they needed to know where to land their ships, and the coastlines of the continent had not been mapped. The Admiralty also thought it made sense to send 'some well-educated and scientific person' to examine the soil and wildlife of the far side of the world.[15] For Robert FitzRoy, the aristocrat who had assumed the captaincy of the *Beagle*, there was another good reason to hire such a man. The expedition would take years and, given that social decorum would prevent the captain from conversing with his crew, he needed another gentleman on board: in the deep loneliness of the southern Pacific, the previous captain of the *Beagle* had gone mad and shot himself.

There was not much to recommend Darwin to FitzRoy. On the one hand, the aspiring naturalist was an avowed Whig and a liberal who supported the major causes of the day, such as the reform of Parliament and the eradication of colonial slavery; it was well known that Darwin's wider family, which had married with the Wedgwood potters, was ardently abolitionist. FitzRoy, on the other hand, was staunchly conservative. He had been a Tory candidate at the last general election, campaigning to safeguard the Church of England and other 'long-established institutions' against 'the innovations which now threaten them' and, in Darwin's telling, believed that enslaved people were perfectly content, undesiring of emancipation.[16] Beyond these political differences, Darwin promised little as a sailor: at six feet tall, he worried about occupying 'so small' a cabin for years on end and went through 'ludicrous difficulty' when getting into his hammock. Nor did he have the stomach for the sea. As the *Beagle* pitched and rolled upon leaving port, Darwin 'suffered most dreadfully' and 'was soon made rather sick'. It was a problem he could not shake. In 'boisterous weather', he would always look 'forward to sea-sickness with utter dismay'.[17]

Despite these problems, FitzRoy accepted Darwin as a companion and, by the early weeks of 1832, the *Beagle* had left the English Channel for the waters of the African Atlantic. Using the Portuguese island of Madeira to gain its bearings, the ship's first scheduled port of call – to Darwin's delight – was his cherished Tenerife. Twelves miles from the coast, he observed a 'Mass of lofty rock with a most remarkably bold & varied outline', a 'sugar loaf' towering in the sky, and 'brown & desolate hills . . . spotted with patches of a light green vegetation'. But so close to realising his dream, to exploring the valleys and mountains of the island, the Spanish consul delivered unwelcome news: because of the ongoing cholera epidemic in England, he would impose twelve days of quarantine on the *Beagle*. FitzRoy would not wait, so they sailed on: 'Oh misery, misery', lamented Darwin. The disappointment was short-lived. By mid-January the *Beagle* had anchored at Cape Verde, where 'the Volcanic fire of past ages' had rendered the soil 'sterile & unfit for vegetation' but exposed something wonderful.[18] Here, exploring the sands and palm trees, Darwin deduced that 'a stream of lava [had] formerly flowed over the bed of the sea, [which was] formed of . . . recent shells and corals'. Since then, 'the whole island ha[d] been upheaved' and there was further evidence of subsidence around the craters.[19] In other words, it looked as if volcanic forces had worked changes *after* the initial creation of the island.

As the *Beagle* sailed further south, the air grew 'damp & oppressive', the thermometer stuck 'night & day between 75 and 80'.[20] On crossing the equator, the crew observed the usual ceremonies: dressing as Neptune, they blindfolded, soaked, and shaved the first-timers, Darwin among them. 'Diversion is often the best medicine', reflected FitzRoy, who knew the dangers of the doldrums.[21] There followed moorings at Bahia and Rio de Janeiro; at the former, Darwin lost himself in the rainforest, collecting as many specimens of foreign flora and fauna as he could ship back to the *Beagle*. Months later, at Montevideo, he and his shipmates aided the local consul in quelling a mutiny: 'We immediately armed & manned all our boats', he reported to his sister, '& . . . occupied the principal fort. It was something new to me to walk with Pistols & Cutlass through the streets of a Town.'[22] He then spent October touring the outskirts of Buenos Aires. There were plenty of rheas, those large, ostrich-like birds, but the fossilised remains of the *Megatherium* were more exciting: 'As the only specimens in Europe are at Madrid', Darwin wrote home in glee, 'this [discovery] alone is enough to repay some wearisome minutes'.[23]

By December 1832 they had reached Tierra del Fuego, the ends of the Earth, where three native Fuegians, who had spent several years in Britain before sailing on the *Beagle*, went home. Meeting the people of the archipelago, Darwin baulked. 'I would not have believed how entire the difference between savage & civilized man is', he wrote in his diary: 'It is greater than between a wild & domesticated animal, in as much as in man there is greater power of improvement.'[24] It was a moment that would stay with Darwin in later years. For much of 1833, the *Beagle* toured the eastern coastline of the continent, navigating and mapping the sounds and waterways of the South Atlantic, and visiting the Falkland Islands that Britain had only recently reclaimed. That Christmas, the crew staged their own, drunken version of the Olympic Games outside the Patagonian fishing village of Puerto Deseado before rounding the Straits of Magellan.

At this point, the whole mission fell into doubt. The Admiralty had written to FitzRoy, rebuking him for the purchase of a second ship to accompany the *Beagle*, and so the captain slumped into depression. He even threatened to sail home without exploring the Chilean coastline or completing the circumnavigation: 'the cold manner [in which] the Admiralty . . . have treated him', Darwin noted, had made the captain 'very thin & unwell' and 'his mind . . . deranged'.[25] Yet on discharging the second ship and paying off its crew, FitzRoy recovered his sense of optimism, and the *Beagle* resumed its survey. They spent the next weeks and months among the bays and inlets of the south Pacific, and in September 1835 – after witnessing the destruction of the earthquake at Concepción – they arrived in the waters of the Galapagos Islands.

Scattered either side of the equator some 500 miles off the coast of Ecuador, the islands made little impression on first landing. For Darwin, these volcanic plugs 'compared to what we might imagine the cultivated parts of the Infernal regions to be'.[26] FitzRoy was equally unconvinced, writing of 'black, dismal-looking heaps of broken lava, forming a shore fit for pandemonium'.[27] Yet as the crew roamed across the black sand and ancient lava, they took note of an extraordinary environment. There were turtles in the bays, enormous tortoises pawing over the rocks, and mockingbirds flitting through the air. None of the creatures seemed to know friend from foe, so the hunting was easy game. Even more curiously, as Darwin hopped across the archipelago, it became clear that the animals on each island differed ever so slightly from one to the next, though the English-speaking governor of the prison colony on Charles Island believed he could 'tell the island by looking at the tortoise'.[28] Darwin killed and collected his samples – birds, lizards, plants, and flowers – and pondered the facts: 'Each variety is constant in its own Island', he noted, and they sailed to the west, across the great expanse of the Pacific.[29]

The *Beagle* made good time across the open water, stopping only when provisions demanded. Darwin marvelled at the miracles wrought by Christian missionaries on Tahiti, then at the explosive growth of Sydney: the Australian city, he thought, was one of the '100 Wonders of the world' and 'a most magnificent testimony to the power of the British nation'. Across the Indian Ocean, Darwin began 'totally rewriting' his ideas on geology before the *Beagle* docked at the Cape of Good Hope, where he and FitzRoy dined with the astronomer John Herschel. Under Table Mountain, walking through Herschel's 'pretty garden full of Cape bulbs', the trio talked of volcanoes and the movement of continents. Returning to sea in July 1835, Darwin longed for home, but FitzRoy insisted on checking his old measurements; back at Bahia in northern Brazil, Darwin feasted on mangoes and fixed in his mind his last sight of the 'great wild, untidy, luxuriant hot house' of the rainforest.[30] Now there were only two stops left before home – Cape Verde, then the Azores – and in the first week of October the *Beagle* hove into Falmouth in Cornwall. As Darwin disembarked, so came the evidence of his five years around the world. There was a diary running to 770 pages, 1,383 pages of notes on geology, and another 368 on zoology. It took much greater effort to unload the physical record of the voyage. There were 1,529 species in spirits, 3,907 pieces of skin and bone, birds and coral in boxes, the half-eaten carcass of a rhea, and a young Galapagos tortoise. It was with 'a comical mixture of dread & satisfaction' that Darwin, still a young man, looked ahead 'to the amount of work, which remains for me in England'.[31]

Since announcing the *Iguanodon* to the world and claiming fellowship of the Royal Society, Gideon Mantell had fallen back into the rhythms of provincial life. He maintained his medical practice at Lewes, palliating the agues and ailments which abounded in the Sussex countryside. In one gruesome incident he 'removed several portions of skull that had been forced into the brain' of a sixteen-year-old boy who had fallen beneath a horse, but Mantell's 'trephining' could not save the patient. He was also invested in the fate of Hannah Russell, a widow from Horsham who had been convicted of poisoning her husband with arsenic. Deducing from reports that the evidence against Russell was 'founded on incorrect experiments', Mantell had conveyed his doubts to the Home Secretary and, by showing that Russell's husband probably died from nothing more suspicious than angina, he secured her pardon. He was too late, alas, to rescue her alleged accomplice: 'for this supposed crime', Mantell regretted, 'a poor lad of 18 was executed'.[32]

The good doctor was immersing himself in civic life as well, drafting resolutions for town meetings and taking 'pleasure in moving the public mind and guiding it unseen'; he took greater pride in complaining that 'neither profit, credit, nor thanks' attended his work. One notional reward, however, was meeting the new king, William IV, who visited Lewes with Queen Adelaide in 1830. 'Various ceremonies' prevented the royal couple from inspecting Mantell's museum, but the doctor had his audience. Kneeling before the king, Mantell thanked him for 'deigning on a former occasion to accept of my Geological Works' and presented the gift of a history of Lewes. The king was underwhelmed. He brushed off Mantell with 'Much obliged, much obliged' before instructing an aide, 'in his usual hasty manner', to take away the book.[33]

Mantell kept up his research too. In 1831 he published a well-received article on 'the geological age of reptiles' where, in comparing the *Megalosaurus* to the monitor lizard, and his own *Iguanodon* to the reptiles of the Caribbean, he marvelled that he and his fellow geologists had proved there 'was a period when *the earth was peopled by ... quadrupeds of a most appalling magnitude*, and that [such] reptiles were the *Lords of Creation* before the existence of the human race!'[34] As for fieldwork, there had been a little local difficulty. Until 1830, the quarrymen at Cuckfield, north of Lewes, had given Mantell first refusal on every interesting find in the limestone, but now they informed him that 'a customer on the spot, a gentleman new to collecting fossils', had purchased their favour. 'When there are thousands of quarries in England which would be equally productive', Mantell mourned, 'I must be robbed of the fruits of my industry and trouble!'[35] Still, some loyalty endured: in 1832, when the quarrymen's explosives exposed 'a great consarn of bites and boanes' that nobody could identify, it was Mantell they called.

He arrived at the pit to find 'a mass of stone blown into fifty pieces' and, 'with much difficulty and great labour', he reconstructed a slab exhibiting '12 vertebrae ... with many ribs, coracoid bones, omoplates, chevron bones, and several of those curious dermal bones which support the scales'.[36] These were unquestionably the remains of an ancient reptile, but when Mantell had the slab transported to Lewes, and as he spent the next three months examining the fossils, it became clear that it was 'a new genus of saurians' entirely. The most striking feature was 'a series of bony processes' in the shape of

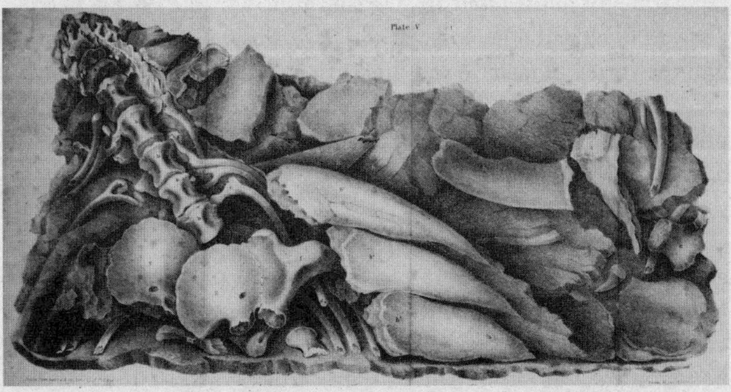

The fossilised remains of the *Hylaeosaurus*, which Mantell announced in 1833

triangles that ran in parallel to the creature's backbone. Between five and seventeen inches in length, they formed 'a formidable row of spines' – an awesome piece of armour – and Mantell had seen nothing like it before.[37] By November he had proclaimed the discovery of an 'osteological organisation [that was] peculiar and important', a creature that he christened *Hylaeosaurus armatus*, meaning 'the armoured lizard of the forest'.[38] The formal presentation of the paper and the *Hylaeosaurus* would take place in January 1833 and, despite the surprisingly 'unhandsome conduct' of William Buckland, who was 'unmerited' in questioning the surgeon's analysis of the spinal plates, Mantell had 'every reason to be highly gratified with the reception my discovery met with'.[39]

But Mantell was restless. Only days before his presentation of the *Hylaeosaurus*, he had grieved in his diary that 'so many trials and annoyances' had beset him that he had 'neither inclination nor time to note the passing events in [his] dull and wearisome existence'. The truth was that, despite making yet another significant discovery, Mantell had grown bored of Lewes and its basic, bucolic routines. Instead, his eyes had wandered south to Brighton, where he had been gossiping backstage with the fast-living actor Edmund Kean. More to the point, Brighton's Royal Pavilion was the favoured residence of the new king, and the seaside resort now figured as the kingdom's second capital, at least in terms of high society. Having secured a degree of reputation for himself, but always aspiring to a higher quality of company, Mantell had to choose. 'Shall I leave this dull place and venture into the vortex of fashion and dissipation at Brighton,' he asked himself, 'or shall I not?' With four children to raise and the medical practice at Lewes providing reliable income, Mantell knew that 'Prudence' said 'stay where you are'; but ambition, the devil on his shoulder, said 'go and prosper!'[40]

In 1833, ambition won the debate. In the same week that Darwin and the *Beagle*'s crew re-created the Olympic Games in Patagonia, Mantell left Lewes for Brighton; funded by a grant of £1,000 from the Earl of Egremont, he moved his family and servants into a six-storey house on the Old Steine within sight of the Royal Pavilion. Within months, Mantell was revelling in fashionable company. 'My reception in this town has certainly been very flattering', he noted in his diary:

'All the principal persons in this place have called upon me, and invited me to their houses.' Moreover, Mantell's museum – his collection of fossils, rocks, and other specimens, which he had reconstructed in Brighton – had been 'visited by nearly a thousand persons' already. Besides all this, there were excursions to Glyndebourne and coach trips to London. But trouble persisted. Despite that gift from Lord Egremont, and though his public lectures turned a profit, Mantell's 'professional prospects [were] not encouraging'. Having left his loyal patients in Lewes, he was struggling to build a new practice in Brighton and, more worrisome, he had failed to sell his old house: 'Not one bidder appeared', he wept. 'What I am to do, I know not!'[41]

Yet if these early months in Brighton put Mantell under strain, they also witnessed another remarkable discovery. It is now a suburban sprawl of sports pitches and cul-de-sacs, but the area to the west of Maidstone town centre in Kent was once the site of several limestone quarries. In one of these pits, in May 1834, workmen in the employ of William Harding Bensted were exploring the 'lowermost strata' of rock when they exposed a 'remarkable substance' that resembled petrified wood. Bensted, who would in time assume membership of the Geological Society, recognised that it was something different. Erecting 'a temporary shed' over the fragments, he worked with mallet and chisel to clear away the stone, soon revealing 'the skeleton of an extra-ordinary animal which had been buried in the bowels of the earth . . . in the earliest ages of its existence'. As the *American Journal of Science* reported, 'the story of the discovery . . . of an antediluvian giant' quickly spread abroad. The 'gentry of the neighbourhood' and London's naturalists descended upon Maidstone to inspect these fossils, but none knew what they were; at a loss, Bensted summoned Mantell to Kent.[42]

On the first Tuesday of June 1834, Mantell took the afternoon stagecoach from Brighton to London and, after staying overnight at the Golden Cross tavern on the Strand, he set off for Maidstone. Arriving at five o'clock, he met and dined with Bensted at the Mitre Inn before the pair decamped to the quarry. In the evening light of high summer, Mantell identified the bones at once: 'they [were] the lower extremities of the Iguanodon'. He knew that it was more than just another specimen, too: because here the limestone of Kent was presenting a more complete, better-preserved *Iguanodon* than anything he had found in Sussex. He had to have it, though Bensted would drive a hard bargain. Mantell's first offer was £10, and this was dismissed out of hand. £20? No, only £25 would do. But this was a price – almost £3,500 today – beyond the means of the struggling surgeon. Happily, Mantell's benefactors intervened: 'My *very very* kind friends Horace Smith and Mr Ricardo', he noted, 'took it upon themselves to obtain [it] and present it to me'. By August the slab had arrived in Brighton and, within another few weeks, Mantell had chiselled out the *Iguanodon* and placed it prominently within his museum. It was a display that attracted the architect George Basevi and the painter John Martin to the Mantells' home; as for Maidstone, the town remembers the *Iguanodon* in its coat of arms.[43]

As palaeontologists took note of the new *Iguanodon*, Mantell's reputation bloomed again. He would publish an exposition of his career, *The Wonders of Geology* – for which John Martin provided a terrifying depiction of the new *Iguanodon* – while an honorary degree from Yale confirmed his transatlantic renown. 'If fame and reputation could confer happiness', he reflected, he would surely be happy. Yet still he was not. As before, celebrity had not brought security. In the world of nineteenth-century medicine, where surgeons depended upon the loyalty and custom of their patients, the people of Brighton had proved unwilling to leave their trusted doctors for Mantell: 'hosts of visitors' might have been 'murdering' his time, but none of them were patients. By early 1836, it was obvious that Mantell's financial situation demanded drastic action. Reluctantly, he moved his family from their Brighton home, which by the spring had become the commercial, fee-charging Sussex Scientific Institution and Mantellian Museum. Writing in his journal from lodgings, his family having returned to Lewes without him, Mantell proclaimed himself 'sick of the cold-blooded creatures' that surrounded him, sick even 'of existence'. He was in turmoil. And as the next few years would prove – following the rise of a vicious rival – even his reputation, his one consolation, was vulnerable.[44]

8
The Goodness of God

> He gave very clear details of the gradual formation of our earth, which, he is thoroughly convinced, took its rise ages before the Mosaic record. He says that Luther must have taken a similar view, as in his translation of the Bible he puts '1st' at the third verse of the first chapter of Genesis, which showed his belief that the two first verses relate to something anterior.[1]
>
> Lady Caroline Fox on William Buckland, September 1836

Francis Egerton, the eighth Earl of Bridgewater, was one of the great oddities of late Georgian society. An ordained clergyman, he had 'assiduously neglected his parish' and fathered a slew of illegitimate children; an otherwise competent historian, his life of an Elizabethan Lord Chancellor was described by one reviewer as 'the worst piece of biography I have ever had the misfortune to be condemned to read'; and though he was heir to Ashfield House in Hertfordshire, he preferred to live in his Parisian townhouse, where he dressed his dogs for dinner and shot game in the garden.[2] Yet when the Earl of Bridgewater died in 1829, he left two major bequests to the British nation. The first was a prize collection of historical manuscripts which came with a fighting fund for the acquisition of more; even now, many histories of the eighteenth and nineteenth centuries will cite something from the Egerton Papers at the British Library. The second was a gift of £8,000 that he entrusted to Davies Gilbert, the president of the Royal Society and an early sponsor of Gideon Mantell, for sponsoring a series of works 'on the Power, Wisdom, and Goodness of God, as manifested in the Creation'.[3] It was a means of atoning, some said, for a life of iniquity.

The Bridgewater Treatises, as they became known, were intended as the solemn defence of natural theology, the idea that divine design was inherent within nature and therefore revealed by scientific inquiry. A panel of trustees comprising Gilbert, the Archbishop of Canterbury and the Bishop of London would confer the honour of authorship upon only the most 'eminent persons'.[4] They chose the Scottish moral philosopher Thomas Chalmers to write on mankind's 'intellectual and moral faculties'; John Kidd, the first Oxford lecturer in mineralogy, took on medicine; and the Cambridge mathematician William Whewell considered astronomy and physics. With the panel commissioning other treatises on the human hand, the 'instincts of animals', chemistry, and 'vegetable physiology' – copies of which reached Charles Darwin during his voyage on the *Beagle* – one major and controversial topic remained.[5]

The trustees had commissioned William Buckland to justify the ways of God to man as early as 1830, but the Oxford don struggled for years to complete his work on 'geology and mineralogy considered with reference to natural theology'.[6] Some part of this was an admittedly poor work ethic: 'I have about as much command of time here', he confessed to Roderick Murchison, 'as a turnpike man, and as I have not your valuable military talent of early rising, I cannot steal a march upon the enemy by getting over the ground before breakfast'. Another part was the sheer scale of the project, an all-encompassing account of human knowledge of the Earth, as it then stood. Buckland lamented having to engage 'five or six different artists' to make as many as 705 engravings, and then that 'all their errors had severally to be corrected'.[7]

As a man ever anxious to reconcile his geology to prevailing opinion, it did not escape Buckland's notice that the early 1830s had seen a conservative backlash against the geology of Lyell's *Principles*. In 1832 the popular historian Sharon Turner, who had written extensively about the Anglo-Saxons, published the first of three volumes on *The Sacred History of the World*. Declaring that mankind could never properly understand nature 'if its creation by the Deity be excluded', Turner embraced 'the great Newtonian principle of the Divine causation of all things'. As for the 'saurian animals' whose fossils had emerged from the earth, animals which had 'no similitude' with any extant species, Turner had a simple explanation: because Moses had recorded only 'one general creation', all the plesiosaurs and iguanodons necessarily came into existence in the beginning; afterwards, during the 1,656 years that in James Ussher's calculation separated Creation from the Flood, the 'spirit of ELOHIM' – that is, God – remained in operation, raising up and laying down the rocks which thus entombed these lizards. Reprinted by Harper's in New York, the *Sacred History* was a thunderous expression of literalist Christianity.[8]

Even more troubling for Buckland, Oxford's antiquated establishment was fortifying its opposition to innovative science. Giving the Bampton Lectures on theology, the cleric Frederick Nolan railed at the impiety of geological research. He mocked the 'arbitrary system' and 'vain imaginations' of the geologist, deplored the notion that a 'day' in Genesis could have meant anything other than twenty-four hours, and abhorred the idea that geologists knew any more than 'the Hebrew lawgiver', who 'could not have been unversed in the general principles' of science.[9] The ferocity of Nolan's attack astonished Mary Buckland: 'Alas! My poor husband', she wrote, 'Could he be carried back a century, fire and faggot would have been his fate, and I daresay our Bampton Lecturer would have thought it his duty to assist at such an Auto da Fé'.[10] It was a solemn lesson that, for all of Charles Lyell's commercial success, there endured passionate hostility towards the more radical implications of this new geology.

In the wake of wider tumult, Oxford was the site of other theological developments. Since 1828, in what historians have called 'the collapse of the *ancien régime*', the political face of the United Kingdom had changed dramatically. The Great Reform Act had swept away dozens of 'rotten' boroughs and expanded the electoral franchise, Parliament had finally abolished colonial slavery, while laws of 1828 and 1829 had lifted restrictions upon Dissenting Protestants and Catholics.[11] For some High-Church Anglicans, the final straw was the Irish Church Temporalities Act, which eliminated ten dioceses in Ireland. They now believed that Earl Grey's Whigs, who had replaced the Tories after sixty years of almost unbroken conservative rule, were the mortal enemy of established religion. In one furious pamphlet, Thomas Elrington, the Bishop of Ferns and Leighlin in south-east Ireland, railed against 'invasions of the fundamental principles of our spiritual Church', the 'very existence' of which he – and others – thought was under threat.[12]

For John Keble, an Oxford graduate and Gloucestershire vicar, all this amounted to 'national apostasy'. Preaching before Oxford's judges on the portentous date of Bastille Day in 1833, he declaimed against a modern Britain where 'liberties are to be taken, and the intrusive passions of men to be indulged, [and where] precedent and permission, or what sounds like them, may be easily found and quoted for everything'. Lambasting both 'public measure' and individual men, Keble painted the 'fashionable liberality' of the 1830s as the same deplorable force which had 'led the Jews voluntarily to set about degrading themselves . . . with the idolatrous Gentiles'. And if such atrocious developments were to be 'forced on the Legislature by public opinion', was 'APOSTASY', Keble asked, 'too hard a word to describe the temper of th[e] nation'?[13]

Apocalyptic warnings about moral decline were not unique to Gloucestershire vicars or to the summer of 1833, but what set Keble's sermon apart was the impetus that it gave to wider change. Several Oxford academics, most of them past or present fellows of Oriel College, joined in the belief that the government and the wider nation were betraying the true nature and historical mission of the Church. Among them were the Hebrew scholar Edward Bouverie Pusey, the conservative priest Richard Hurrell Froude, and John Henry Newman, a former evangelical who now believed that the Church should be entirely 'independent of the State . . . endowed with rights, prerogatives, and powers of its own'.[14] Founding *Tracts for the Times* as a means of publicising their new theology, this Oxford Movement would argue in time that, if the Church of England wished to recover its past glories, it should resurrect the ceremony, symbolism, and appeals to emotion which characterised Catholicism and Eastern Orthodoxy.[15] Although many 'Tractarians', as they were also known, preached in British slums, interceding among the poor where the government would not, there was little room for liberalism in the Movement.[16] Newman and Froude, for instance, were both aghast at the abolition of colonial slavery, the latter describing Africans to

Keble as 'so horridly ugly [that] . . . one can distinctly trace the differences of caste in all shades from man to monkey'.[17] All this meant, as Buckland struggled with his treatise, that the intellectual tenor of Oxford life was ever less welcoming to innovation.

Whatever progress Buckland made with his work owed much to the patience of his wife. Besides travelling with him on many of his field trips, and then tolerating the wildlife that decorated their rooms at Christ Church, Mary Buckland had often played a key, practical role in her husband's research. On one occasion, when he was struggling to identify which creature had left a set of fossilised footprints in sandstone, it was she who rolled a slab of dough, laid it upon the kitchen table, and set their pet tortoise upon it; to their delight, the footprints matched. During the arduous writing of the Bridgwater Treatise, she rendered a greater service. 'My mother sat up night after night,' recalled the Bucklands' son Frank, 'for weeks and months consecutively, writing to my father's dictation; and this, often till the sun's rays, shining through the shutters at early morn, warned the husband to cease from thinking, and the wife to rest her weary hand'. In time, this working pattern produced results. By the autumn of 1835, Buckland had shown early drafts to several bishops and the Tractarian scholar Pusey, and all of them were 'perfectly content' with Buckland's thesis. By the time of the British Association's 1836 meeting in Bristol, the labour was almost complete. 'I am astonished it has been finished so soon', he told his colleagues: 'Such is the intricacy of the subject, such the tiresomeness of the details, that were the work to be done over again, no power on earth should induce me to undertake it.'[18]

When Buckland finished the treatise in May 1836, he made clear that geology, not theology, was the junior subject. (This was the same position that he had adopted for his inaugural lecture back in 1819.) In the preface, he declared that the purpose of 'the Materials of the Earth' was to furnish 'abundant proofs of wise and provident Intention', that geological phenomena would disprove heretical ideas about the 'gradual transmutation of one species into another', and that fossilised remains – rather than calling biblical history into question – in fact demonstrated 'the exercise of stupendous Intelligence and Power'. The collected weight of geological evidence would 'afford an argument of surpassing force against the doctrines of the Atheist and

Polytheist' and illustrate 'the highest Attributes of the One Living and True God'.[19]

Properly speaking, Buckland's treatise concerned palaeontology more than geology, for most of its nearly 1,000 pages addressed the fossil lizards that were so prominent in contemporary imagination. And because of his lateness in publication, Buckland was not even the first Bridgewater author to discuss them. In 1835, the entomologist William Kirby had speculated, just like Robert Hooke in the seventeenth century, that some of these ancient beasts could still inhabit the deeper reaches of the oceans. 'It would not be wonderful', he suggested, 'that some of the Saurian race, especially the marine ones, should have their station in the subterranean waters, which would sufficiently account for their never having been seen except in a fossil state'.[20] Buckland would take the apology further. In his view, the *Plesiosaurus* was 'one of the most anomalous and monstrous productions of the ancient systems of creation'; the *Pterodactyl*, 'Milton's fiend', was so peculiar that it could only have sprung from 'some intelligent First Cause'; and the teeth and claws of the *Megalosaurus* were no abomination, but instruments 'adapted to effect the work of death most speedily'.[21] The fossils of southern England were not aberrations which undermined sacred history, but the careful work of the Lord.

As ever, the opinion of John Murray's *Quarterly Review* mattered most, but Buckland need not have worried: the leading conservative magazine in the English-speaking world praised the treatise as 'a most instructive and interesting volume, of which every page is pregnant with facts inestimably precious to the natural theologian'. Pointedly, the reviewer commended Buckland for swelling 'the chorus in which all creation "hymns His praise"' and for testifying 'to His unlimited power, wisdom, and benevolence'.[22] Likewise, the *Edinburgh Review* found much to admire, writing that Buckland's work was 'calculated to inspire the most affectionate veneration for that Great Being who has made . . . the material world'; as for the fossil creatures, it urged readers to think of them as belonging to a 'series of creations which the Almighty has successively extinguished, and successively renewed'.[23] Charles Lyell, whose geological theories of long-term, gradual change could otherwise have conflicted with Buckland's theology, recognised the prestige that Buckland's treatise had given to their work: addressing the Geological Society in February 1837, he declared that Buckland's work had gone some way 'towards the filling up one of the greatest blanks which existed in the literature of our science'.[24]

As Buckland basked in critical acclaim and turned to new projects – he would search the Alps and Scottish Highlands for evidence that glaciers had carved their own valleys – Charles Darwin was resuming life in England. Having landed at Falmouth in October 1836, he had raced from Cornwall to Shrewsbury in just two days, slipping into his old room at night, then surprising his father and sisters at breakfast. 'Dead and half alive', exhausted from the voyage, he began writing to the friends and family he had not seen for years.[25] The most urgent letters, however, went to the naturalists who could explain the specimens that served as souvenirs of the *Beagle*. Within a fortnight, Henslow at Cambridge had agreed to look at the plants; Darwin began to discuss the geology with Sedgwick and Lyell; and the anatomists of London embraced him as a source of abundant research material. Quickly, a plan came together for a multi-volume, multi-author survey of all that Darwin had seen and collected during his time at sea.

No longer desiring a quiet country parish in the Church of England, Darwin went down to London. Living with his brother Eras in Soho, he spent his days walking to the learned institutions of the metropolis: to the British Museum in Bloomsbury, the Geological Society at Somerset House, and the Zoological Society in Regent's Park. In these august halls of science, Darwin's reputation preceded him: his Cambridge mentor Henslow had already published several letters from the *Beagle*, presenting them to an excited readership as 'the first thoughts which occur to a traveller respecting what he sees'.[26] Darwin kept more radical company too: Eras was 'driving out' with the Dissenter and reformer Harriet Martineau, whose *Illustrations* were introducing and explaining political economy to the wider public.[27] Charles found Martineau 'very agreeable' but a 'little ugly' and 'overwhelmed with her own projects, her own thoughts and own abilities'; Eras apparently got over these unfeminine obsessions by 'not looking at her as a woman'.[28]

As Darwin walked the streets of London, money was of no concern. His father, 'the Doctor', had amassed one of the largest fortunes in Shropshire through judicious investment and moneylending, and he guaranteed Charles a yearly allowance of some £400, more than £32,000 today. It was enough to live independently, maintain a servant, and afford the services of the fossil-trading artist George Sowerby, who would illustrate many of the *Beagle* specimens. Family was another matter. Though he enjoyed the intellectual energy of the capital, Darwin – like so many people spending their first winters in the city – despaired at the noise, dirt, closeness, and loneliness of life in London. How much more time and space to think would he have in the countryside? How much better to take a wife, to raise a family? The idea of domestic comforts was all the more poignant for having spent so long in the company of sailors, and on a stray blue sheet of paper he listed the costs and benefits of marriage. He weighed the 'Charms of music & female chit-chat' against the 'Conversation of clever men at clubs', the warmth of a 'Constant companion, & friend in old age' against the 'freedom to go where one liked'. On balance, marriage won out: 'Imagine living all one's day solitarily in [a] smoky dirty London House', he concluded. 'Only picture to yourself a nice soft wife on a sofa with good fire, & books'.[29] In 1838, Darwin bound his family even more tightly to the Wedgwoods when he proposed to his first cousin, Emma; they married two months later, moving into a house on Gower Street in Bloomsbury.

By the spring of 1839, Darwin had become an author, his diary from the *Beagle* appearing alongside the journals of Robert FitzRoy and Philip Parker King, a former captain of the ship. Preparing and publishing these volumes had been excruciating, for Darwin complained repeatedly of something close to writer's block, looking enviously at the sentences, chapters, and books that flowed fluently for Miss Martineau. Coaxing a manuscript from FitzRoy was even harder. The sailor, who took great offence that Darwin's first draft had omitted 'any notice of the [*Beagle*'s] officers', was a self-confessed 'slow coach'.[30] Even so, the journals impressed most readers. One colleague thought Darwin's work 'as full of good original wholesome food as an egg', while praise from Alexander von Humboldt, the inspiration of Darwin's travel, brought the ultimate reward.[31] Yet if

Darwin ever doubted that 'bold generalisations' about sensitive issues could lead him into trouble, the *Athenaeum* literary magazine convinced him of those dangers. It bridled at the young man's suggestion that, if geology could explain the presence of fossil shells in the Andes, 'at least one million of years must have elapsed since ... the sea washed the feet of the Cordillera'.[32] There was also the coda that FitzRoy had appended to his journal, a burst of religious enthusiasm where he discoursed on the lifespan of the patriarchs, 'the landing of [Noah's] ark on a mountain of middle height', and the miracle the Lord performed by preventing the animals on the Ark from devouring one another.[33] It was further notice, as much as Buckland's treatise, that geologists and other naturalists still operated in a world of sincere religious conviction.

Indeed, even as London science feted his reports from the *Beagle*, a crisis of conscience was creeping upon Darwin. Chief among his concerns were the differences that he had observed among the animals of the southern hemisphere, and the conclusions that he was beginning to draw. For instance, when the ornithologist John Gould inspected the finches and mockingbirds that Darwin had collected from the Galapagos, he confirmed that each island had hosted not just different varieties but different *species*; the tortoises, too, appeared to be unique to their islands. But why and how had this happened? Could Darwin explain these differences as the result of the immigration of particular species to particular islands? He could not. Rather, he was turning over in his mind the idea that all the finches had once been the same, that they had changed in their constitution only *after* arriving on the islands. Forty years before, his grandfather Erasmus had written of an animal 'acquiring new parts' and 'continuing to improve by his own inherent activity', the Lamarckians had long promoted the idea of the transmutation of species, and now Darwin was contemplating the idea of 'descent'.[34] It was perilously close, he knew, to what the geologist Adam Sedgwick had damned as 'gross and ... filthy' French thinking.[35]

For the most part, therefore, he confined these thoughts to a secret notebook, for he knew the penalty for apostasy. When Darwin was a student at Cambridge, the radicals Richard Carlile and Robert Taylor had arrived in town to preach a gospel of subversion. Carlile had already served six years in prison for publishing Paine's *Age of Reason* and his comrade, Taylor, was a former priest who had suffered prosecution for blasphemy; the latter had written his *Diegesis*, a searing depiction of Christianity as a pastiche of pagan mythologies, from a prison in Rutland.[36] Together, Carlile and Taylor had embarked upon an 'infidel home missionary tour', striving to prove to open-minded Britons that 'Jesus Christ, alleged to have been of Nazareth, never existed'. Upon coming to Cambridge in 1829, they established 'Infidel Head-Quarters' above a print shop at 7 Rose Crescent and, each day, they set forth in full academic dress to dissuade freethinking students of 'the merits of Christian religion'. Carlile reported that perhaps fifty 'young collegians' had proven 'bold in avowing Infidelity among each other', but the Cambridge authorities had acted decisively to curtail their mission: the landlord at Rose Crescent found that his licence to accommodate lodgers had been revoked, and rumours abounded of vigilantes looking to dispense godly justice more directly. As Taylor and Carlile swept up their belongings and made haste for Wisbech to the north, their persecution made a lasting impression on the young Darwin: even decades later he would fear condemnation, like Taylor, as 'the Devil's Chaplain'.[37]

The fate of radical naturalists was no more appealing, and the treatment of Robert Grant was a case in point. At Edinburgh, Darwin had learned at the side of Grant, who was a leading expert on marine invertebrates. Yet in seeking to connect simple, undeveloped creatures such as sponges and polyps to more sophisticated organisms, Grant had embraced the transmutational zoology of Lamarck and Geoffroy, with whom he had studied in Paris. So long as he remained at Edinburgh, simple distance kept Grant safe from censure; in 1827, however, when he took the first chair in zoology at London, the patrician guardians of British science paid closer attention to his work. For a while, they and the university authorities tolerated his teaching that animals changed in response to 'climate, domestication, and other external circumstances', but by 1838 they had resolved to destroy his reputation.[38] Inviting Grant to speak before the Geological Society on a fossilised 'opossum' from Oxfordshire, Britain's leading naturalists – including William Buckland – baited him into describing that fossil as reptilian and declaring that all such reptiles, having been found in the oldest rocks, would transmute into mammals. This was the heresy that the Geological Society had anticipated and, with Buckland having invited the Lord Chancellor 'to witness our Skirmish', the whole evening amounted to an intellectual assassination. Darwin was there to see it. He agreed with Grant, writing in his diary that the opossum was 'the father of all Mammalia in ages long past', and duly feared the same ostracism.[39]

But Darwin could not shake the idea of 'descent', nor the ideas of the clergyman Thomas Malthus, who had supposed an eternal struggle for existence in a world of finite resources. And as Darwin spent the late 1830s talking to farmers and dog-handlers about the qualities and differences that they could breed into their animals, there was the further matter of Jenny, the orangutan who since November 1837 had lived in the giraffe enclosure at London Zoo. Dressed in children's clothing and trained to drink tea from a cup, Jenny had amused the patrons of the Zoo, but she fascinated Darwin. Returning time

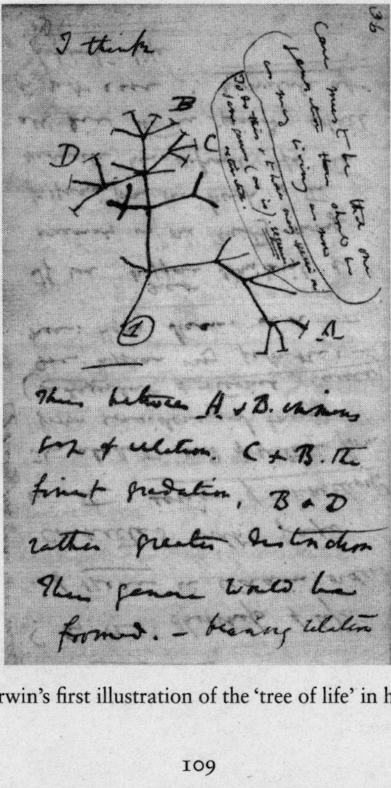

Charles Darwin's first illustration of the 'tree of life' in his notebook

and again to see her in Regent's Park, he noted how the orangutan would learn to follow instructions, to hold open doors, and to rejoice in eating apples 'with the most contented countenance imaginable'.[40] At other places in his notebooks, Darwin marvelled at her 'untying a very difficult knot', her expressions of 'fear or shame', and her apparent readiness to play the mouth organ when guided to do so.[41] It would be some time before Darwin collected his thoughts fully but, as early as 1838, there were questions and doubts: 'Let man visit Ouranoutang in domestication', he wrote in Notebook 'C'; let him 'see its passion & rage . . . and then let him boast of his proud preeminence . . . Man in his arrogance thinks himself a great work, worthy the interposition of a deity. [It would be] more humble and I believe true to consider him created from animals.'[42] In Notebook 'B', meanwhile, under the words 'I think', Darwin had sketched something which looked like a tree, its branches sprawling every which way from a common origin. These were dangerous ideas indeed.

9
Terrible Lizards

> [Religious] ministers have been, through all recorded time, and are, at this moment, from pole to pole, the legalised prime demoralisers of our species. They pour their poison of lies into the ear of cradled infancy – they debauch reason in the very womb . . . If every priest was at the bottom of the Red Sea, society would be infinitely more happy than it is at the present moment.[1]
>
> *The Oracle of Reason* (1842)

Across the country but especially in the south-west counties, several factors had combined to make Britain the global frontier of palaeontology. There was the sheer good luck that so much of the landscape was composed of the relatively soft, sedimentary rocks that trapped and preserved prehistoric fossils; there was the pulsing drive of the Industrial Revolution, which had sunk numberless quarries and driven countless canals through those same rocks; and there was a concentration of universities and learned societies where amateurs and professors could discuss the fossils that they found.[2] By 1841, the Annings had excavated the *Ichthyosaurus*, the *Plesiosaurus*, and the flying *Dimorphodon* from the cliffs of Lyme Regis; William Buckland had pieced together the *Megalosaurus* from the fields of Oxfordshire; and Gideon Mantell had announced the *Iguanodon* and the armoured *Hylaeosaurus*. British science had also taken note of a series of foreign discoveries. There was the *Mosasaurus* from the Netherlands, the *Pterodactyl* of Bavaria, the *Megatherium* from South America, and the curious three-toed footprints that Edward Hitchcock, one of

Emily Dickinson's mentors at Amherst College, had found in the flagstones of western Massachusetts. The last of these discoveries, at first mistaken as 'turkey tracks', would inspire lines of verse from Henry Wadsworth Longfellow: 'As once by the margin of rivers / Stalked those birds, that have left only their footprints'.[3]

There was now an established vocabulary for describing these monsters. They were 'terrific', 'wonderful', 'appalling', 'fearsome', and 'antique'; in more fantastic prose, the fossil-hunter Thomas Hawkins raved about 'carnivorous automata' spawned by Satan, machines 'contrived for none imaginable End worthy of a god'.[4] Nonetheless, and despite William Conybeare suggesting 'enaliosauria' to describe the maritime creatures, there was no general term that went beyond the evocative, almost mythological language of 'dragons', 'beasts', and 'fossil-Lizards'.[5] So, what should they call the *Iguanodon*, the *Megalosaurus*, and the *Hylaeosaurus*, all of which dwelled on land, collectively? It was in the spring of 1841, at the Plymouth meeting of the British Association for the Advancement of Science, that the new leading man of British zoology – and one of its most reactionary figures – gave the persuasive, binding answer.

Richard Owen was born in 1804 in Lancaster, the son of a merchant who traded with the West Indies, and from an early age he displayed an uncommon belligerence. When a new student at school exhibited such intelligence that Owen's teachers 'considerably raised the standard of work', he was 'loud in his taunting expressions of disgust'. But in the fight that followed between Owen and his new classmate, it was the latter who administered 'a couple of black eyes'; by coincidence that student was William Whewell, who became a leading Cambridge mathematician and an author of a Bridgewater Treatise. Unlike many among the first generation of palaeontologists, who spent their childhoods finding fossils among the fields and hills of England, Owen at first took little interest in the natural world, preferring to read of heraldry. Still, at the age of sixteen, he followed Gideon Mantell into a medical apprenticeship, his parents contracting with a surgeon in Lancaster that he should receive 'meat, drink, washing and lodging, and also decent suitable cloathes' [*sic*] while learning the 'arts, businesses, professions, and mysteries of a surgeon apothecary and man midwife'. Most of Owen's studies took place at

Lancaster Castle, which served as the town's gaol and where, in the highest room in 'Hadrian's Tower', medical students learned from their masters' dissection of dead inmates, whom Owen described as 'the unfortunates who by natural death [were] liberated from prison'.[6]

The castle was also the site of two incidents that Owen would often recount. In the first, on climbing the tower late at night to retrieve some instruments, he perceived a 'tall and thin [figure] leaning against or clasping the central stone pillar of the staircase'; more alarmingly, he also saw a pair of 'half-opened glassy eyes'. Scared witless, and making his escape by 'precipitate descent', Owen saw a 'second figure in white' appear before him 'as if to intercept my passage'. Had he seen ghosts? Of course not, but the shock was enough that he vowed 'never, never again to desecrate the Christian corpse, and to quit a profession that could only be learnt by such practices'. He did not keep this promise. Only weeks later, re-enthused by a treatise on the 'Varieties of the Human Race', Owen stole back into the tower on the night that a black prisoner had died in the cells. His mission was to sever the dead man's head from his body and to carry it home, where he could study the skull in private and satisfy his 'craniological longings'. Taking up the saw that lay beside the corpse, Owen secured what he called his 'specimen of the Ethiopian race' and placed it in a 'strong brown-paper bag'. But it was 'a fine frosty evening in January', and on leaving the castle Owen slipped on the ice; he fell forwards, the bag with him. In Owen's telling, this 'jerked the negro's head out of the bag and sent it bounding down the slippery surface' of the road. At the foot of the hill, the dead man's head bounded through the front door of a cottage belonging to the widow and daughter of a former slave-ship captain. Its 'white protruding eyeballs' struck terror into the women: 'What could this be', mused Owen, 'except an apparition' from the life of the departed slaver? Perversely, Owen would relish telling this story for laughs, but it was an early indication of his intellectual obsession with the human skull.[7]

Later spending two terms at Edinburgh – where he indulged with fellow students in 'Het Pints' of boiled ale, whisky, and spice – Owen completed his apprenticeship in 1826 and moved to London, taking up a position at St Bartholomew's Hospital. His nocturnal chicanery at Lancaster now paid dividends. With his proficiency in dissection attracting the attention of senior medical figures, William Clift, the conservator of the museum at the Royal College of Surgeons, appointed Owen as his assistant. Over the next five years, in the bowels of the museum and in the shadow of vast 'stoppered bottles of alcohol' filled with exotic specimens, Owen worked through the remains of John Hunter's collections. It was difficult, tedious work, for there was no 'adequate catalogue', but the counsel of his mother kept Owen focused: 'I am most anxious, my dear boy,' she wrote, 'for your improvement and success'. In time, the son honoured the mother. Celebrated papers on the orangutan, the platypus, and the pearly nautilus mollusc built his reputation and, with Clift recommending Owen as 'sober and sedate very far beyond any young man I ever knew', the Birmingham Hospital offered him the job of house surgeon. He took it, but on finding the Midlands drab and poor, Owen returned to London within days.[8]

There was another reason that Owen gave up on Birmingham. In 1827, William Clift's daughter had been hanging a pair of bellpulls in her mother's bedroom when she fell from the stepladder, knocking herself unconscious; in the emergency, the family had summoned a surgeon, and Owen had answered the call. Miss Caroline Clift enchanted him at first sight, but even in the firmness of youth there was not much to commend Owen to her. Though tall, he was painfully gaunt and, in the words of one historian, his face was 'very large, with a prodigious forehead, and very large eyes, which seemed highly speculative and pondering in thought'. Thomas Carlyle was less charitable still: when he first met Owen, he described 'a man of a huge coarse head, with a projecting brow and chin (like a cheese in the *last* quarter)'.[9] Despite this, Owen and Miss Clift hit it off and, by the end of 1827, they were engaged. The only impediment to marriage was Owen's penury: though he was an esteemed anatomist, Mrs Clift insisted that he 'should have sufficient means to provide for her daughter before she would hear of the [ceremony] taking place'. The couple therefore embarked upon extensive correspondence, a courtship in print, though it would be difficult to describe Owen as a romantic: 'You have now, my dear Caroline,' he wrote to his fiancée, 'effected what I long wished [and] directed your thoughts in a definite channel on the subject of our approaching union, a thing certain and fixed'.[10]

They married at last in 1835 in St Pancras Church at Euston and, after a brief visit to Oxford under the supervision of the indomitable Mrs Clift, the newlyweds moved into Owen's new apartments at Holborn. This would be home, but it was a laboratory too: when Owen became the Hunterian professor of anatomy at the Royal College of Surgeons, he gained the right of first refusal to any animal which died at London Zoo. It followed that, besides accounts of polite society – of arrangements from *Don Giovanni*, late suppers, and readings from Dickens' 'Boz' – Caroline Owen's diaries from these early years of marriage describe the constant receipt of specimens and carcasses. On one occasion, she came home to find a dead rhinoceros in the hallway; on another, she wrote: 'To-day Richard cut up the giraffe which died at the Zoological Gardens. Afterwards he went to the Royal Institution to dissect a snake.' And as with Mary Buckland and Mary Mantell, Caroline Owen often played an important role in her husband's work. She translated scholarly treatises from German, drew 'outlines of shark's teeth', and in December 1836 dutifully wrapped a tortoise in flannel and 'put it in the front cellar' before retiring for the evening. Of course, this could be a messy business, and one of its less salubrious aspects was the delivery of the putrefying flesh that Owen needed for his investigations into worms: 'I could not get over the smell', she complained, 'of the decaying piece of muscle for hours'.[11]

As the crown prince of British anatomy, Owen was by the mid-1830s moving in the most fashionable circles. He was dining with Charles Lyell and the mathematician Charles Babbage, offering counsel to Charles Darwin on zoological specimens from the *Beagle*, and establishing friendly terms with the aristocratic guardians of society: George Grey, the son of the earl and former prime minister, would write to Owen from Tenerife, where he had procured 'the skull of a guancho (an aboriginal of the Canary Islands)' for his collection. Even William Whewell had forgiven Owen for their schoolyard fisticuffs, writing in 1837 that he was 'much pleased' to hear of his forthcoming editions of John Hunter's zoology. Owen relished this life, and he married technical brilliance to the masterful administration of scientific society: he sat on the right committees, procured the right patronage, and cultivated the right friendships. Honours came in a stream from the Royal Institution, the Geological Society, and Russia's Imperial Academy of Sciences; even allegations of plagiarism could not stop his ascent.[12]

Although this early eminence was in the study of the living, Owen was also a student of the ancient reptiles that had once stalked the land and waters of the British Isles. In 1825, the fossil-hunter John Kingdon had found several peculiar vertebrae in a quarry outside Chipping Norton; William Buckland had collected similar remains from the Oxfordshire stone; and as the workmen of the London and Birmingham Railway drove through the hills of Northamptonshire – an exercise that one engineer described as 'the greatest public work ever executed' – they exposed arm-bones and shoulder blades of 'gigantic proportions'. In the early 1840s, Owen brought these fossils together as evidence of a 'saurian genus distinct from' any already known. He called the new creature *Cetiosaurus*, from the Greek for 'whale lizard', and supposed it was a vast, maritime creature of 'carnivorous habits'.[13] He was wrong – the *Cetiosaurus* was in fact a four-legged, land-dwelling herbivore – but it was a major discovery all the same.

Among his gifts, however, Owen had developed a special faculty for making enemies. One of his motives could be the merely personal, especially where the achievements of a potential rival exposed any apparent deficiency in his own record. Gideon Mantell, for instance, could not hope to match Owen for social cachet or political influence, but he had discovered two ancient reptiles whereas Owen had not, and that stoked a fearsome jealousy. Therefore, when Mantell submitted papers to the Royal Society, Owen manoeuvred behind the scenes to prevent their acceptance; and when Mantell suggested that the *Iguanodon*'s 'spike' belonged to its thumb, Owen placed it instead on the creature's skull, ridiculing the provincial surgeon (although time would prove Mantell correct).[14] For the most part, though, Owen persecuted ideological enemies, especially those who advocated theories of transmutation. For Owen and other orthodox Anglicans such as William Buckland, who had become a fast ally, these ideas were heretical because they supposed that the Lord had erred at the moment of Creation: how could animals improve on what was already perfect?

Moreover, these theories were politically explosive: by implying that lowly creatures could rise independently in the great chain of being, they provided an unwelcome natural analogy for the transformation that Chartists, among others, were pursuing in British politics. Owen did what he could to arrest the spread of these ideas, not least by vetoing the zoologist Robert Grant's admission to the Geological Society, but it was in fact by naming the term 'dinosaur' that he took a more precise and considered aim at what Darwin was privately calling 'descent'.[15]

On that day in Plymouth, in a speech which ran to 144 pages in print, Owen coined the term '*Dinosauria*', meaning 'terrible lizards', to describe the *Iguanodon*, the *Megalosaurus*, and the *Hylaeosaurus*. In his view, these dinosaurs exhibited 'superior adaptation to terrestrial life' and had 'a highly organized centre of circulation'; the *Megalosaurus* and *Iguanodon* in particular represented the 'most perfect modifications of the Reptilian type, attained the greatest bulk, and must have played the most conspicuous parts . . . that this world has ever witnessed in oviparous and cold-blooded creatures'. Most importantly, they were 'as superior in organization and in bulk to the Crocodiles that preceded them as those which came after them'.[16] This was crucial, for although it might appear counter-intuitive to the modern reader, Owen had created the category of dinosaur not to encourage developmental theory but to rubbish it. In fact, by situating the dinosaur, the 'pinnacle of reptilian perfection', at the very beginning of knowable history, he was seeking to explode the profane notion that species had changed over the generations from less complex to more intricate organisms.[17] All through the speech, therefore, he teased a leading question: how could any theory of progressive transmutation be correct when the dinosaurs, the most stupendous reptiles ever known, had existed only in the beginning? If there was any merit to evolutionary theory, he asked, would dinosaurs not be alive and well today? In making these arguments, Owen sought to strike a deliberately conservative tone and to wrest control of palaeontology from heathens such as Lamarck, Geoffroy, and Grant. And while this description of dinosaurs and Owen's legitimisation of their study would later attain a radical significance that he could not have foreseen in 1841, the speech landed perfectly at the time. On returning to the College of Surgeons, Owen found a letter from the prime minister, Robert Peel, advising that his distinguished public service had earned an annual pension of £200 from the Civil List.

Not everyone enjoyed such generous favours. It was ninety miles from Birmingham to Bristol, but George Jacob Holyoake could not afford even third-class passage on the coach, so he was walking. Holyoake was twenty-five and a former devotee of Robert Owen, the early socialist thinker who advocated the formation of utopian, cooperative communities as alternatives to the dark, Satanic mills of the Industrial Revolution. In the late 1830s, therefore, as nascent trade unions and 'combinations' sought support for workers' rights, and as the People's Charter inspired widespread agitation for political reform, Holyoake had been preaching the socialist gospel in Birmingham and Sheffield. Yet when Robert Owen – no relation to Richard – decided that his 'social missionaries' should take a religious oath that would allow them to preach on Sundays, Holyoake and several others had quit the fold in disgust. They were not Sabbatarian; they just would not countenance any sort of religious profession. One of Holyoake's fellow rebels was Charles Southwell, the youngest of thirty-three children and the editor of Britain's first avowedly atheistic periodical, *The Oracle of Reason*, which in late 1841 provoked the fury of Church and state alike when it described the Bible as a 'revoltingly odious Jew production'.[18] Magistrates in Bristol charged Southwell with blasphemy and secured a conviction; he received a prison sentence of twelve months, and it was to visit him that Holyoake was walking to Bristol in early 1842.

Despite falling 'among thieves by the way', Holyoake had made it as far as Cheltenham, where he agreed to speak to the local Mechanics' Institute, one of those pioneering establishments which since the 1820s had provided education in the arts to working men.[19] The subject of Holyoake's lecture was home colonisation, which was the policy – extremely popular in the early nineteenth century – that 'surplus' populations should move to underdeveloped parts of the British Isles instead of emigrating overseas. The clergy of Cheltenham, however, had listened to Henry Phillpotts, the strident Bishop of Exeter, who was imploring his brethren to make 'inquiries ... into the diffusion of blasphemous and immoral publications'; accordingly, they laid a trap for Holyoake, who already had a reputation as an infidel.[20] As he spoke to roughly one hundred Chartists, freethinkers, and a 'sprinkling of adventurous ladies', Holyoake stuck to his topic and said nothing of religion, nothing to provoke. But when he invited questions from the audience, 'a local preacher of a darkly zealous order' stepped forward: Holyoake had spoken 'a good deal about their duty to man', he said, 'but what about their duty to God?'[21]

He could not resist. Holyoake proclaimed to the room that 'religion . . . has ever poisoned the fountain-springs of morality'; he compared the history of religion to 'mental degradation and oppression'; and he damned the clergy as 'fierce and inhuman myrmidons'. At the peroration, he declared: 'I wish not to hear the name of God, I shudder at the thought of religion, I flee the Bible as a viper, and revolt at the touch of a Christian.'[22] The crowd laughed and many applauded, but the questioner's work was done. Two days later, the *Cheltenham Chronicle* printed a forthright condemnation of Holyoake's remarks, and news of the furore reached him at Bristol, where he had finally seen Southwell in prison.[23] Now, Holyoake walked back to Cheltenham to meet his accusers and to speak again, this time on the more incendiary topic of 'Civil and Religious Liberty'. After an hour of this second lecture, the superintendent of the police entered the hall, armed with 'all the available force', and the constables lined themselves along the walls, their 'shining hats form[ing] a picturesque background'.[24] Having waited for Holyoake to finish, they approached him, arrested him, and escorted him from the Mechanics' Institute. In a touching display of support, Holyoake's audience went too, walking with him to the gaol.

Remanded in custody, Holyoake expected no favour from the authorities in Cheltenham; he thought the town so hostile to radical thinking that any jury would find a 'man guilty of blasphemy who boiled his tea-kettle on a Sunday'.[25] Even so, the charge that magistrates laid against him was extravagant: that he 'maliciously, unlawfully, and wickedly did compose, speak, utter, pronounce, and publish with a loud voice' such things as were 'to the high displeasure of almighty god, to the great scandal and reproach of the Christian religion . . . and against the peace of our lady the queen'.[26] The treatment that

Holyoake received in gaol was no better. He subsisted in a filthy room with 'verminous' company and, when the police surgeon called, it was not to examine Holyoake's health but to interrogate his beliefs: the surgeon was sorry, he said, that 'the days were gone by when we could send you . . . to the stake instead of to . . . Jail'. The trial would take place at Gloucester, a few miles to the west, and so the authorities bundled Holyoake into an early railway carriage, his hands bound in irons that tore the skin from his wrists. When he disembarked, the guards searched his pockets, seized his notebook and papers, and threw him into another lice-ridden cell where 'the fetor of death' haunted all corners. He stayed there a fortnight, waiting for the courts to open, with the other inmates offering him mint-tea 'to help him swallow his coarse bread'.[27] By the time that two friends had posted bail of £100 – some £8,000 today – the Home Secretary had spoken in the House of Commons about 'serious irregularities' in the proceedings against Holyoake but declined to intervene in the case.[28]

On Monday 15 August 1842, an 'excited audience' of country gentry, ladies, clergymen, and freethinkers streamed into Shire Hall in the shadow of Gloucester's cathedral. They watched as a jury of seven farmers, a miller, a maltster, a grocer, a poulterer, and a shopkeeper swore their oaths and took their seats.[29] Holyoake appeared, climbing into the dock to arrange his papers and books, and he did not look well: one lawyer described him as 'a wretched-looking creature . . . whose wiry and dishevelled hair, [his] lip unconscious of the razor's edge, and dingy looks, [gave] him the appearance of a low German student'.[30] Mounting his own defence, Holyoake cross-examined the witness for the prosecution, a printer at the local newspaper, who had accused him of reviling 'the majesty of heaven'; he also resisted repeated entreaties from the judge, Mr Justice Erskine, to plead that he had spoken merely from impulse, or that he meant only to say that 'the incomes of the clergy ought to be reduced'.[31] Over six long hours, he refused to compromise, quoting from William Godwin and Goethe, and arguing only 'for free speech'. Conviction and a sentence of 'the Common Jail for six calendar months' were the reward and, by the time that Holyoake emerged from prison, his daughter Madeline had died of malnutrition.[32]

*

It might be tempting to dismiss the case of George Jacob Holyoake as a merely provincial tragedy, as the story of a principled man who fell victim to the connivances of reactionary clerics in the holy heartlands of the West Country. But the response of the press to Holyoake's conviction spoke of a broader hostility towards irreligious thinking. For *The Times*, his defence consisted of 'sophistries and absurdities', and the paper of record declared that six months in prison would give that 'miserable-looking lad ... ample opportunity of reflecting upon the enormity of his guilt, and of discovering the futility of the impious doctrines which he has the effrontery to avow'.[33] Fears of prosecution for blasphemy and of persecution for infidelity were real and pervasive, and they were chief among the reasons why Charles Darwin had not yet committed to paper the ideas which had vexed him since returning from sea: 'I was so anxious to avoid prejudice', he confessed, 'that I determined not for some time to write even the briefest sketch'.[34]

Now father to two young children, William and Annie, Darwin had spent the early summer of 1842 out of London, first in Staffordshire and then at Shrewsbury. And it was there, in the peace of the countryside, over thirty-five pages of 'bad paper with a soft pencil' and with many of his words 'ending in mere scrawls', that he drew together his ideas about species and descent. Here, in a manuscript that his son typed up and published at the turn of the twentieth century, we can see Darwin express so many of his key phrases and concepts for the first time. He writes that 'each parent transmits its peculiarities' to its children and that 'variations ... evince a tendency to become hereditary'; he ponders the 'concealed war of nature' which has bequeathed 'the creation of the higher animals'; and he makes notes under the heading of 'Natural Selection'. There was even room for the *Plesiosaurus*, whose changes in structure – that is, the 'morphology' of its extended neck – had been so extreme that they were now unknown in the world, its variation in form confined alone to what Darwin would call 'extinct gigantic sea-lizards'.[35]

And what of God? He is there, the 'creator of this universe', but only in the beginning. There was no role for the Lord beyond impressing his laws upon matter, for in Darwin's view it was ridiculous – 'derogatory' – to suggest that He had attended the creation of 'the myriads of creeping parasites and [slimy] worms which have swarmed each day of life on land and water (on) [this] one globe'. Was it part of a divine plan, Darwin asked himself, that certain animals had been created 'to lay their eggs in [the] bowels and flesh of other[s], that some organisms should delight in cruelty . . . that annually there should be an incalculable waste of eggs and pollen?'[36] He did not think so, but he would keep that conclusion to himself. These thirty-five pages, this origin of *The Origin of Species*, would remain a secret for now. For a man so concerned with social propriety, whose anxieties were now finding painful expression in persistent problems with his stomach, it was perhaps the judicious course of action. Indeed, when in 1844 another author struck forth in a best-selling book, detailing a history of the universe from which God was almost entirely absent, he needed to do so under conditions of the strictest anonymity. The identity of the author of *Vestiges of the Natural History of Creation* would become one of the great mysteries of the mid-nineteenth century.

10
Abominable Books

This world was once a fluid haze of light,
Till towards the centre set the starry tides,
And eddied into suns, that wheeling cast
The planets: then the monster, then the man;
Tattooed or wooded, winter-clad in skins,
Raw from the prime, and crushing down his mate;
As yet we find in barbarous isles, and have
Among the lowest.[1]

Tennyson, *The Princess* (1847)

In the 1840s, British literature was on the cusp of a golden age. Tennyson had published 'The Lady of Shalott', 'Ulysses', and 'Break, Break, Break' in the collection *Poems*, while Robert Browning's 'My Last Duchess' and 'The Pied Piper of Hamelin' appeared in *Dramatic Lyrics*. The novel was emerging as the dominant literary form, and the British were its masters: the Brontë sisters were writing *Jane Eyre* and *Wuthering Heights*, and William Makepeace Thackeray – in whose work the prehistoric *Megatherium* would appear with frequency – told of *The Luck of Barry Lyndon*.[2] Charles Dickens, who was taking 'unspeakable interest and pleasure in natural history', published *The Old Curiosity Shop*, *A Christmas Carol*, and *Dombey and Son*, where a 'great earthquake' churned the earth and exposed the rocks and soils and 'fragments' of ages past.[3] And in the weekly press, the *Illustrated London News* opened its doors, while Henry Mayhew founded the satirical magazine *Punch*. All were available to read at the new

London Library, which Thomas Carlyle, whose lectures on the heroic were shaping the 'great man' theory of history, had helped to found as a refuge from the cramped and damp conditions at the British Museum. There was also a thriving readership for popular science. Yet rather than geology, to which Charles Lyell and Gideon Mantell had now published popular and successful introductions, it was often the 'science' of phrenology that commanded the greater share of public attention.[4]

Put simply, phrenology was the means of inferring a person's intelligence and character from the size and shape of the skull.[5] It was the invention of Franz Joseph Gall, a German anatomist who believed that the brain was the aggregate of multiple organs, all of which had precise locations and specific functions; the size of those organs, which phrenologists could assess by applying callipers to the cranium, then determined the strength of that person's mental faculties. Among the twenty-seven different functions that Gall identified were 'the gift of music', 'comparative sagacity', and 'educability'. The last of these he defined as 'the faculty of being instructed by means of external objects', and he claimed it was strongest in people who had a 'prominent and perpendicular' forehead.[6]

The leading phrenologist in Britain was George Combe, the son of an Edinburgh brewer and one of thirteen children who had grown up in an atmosphere of stultifying Calvinism. Bored witless by sermons on the Sabbath and quickly estranged from his devout family, Combe was a conscious rebel against the religious strictures that, in his view, inhibited creativity. From an early age, therefore, he had set himself 'to write some useful book on human nature' and there was no doubt – at least in his own mind – that he would achieve greatness in doing so: 'A desire of fame', he noted in his diary, 'may be one mark of a mind that deserves it'.[7] Combe's damascene moment occurred in 1816, when Gall's former assistant, the physician Johann Spurzheim, came to Scotland to defend phrenology's honour against attacks in the Edinburgh papers. On listening to Spurzheim and witnessing his demonstrations, Combe became convinced that this new science could unlock the secrets of the soul.

Combe decanted his views on phrenology into *The Constitution of Man*, which first appeared in print in 1828. Here, he promised to illuminate the relationship between the laws of nature and 'a theory of [the human] Mind', all with a view 'to the improvement of education, and the regulation of individual conduct'; in doing so, he cast himself as the heir not of medical men, but of moral philosophers such as Francis Hutcheson and Dugald Stewart.[8] Although the *Constitution* flopped at first, selling only a few hundred copies, it became the bible of the science when an expanded, subsidised edition came out in 1832. By then, 'craniological mania [had] spread like a plague ... possess[ing] every gradation of society from the kitchen to the gavel'. Armies of phrenological lecturers, seeking quick money from the gullible, now set forth into the shires on errands of unashamed quackery. At the Mechanics' Institute in Banbury the Reverend Dr Eden, who had 'given hundreds of Analyses and Sketches of the Inhabitants of the Principal Towns', promised to demonstrate 'the Physiology of the Brain and other leading principles of Phrenology [and] its harmony with Sacred Scriptures'. In Leicester a handbill promoted the services of Dr Bushea, who between the hours of eleven and five o'clock and then again between seven and ten could provide answers to the all-important questions of the age: 'Are we about to marry?' 'Have we children?' 'Do we desire to know the true Characters of Clerks, Shopmen, or Domestic Servants?' With the 1836 edition of the *Constitution* selling more than 80,000 copies, Combe was the great champion of popular science. One magazine described phrenology as a 'species of intellectual mushroom', while the *Spectator* wagered its readers that 'if in a manufacturing district you meet with an artisan whose sagacious conversation and tidy appearance convince you that he is one of the more favourite specimens of his class, enter his house, and it is ten to one but you find COMBE's *Constitution of Man*'.[9]

Besides tapping into the prevailing Victorian spirit of 'progress' and self-improvement, Combe's great advantage was in promoting a scientific discipline in which ordinary people could participate. To engage properly in geology, for instance, a person required the time and money for fieldwork. Roderick Murchison had just announced the discovery of the Permian system of rocks that was formed before the Triassic period, but this work had obliged him 'to invade Russia' and explore the Urals with the support of the tsarist regime;

accordingly, he had named the new system after the Russian region of Perm.[10] In a similar vein, any aspiring zoologist might require access to the exotic specimens which Darwin procured in South America, or which Richard Owen received from the outposts of the British Empire. Conversely, anyone could be a phrenologist so long as they could afford a copy of the *Constitution* and a pair of callipers, and so it brought scientific inquiry into the home in a way that zoology never could. This did not sit well with the patrician guardians of science. Although some organisations, such as the Society for the Diffusion of Useful Knowledge, had been promoting cheaper copies of 'instructive' books for years, they thought there was real danger in allowing the wider population to consume such literature without appropriate guidance. Indeed, with Robert Owen and George Holyoake preaching a secular gospel to Chartists and socialists, the advance of infidelity among the lower orders was a source of growing anxiety: 'If he chances to have any religion,' observed Friedrich Engels of the British worker, 'he has it only in name, not even in theory. Practically he lives for this world, and . . . among the masses there prevails almost universally a total indifference to religion.'[11]

Combe's critics had further problems with the religious implications of phrenology. This was because, by ascribing moral and emotional qualities to anatomical features of the human skull, the *Constitution* was flirting with materialism. This was the belief that physical matter, such as atoms within the human body, had an energy of its own and did not require activation or direction by the Lord. But if removing the divine from the physical world was sacrilegious in itself, there was also a disturbing political analogy: if the lower orders of British society were like physical matter and possessed an inherent agency, this was an implicit threat to the station of the ruling classes and to a social order that many assumed to have been divinely organised. All this meant that many reactions to the *Constitution* were brimming with invective which mixed the religious with the political. One of Combe's opponents sought to measure phrenology against 'the word of God' and found it wanting. His pamphlet, which concluded that phrenology was 'antichrist and injurious to individuals and families', featured on its frontispiece a burning skull and the warning from Paul's first letter to the Corinthians that 'the fire shall try every man's work'.[12] On another occasion, the mere sight of the *Constitution* sent a lady into a fit: 'She threw it down as if it had been a serpent, and exclaimed – "Oh, . . . how *can* you print that abominable book?"'[13]

By creating this marketplace for controversial science, Combe and the phrenologists had ploughed a furrow for others, and especially for the author of *Vestiges of the Natural History of Creation*, a book which appeared in late 1844, shrouded in mystery. Published anonymously, and with a title harking back to James Hutton's aphorism about the history of the Earth having 'no beginning' and 'no prospect of an end', the *Vestiges* described a world which had coalesced from the gas and dust of stellar explosions, and where species of animal had emerged, died out, and changed in response to the demands of their environment. Of course, neither idea was born new in the *Vestiges*: Cuvier had long since established the soundness of extinction, while theories of 'transmutation' had circulated – albeit under contempt – for decades. The *Vestiges*, however, put them together in a coherent history of the world; moreover, its author rejected the book of Genesis entirely, condemning 'ideas about the organic creation [that] appear only as a mistaken inference from the [scriptural] text'. In his view, any schemes which put faith in that book belonged to 'a time when man's ignorance prevented him from drawing . . . a just conclusion'.[14]

Even more controversially, the author of the *Vestiges* suggested that 'the Divine Author', whom he conceded was present at the original creation, had not intervened subsequently to bring about changes in the world. 'How can we suppose that the august Being who brought all these countless worlds into form', he asked, 'was to interfere personally and specially on every occasion when a new shell-fish or reptile was to be ushered into existence . . . ?' As we have seen, this thought had occurred to Charles Darwin, but he was keeping it to himself, safe in his notebooks. The author of the *Vestiges*, however, would state it plainly: the notion of divine intervention recurring throughout history was 'too ridiculous to be for a moment entertained', mostly because it undermined any deity's pretence to omniscience. After all, if the Divine Author had made mistakes which demanded correction, would this not 'greatly detract from his foresight' and 'lower him towards the level of our own humble intellects'? It would be 'the narrowest of all views of the Deity', the author considered, to expect the Lord to act 'constantly in particular ways for particular occasions'; the truly sacred had no business with the minutiae of the profane.[15]

There was no question that dinosaurs and other fossil reptiles had informed the author's views of the world and its history. In discussing the 'era of the oolite' – that is, the limestone of southern England – the author referred to the 'huge saurian carnivora . . . [who] plied . . . their destructive vocation' in the seas of past ages; on land there were megalosaurs and the *Iguanodon*, 'a creature of the character of the iguana of the Ganges, but reaching a hundred feet in length'; and in the ancient air flew 'at least six species of the flying saurian, the pterodactyle'. Beyond recounting the discoveries of British palaeontologists, the author also evoked Lamarck and Grant in suggesting that the dinosaurs were well suited to their time and place: 'The huge saurians', he wrote, 'appear to have been precisely adapted to the low muddy coasts and sea margins of the time when they flourished'. But what time was that? The author does not mention dinosaurs until almost one hundred pages into the *Vestiges* and, before that, the book teems with astronomical speculations, accounts of the 'commencement of organic life' from which all other life had flourished, and descriptions of prehistoric corals, sandstones, and carboniferous formations. So how old was the Earth? And how wrong was biblical history?[16]

In one of the great studies of any single book, the historian James Secord has dissected the 'Victorian sensation' that the *Vestiges* became. First there was a clever marketing campaign. On a scale unusual for the time, and in addition to taking out dozens of advertisements, the publisher John Churchill distributed 150 free copies – fully one-fifth of the initial print run – to newspapers of all political persuasions, intellectual journals, libraries, and leading naturalists, thereby guaranteeing the attention of the men who shaped public opinion. This did not always procure the desired reviews, and many prestigious journals did not notice the *Vestiges*, but commercial success followed all the same. October 1844's first edition of 750 copies sold out almost immediately, as did December's second edition of 1,000; by January

1846, three more editions and a supplementary 'sequel' had appeared, with readers snapping up all 6,500 copies at the retail price of 7*s* 6*d*. Besides the simple volume of business, it was the quality of its readership which set apart the *Vestiges*, which was 'being extensively read in the highest circles'. The poet Tennyson professed himself 'quite excited' by the book, while Benjamin Disraeli thought it was 'convulsing the world' and causing 'the greatest sensation & confusion'. And as she sat on a sofa, a table and lamp beside her, Queen Victoria listened as Prince Albert read the *Vestiges* aloud.[17]

Not everyone liked what they read. The Bridgewater Treatise author William Whewell thought the *Vestiges* was 'unscrupulous and false' and that it appealed only to ignorant readers who had 'no power or habit of judging scientific truth'; the geologist George Featherstonehaugh, who had once bought fossils from Mary Anning, regarded the book as 'the signal of the Revolt against the Church'. Another critic damned the *Vestiges* as the despicable consequence of 'unwashed Radicals' having intruded upon national life since the passage of the Great Reform Act.[18] And in a furious essay for the *Edinburgh Review*, one of the longest ever published in the journal, Adam Sedgwick ranted against an author who told readers that 'their Bible is a fable when it teaches them that they were made in the image of God [and] that they are the children of apes and the breeders of monsters'. The *Vestiges*, he concluded, represented 'the progression and development of a rank, unbending, and degrading materialism'.[19] For all the free-thinking radicals who welcomed its daring vision, the men whom Secord calls the 'clerical magistrates of Nature' attacked the impiety that the author had embedded within the *Vestiges*.[20] But the author did not mind. In fact, he relished the controversy: in a later edition he noted that even if 'obloquy has been poured upon the nameless author from a score of sources', the *Vestiges* had provoked 'professing adversaries [to] write books in imitation of his'.[21] It was flattery of sorts.

But who was the author? Publishing anonymously or under a pseudonym was, of course, quite common for the time, even for successful authors. But unlike Charlotte Brontë, who had quickly revealed herself as 'the unknown power whose books ha[d] set all London talking', the author of the *Vestiges* was truly a secret.[22] Thomas Choate Savill, who operated one of London's largest printing shops near Charing Cross, had no idea whose words were going through his press. Nor did the publisher John Churchill, for the *Vestiges* had gone from the author's mind to the bookshelf in a manner worthy of the most intricate espionage. First, the author wrote each section of the *Vestiges* by hand; then, lest anybody down the line recognise his writing, his wife copied each word of the manuscript onto fresh pages. Now, an envoy intervened: the Manchester-based journalist Alexander Ireland took each section of the book from the author's wife to the publisher, Churchill; thereafter, when each page went through the press, Savill delivered the proofs not to the author but to Ireland, who conveyed them to the author. Beyond the author's wife, brother, lawyer, and envoy, *nobody* knew the truth about his identity.

Trying to divine the authorship of the *Vestiges* became one of the great parlour games of the 1840s and, for Gideon Mantell, it was *because* of its anonymity that the book became so controversial: 'A little volume of 390 pages', he wrote to a friend, 'has made a great sensation, chiefly I believe because the author cannot be detected'.[23] But who was it? There were rumours at Cambridge parties that it was, in fact, Mantell; in London, there were whispers about Richard Vyvyan, an MP whose writings on natural history had earned him fellowship of the Royal Society. Others thought that it might have been George Combe's second great work, while another theory held that all the book's moral failings were explicable by female authorship: Harriet Martineau and the mathematician Ada Lovelace were among the suspects. Others still were moved to deny that the book was theirs. Writing to his cousin, Charles Darwin described the *Vestiges* as 'that strange unphilosophical, but capitally-written book . . . [that] has made more talk than any other of late'.[24] As for rumours that he was the author, Darwin was 'much flattered & unflattered', and he read Sedgwick's review of the *Vestiges* – where his old Cambridge mentor had argued fiercely against 'the mutability of species' – with 'fear & trembling'.[25]

As the guessing continued, the suspicions of Britain's more perceptive commentators fell upon the Edinburgh-based publisher Robert Chambers. The author of dozens of books on Scottish history and the editor of the *Edinburgh Journal*, Chambers had also dabbled in phrenology and geology, and by the spring of 1845 sharper eyes had noticed too many similarities between the *Vestiges* and his other works for coincidence alone to explain. With his authorship becoming an open secret among the cognoscenti, the conservative critics of the *Vestiges* at last found an object for their fury. At the 1847 meeting of the British Association, where Chambers presented a paper on 'raised beaches' and 'ancient sea-margins', the old guard of Buckland, Murchison, and Lyell made sure that he was 'roughly handled'; it seems that Lyell did so 'purposely that C[hambers] might see that reasonings in the style of the author of the *Vestiges* would not be tolerated among scientific men'.[26] The next year, even the rumour of association with the *Vestiges* was enough to scupper Chambers's nomination for lord provost of Edinburgh. It was on account of such reactions – of the enduring hostility to works which denied the role of God in the natural world – that Chambers never once confessed publicly to writing the *Vestiges*, not even in his posthumous memoir. It was a book which had embraced and promoted ideas that remained heretical and, as a man with responsibilities, he could not risk the ruin that would follow acknowledgement. Later asked by his son-in-law why he continued to deny title to his great work, he pointed to his house and eleven children and said simply, 'I have eleven reasons'.[27] For Chambers as for other radical authors of the day, anonymity remained a sanctuary, a defence against the worst of the allegations that attended the profession of heresy. The case of one young academic at Oxford, who in 1849 dared put his name to a work of similar profanity, was proof of that.

11

Nemesis

Are God and Nature then at strife,
That Nature lends such evil dreams?
So careful of the type she seems,
So careless of the single life; . . .

Who trusted God was love indeed
And love Creation's final law–
Tho' Nature, red in tooth and claw
With ravine, shriek'd against his creed.[1]

Tennyson, *In Memoriam, A.H.* (1850)

James Anthony Froude was only thirty when his world collapsed.[2] His family had expected much of him: his censorious father was an Anglican archdeacon, one of his brothers had been a stalwart of the Oxford Movement, and another would become a leading naval architect. There was a degree of relief, therefore, when Anthony, as they called him, overcame a sickly childhood and merciless bullying at public school to write a prize-winning essay on political economy at Oxford, where in 1842 he became a fellow of Exeter College. Holy orders came afterwards as a condition of the fellowship and, by the mid-1840s, it appeared as if Froude would lead the same kind of stolid, comfortable, and eminently respectable life at the university that William Buckland had enjoyed since the 1810s; now a tall, athletic man with a ready smile, Froude sat in splendour at high table. The problem was that he had been eclectic in his reading. Far from keeping to economics, or the classics, or the conservative theology that was pervasive at Oxford, Froude had indulged in the 'higher' biblical criticism that was flourishing in Germany.

At universities including Tübingen, Göttingen, and Erlangen, scholars such as Ludwig Feuerbach and David Strauss had pioneered the analysis of the Bible and Christianity by reference to secular evidence. By testing the historical and miraculous claims of the holy books against historical facts and observable phenomena, they had sought to assess the integrity and consistency of the Bible as a whole, and some of their conclusions were sensational. In *The Essence of Christianity* (1841), Feuerbach developed an anthropological approach to religion, which he described as merely 'the dream of the human mind'. Here, the God of justice represented human ideals of justice, and the God of love was the perfection of human ideals of love; it followed that Christ the miracle-worker was 'nothing else than a product and reflex of the supernatural human mind'.[3] In *The Life of Jesus* (1835) Strauss had meanwhile looked at the gospels, striving to separate historical evidence from mythology. Though he did not deny that Christ had lived, Strauss decried the New Testament's 'false facts and impossible consequences which no eye-witness could have related'. Fatally, in his view, 'there was [for a long time] no written account of the life of Jesus', so that 'oral narratives alone were transmitted'; such tales had become 'tinged with the marvellous', growing into 'historical myth[s]'. For Strauss, these stories 'like all other legends were fashioned by degree', only in time acquiring 'a fixed form in our written Gospels'.[4]

Distance and ignorance had once insulated Oxford from the most dramatic of these provocations. In the earlier decades of the nineteenth century, when the syllabus remained staunchly classical, 'only two persons . . . were said to know German', and when the Tractarian theologian Edward Bouverie Pusey returned to Oxford from two spells at Göttingen, he sought only to explain German rationalism, not to promote it.[5] Still, Pusey had understood the power of the criticism that was gaining strength on the continent: 'This will all come upon us in England,' he cried, 'and how utterly unprepared for it we are!'[6] (Pusey later conducted his own exorcism, throwing his German notes into a fireplace at Christ Church.) This higher criticism arrived properly in 1846 when the twenty-six-year-old Mary Ann Evans, who later wrote as George Eliot, published a three-volume translation of Strauss's *Life of Jesus*, where the German theologian had presented a purely historical biography of Christ, stripped of miracles.

J. A. Froude had been among Eliot's readers. As doubt crept into his thinking, a traumatic trip to Ireland called divine benevolence even further into question. The horrors of Irish starvation, combined with serious illness, prompted discussions of becoming a doctor, but the law of the day forbade clergymen from retraining in the learned professions: 'Deacon I was,' he lamented, 'and deacon I must remain, hung like Mahomet's coffin between earth and heaven'.[7] Returning to Oxford, Froude did not settle to a diet of pamphlets and papers; instead, he reviewed for the journals and channelled his uncertainty into novels. First, under the pseudonym of 'Zeta', came *Shadow of the Clouds* (1847), a morose and 'excruciating' romance in which the protagonist runs up awful debts, sees his true love married off to a lacklustre rival, then dies 'of a spectacular disease'. There was nothing here, as one historian has remarked, to suggest that the author had become a rebellious '*frondeur*' but, as Froude reflected on his time in Ireland, where religion had not redeemed the hunger of the people, he grew restless.[8]

His second novel would catalyse these private doubts into public infamy. The title, *The Nemesis of Faith* (1849), was clue enough, and it was not difficult to discern elements of autobiography in the tale of Markham Sutherland, an aspiring clergyman who begins to question his calling. Yet what distinguished *Nemesis* from Froude's previous work, and from almost all other mid-century novels, was the explicit, excoriating articulation that it gave, through Sutherland's letters, to the author's reservations about religion. For Sutherland and for Froude, the Church loomed as a venal, predatory monster whose ministers valued 'the beauty of Christianity' only as 'a means of advancing them into [the wealth] they condemn'. The Bible? It was 'a jumble of arguments', a mausoleum where 'metaphor became petrified into doctrine', and testament more to 'the massacres of women and children' than to the benevolence of the Almighty. There was a cyclical perversity in the authority that Christianity claimed: 'The Church proves the Bible, and the Bible proves the Church', he wrote, 'cloudy pillars rotating upon air – round and round the theory goes whirling, like the summer wind-gusts'.[9]

And what of the Lord Himself? Froude despaired at the 'goodness' of a god who had chosen to bless 'arbitrarily, for no merit of their own, as an eastern despot chooses his favourites, one small section of mankind, leaving all the world besides to devil-worship and lies'. Just why were the chosen people chosen? And how could Sutherland believe the Lord to be 'all-merciful, all-good' when He was 'jealous, passionate, capricious, [and] revengeful, punishing children for their fathers' sin', tempting men 'into blindness and folly' when He knew they would fall, and punishing them eternally in a 'hell prison-house'? This god was not divine. He was 'a fiend'.[10]

By the end of *Nemesis*, when Sutherland is saved from suicide but dies with his doubt enduring, Froude had ventriloquised the most violent English-language censure of Christianity since Thomas Paine's *Age of Reason*. And though some reviewers took aim at the weakness and melodrama of the story, with Thomas Carlyle advising Froude to burn his 'own smoke, and not trouble other people's nostrils', there was no question that debate turned and turned fiercely on the issue of religion.[11] There was support in some quarters for Froude's historicism. The *Athenaeum*, for instance, conceded that Christians could treat the Old Testament as 'open to all kinds of historical criticism' and likened this case to the furore that surrounded Henry Hart Milman and his 1829 history of the Israelites.[12] More typically, however, the *Morning Herald* raged that 'every page is full of poison' without 'even the show of an antidote'; in its view, the *Nemesis* was 'one long series of attacks on Christianity' which amounted to 'a manual of infidelity'.[13]

At Oxford, that bastion of High-Church Anglicanism, the reaction was splenetic. The most notorious incident occurred at Froude's own college, Exeter, where, on the last Tuesday of February 1849, the undergraduate Arthur Blomfield took his copy of the *Nemesis* into a lecture on theology in the college hall. There Exeter's dean, the conservative churchman William Sewell, was discoursing on devotion and apostasy, taking the *Nemesis* as his subject. When Blomfield disclosed that he possessed a copy of the book, Sewell demanded that he bring it forward. 'No sooner had I complied with his request', the student recalled, 'than he snatched the book from my hands and thrust it into the blazing fire of the college hall'. Sewell stood triumphant by the fireplace, brandishing a poker 'in delightful indignation', consigning Froude's blasphemy to the flames. Within hours, this 'burning of the book was known all over Oxford'.[14]

Whereas Froude had once mixed easily among the brighter minds of the university, befriending Matthew Arnold and the poet Arthur Hugh Clough, he now fell into isolation. 'I am put out of communion altogether here,' he wrote to Clough, 'and must pack my traps and be off somewhere ... I can't read, write, or think, or do anything but groan'.[15] With Oxford growing 'rapidly too hot' for him, Froude resigned his fellowship; the master of Exeter refused even to shake his hand in farewell. This pained the young man, but it was no disaster. The month before the publication of *Nemesis*, he had written to the liberal churchman Charles Kingsley that he wished 'to give up my Fellowship', for he hated the Thirty-Nine Articles which governed the Anglican faith and even 'chapel' itself.[16] He had already planned, like so many disappointed men of the nineteenth century, for a new life in the colonies, far from authorities who believed 'too devoutly in the God of this world'.[17] To that end, he had accepted a teaching post in Tasmania.

Yet as Froude fled the university for the coastal village of Ilfracombe, where he lodged in hiding with Kingsley, the *Nemesis* scandal wrecked even his escape plan. On 10 March, the firebrand bishop Henry Phillpotts, who had been 'accused' of ordaining Froude in holy orders, wrote an open letter to the *Morning Herald* to disavow any responsibility for admitting him to the Church of England. In doing so, the newspaper declaimed against Froude's 'daring impiety' and 'foul libels', demanding that such a work of 'unexplained blasphemy' should 'be the subject of judicial inquiry'.[18] By questioning Froude's fitness to teach, his enemies were 'raising a cry against the Council of University College', which had appointed him to the job in Tasmania.[19] In distress, he withdrew from consideration and chose not to emigrate. Disowned by his family, who cut off his allowance, Froude was alone and without means: 'As I would not do what [my father] wished,' he reflected, 'I must be left in the water to find bottom for myself where I could'.[20]

Froude was not alone in crisis, for ever since leaving London for Kent in 1842, Charles Darwin had suffered in body and spirit. Though a troublesome stomach had plagued him on the *Beagle*, his journals studded with descriptions of torrential vomiting, recent years on land had proved the illness was more than seasickness. Now, when the symptoms came, they came in violence. There was vomiting (of course), but also cramps, gaseous bloating, diarrhoea, and dehydration; exhaustion, fatigue, and headaches joined them. 'Yesterday,' he reported during one trip to Shrewsbury, 'I was not able to forget my stomach for 5 minutes all day long'.[21] Discomfort was one thing, but embarrassment was another and, by the late 1840s, Darwin's stomach was a perpetual excuse for avoiding society. 'If I do not appear by 8 o'clock,' he wrote to Richard Owen, 'you will understand that my stomach has failed me'.[22] He declined another invitation on the grounds that 'my stomach is so particular, that I am afraid to go about before eating'.[23]

Now forty years of age, Darwin was a prisoner of his health; and having exhausted the means and patience of local physicians, the remedy of last resort was hydropathy. This meant an expedition of two months, with 'wife, children & all our servants' in tow, to the Worcestershire spa town of Malvern, where Dr James Gully was purveying an array of water-based treatments for sundry ailments. Even now historians dispute the correct diagnosis – some perceive a hereditary condition, others an exotic infection which lingered from his time at sea – but in 1849, upon consulting his newest patient, Dr Gully concluded that Darwin was suffering from nervous indigestion. He prescribed an exacting schedule. First thing in the morning, the orderlies exposed Darwin to a spirit lamp, heating him until he 'stream[ed] with perspiration', before they took 'towels dripping with cold water' and rubbed him violently until the skin was raw. There followed a cold footbath and another towel that Darwin would sport around his midriff, under his clothes, until bedtime. Complementing this routine were a strict diet and an exercise regime, initially several short walks each day. Over the first week, the treatment 'brought out an eruption all over [Darwin's] legs' but, by the time ten days had passed, he was feeling good enough – 'a little stronger', spared from vomiting – to announce that 'the Water Cure is no quackery'.[24]

Darwin maintained these habits at home, his butler assuming the role of the Malvern orderlies by soaking, rubbing, and wrapping his master as the regime demanded. The Sandwalk that wound around

Down House, alongside hedgerows and oak trees, now figured as the venue for daily exercise. And local tradesmen put up a magnificent contraption in the garden, an elevated cistern which daily drew more than 600 gallons of spring water from the Darwins' well; each morning, Darwin would stand on a stage beneath the tank and pull a string to release the water through a two-inch pipe. There would be no problem in paying for this 'douche', as he called it. Darwin had continued to receive a generous allowance from his father, and whatever Emma did not require for domestic expenses had been invested wisely, especially in the sprawling web of railway lines that were connecting Britain's cities. There was £20,000 in the Great Northern Railway from London to Yorkshire, and thousands more in the line between Leeds and Bradford. By the close of the decade, with both Charles and Emma receiving large inheritances, their investments amounted to more than £7.5 million in today's money. There was enough for everything and for everyone in a growing family: Emma had been pregnant for most of the last decade, blessing their marriage eight times between 1839 and 1850.[25]

Financial comfort had also allowed Darwin to pursue his studies in natural history, for there was never any need to work for wages. When his health allowed, therefore, he had gone over his notes from the *Beagle*, publishing monographs first on the formation of coral reefs and then on volcanic islands.[26] Yet the issues of species and what he had already called 'natural selection' continued to nag. On these more clandestine ideas, Darwin had begun to confide in the young botanist Joseph Dalton Hooker, who had returned a hero from the Ross expedition to the Antarctic: as the sailors mapped the shorelines of the southern continent, Hooker made the first studies of polar plant-life. In January 1844, in perhaps his most famous letter, Darwin had admitted to Hooker his conviction that 'species are not ... immutable', a statement that he linked to 'confessing a murder'.[27] Three years later, when he had worked his ideas on evolution into a 231-page manuscript, it was Hooker who provided notes. This was a 'fair & profitable subject', he thought, with Darwin's arguments against multiple sites of creation deemed 'very good' and the summary of the process of selection 'goodish', but the passage on the domestication of wild species 'not clear at all'.[28]

Hooker's most useful advice, however, was about Darwin's standing. The voyage of the *Beagle* might have made him a celebrity in geological circles, but he had published nothing on species or on zoology generally. In a world where the clerical cabal which ran London science had ostracised Robert Grant for promoting theories of transmutation, and where the Anglican establishment had savaged the *Vestiges*, how could a parvenu such as Darwin expect to have such a radical idea about species accepted? It was a fair point. To prove himself a zoologist and to apply his theories practically, Darwin turned to the remaining specimens from the *Beagle*: the tiny, clinging crustaceans known as barnacles. Though he would do so without Hooker's immediate counsel, the botanist embarking on an expedition to the Himalayas, Darwin would devote the next decade to dissecting and examining hundreds if not thousands of the creatures.

As he worked on the barnacles, Darwin was also reading extensively about religion, and in particular the works of Francis William Newman. The brother of the Tractarian churchman, Newman was a gifted linguist and classicist who, while professor of Latin at University College London, had published three volumes of theology that would shape Darwin's thinking profoundly. In *A History of the Hebrew Monarchy* (1847), Newman applied the new German criticism to the story of the Israelites and its liability 'to produce legends and *mythi*'; in *The Soul* (1849), he deplored 'the dreadful doctrine of the Eternal Hell' and derided Paul's letters to the Corinthians as evidence of the Resurrection and Ascension; and in *Phases of Faith* (1850), by way of the geological rejection of 'the idea of a universal deluge as physically impossible', he condemned 'portions of the Scripture as erroneous and immoral'.[29] Darwin read the last of these on a Sunday afternoon, reclining on the sofa at home and 'dosing himself with tonics'. He thought it was an 'excellent' book, and events would soon give him greater reason to doubt.[30]

Annie Darwin was his eldest daughter and his favourite child. Turning nine in the spring of 1850, she was tall with 'long brown hair and greyish eyes' and a kindly spirit.[31] 'Her cordiality, openness, buoyant joyousness & strong affection', Darwin wrote to a cousin, 'made her most loveable'.[32] But Annie was not well. That summer, she began crying at night and complaining of terrible indigestion: had she inherited her father's infirmity? Headaches and fevers soon followed and, by early 1851, her parents decided that Annie, like her father, needed a trip to Malvern for hydropathy. Dr Gully, so trusted by Darwin, feared that she needed more than water therapy and, within a week, he had referred the young girl to a clairvoyant, who 'described in lurid detail the horrors [that] she saw in Annie's innards'. For a while, however, the water seemed to work and, in confidence that Gully would save his child, Darwin went home to the barnacles. Then, in mid-April, there came an urgent dispatch: a 'bilious fever with typhoid character' had gripped Annie, and she was failing. Charles threw himself and some books into a carriage and, two days later, he was at her bedside. There he stayed for six days and nights, hopes rising and fading with each sign of strength or weakness. On the morning of 23 April, he sat crying by Annie's bed, staring blankly through the window at gathering storm-clouds; at midday the thunder rolled, and Annie breathed her last. The grief was too much, and Darwin broke. As Desmond and Moore observe, 'There was no straw [left] to clutch, no promised resurrection. Christian faith was futile.'[33]

On the last Sunday in March 1851, only weeks before Annie Darwin died, the British government had begun counting God. It was the day of the national census, the sixth such survey that the state had undertaken, and while the previous census of 1841 had asked for the name, age, sex, occupation, and birthplace of every person living in Great Britain, authorities were now enquiring about religious habits too. As theologians and geologists were debating the mysteries of the heavens and the Earth, the state had embraced a culture of numbers, especially as a means of measuring national prowess. At the Board of Trade, a powerful government department which oversaw the economics of the Empire, one civil servant was publishing statistical research on railways, shipping, trade, crime, emigration, education, and even manners: 'If no other indication of the prosperity of the country were to exist,' he wrote, 'we might justify our assertion of that prosperity, by the simple fact, that our numbers have increased'.[34] It was an obsession that Dickens would soon parody through the character of Thomas Gradgrind, the schoolmaster in *Hard Times* who sought to fill the 'little pitchers before him . . . so full of facts'.[35]

Taken every ten years, the census had illuminated several startling trends in British life. In 1811, when the Annings discovered the *Ichthyosaurus*, there were 12.6 million people in the United Kingdom; by 1841, when Richard Owen named the dinosaurs, this number had more than doubled to 27.1 million. The distribution of the population had changed as well. Whereas in 1811 Britain was largely rural, with only one in five people living in towns and cities, technological developments had prompted a parallel process of urbanisation. As the mechanisation of farming made agricultural labour redundant, workers had left their villages to find employment in urban mills and factories; in the space of thirty years, the proportion of Britons living in towns and cities had doubled. The radical journalist William Cobbett had once attacked London as 'the all-devouring Wen', a festering wound which sucked life and labour from the fields, but that process was now repeating in Manchester, Birmingham, Leeds, and many other towns and cities.[36]

This mattered because one of the government's primary concerns, and one of the driving factors behind the census of religious worship, was that British churches were failing to provide for their congregations. The government might have appropriated £1 million for church-building in 1818 and another £500,000 in 1824, and the Established Church Act of 1836 might have created the new dioceses of Ripon and Manchester, but there were widespread fears that the Church of England was floundering. These persistent concerns were thrown into sharper relief by the papal bull of 1850, which restored the Catholic hierarchy that had lain dormant in England since the late sixteenth century; consequently, in one of the last, furious expressions of English 'anti-popery', the Ecclesiastical Titles Act made it a criminal offence for the Catholic Church to name its dioceses after British places.[37] In this context, if the government were to determine how and where to make provision for Anglican teaching, it made sense to work out how many people followed each faith and how often they went to church. As one civil servant explained, 'the religion of a nation must be a matter of extreme solicitude to many minds', for the subjects of the queen were also 'subjects of a higher kingdom'; if taking an ordinary census afforded useful information 'to the temporal guardians of a nation', so would a religious survey assist its 'spiritual teachers'.[38]

Britain's registrar general would organise the secular elements of the 1851 census, but responsibility for the religious survey fell to a young lawyer in his office. Horace Mann was twenty-eight, only recently admitted to the bar, and a natural man of numbers. In a slew of public letters, Mann had already evangelised for the systematic registration of every detail of British life and he now contemplated two methods of 'pursuing a statistical inquiry with respect to the religion of [the] people'.[39] The first involved asking 'each individual, directly, what particular form of religion he professes', but Mann thought this had an overly 'inquisitorial aspect'. The second, less intrusive method involved 'collect[ing] such information as to the religious acts of individuals as will equally, though indirectly, head to the same result'. He thought this observational technique would be more incisive: 'the outward conduct of persons', he advised, 'furnishes a better guide to their religious state than can be gained by merely vague professions'.[40]

In the weeks and months before census day, Mann and his colleagues prepared questionnaires for dispatch to 'all existing edifices or apartments where religious services were customarily performed'. They asked the clergymen of Established churches to record on a black form the number of worshippers at morning, afternoon, and evening services on Sunday 30 March, the number of Sunday schoolers under their care, the date of consecration of their church buildings, the number of seats available therein, and their church's sources of income. Nonconformist ministers, Catholic priests, and Quakers – all those 'not connected with the Establishment' – received a distinct red form which asked the same questions, save for demurring on the matter of money. In time, they received more than 34,000 responses and, by December 1853, Mann was ready to submit his findings to the registrar general and to the prime minister, Lord Aberdeen. Reporting on the 'multiplied diversities' of thirty-five distinct religious communities both 'indigenous' and 'foreign', they calculated that on their chosen Sunday in March 1851 there had been 10.89 million attendances of a religious service in England and Wales.[41] (Scottish data was especially incomplete, so Mann did not prepare a comparable report.) Of these attendances, 5.29 million (forty-eight per cent) were in the Church of England, 2.6 million (twenty-four per cent) belonged to various denominations of Methodism, 1.2 million (eleven per cent) were 'Independents' such as Congregationalists, and only 383,000 (three-and-a-half per cent) were Catholic. The same report gave details of smaller communities of Mormons, Moravians, Jews, Quakers, Plymouth Brethren, and evangelicals who identified with the Calvinism of 'Lady Huntingdon's Connexion'.[42]

But what did this mean? One potential lesson was that the Church of England had lost considerable sway over the English people, just as the reforms of the 1820s had loosened its grip on political life. Another issue for Christian leaders was that the secular census had estimated the population of England and Wales to be 17.9 million, meaning that more than seven million people had not attended a religious service of any description. Moreover, abstention was likely much greater than such a number suggested because the survey had counted *attendances* rather than individuals: the deeply pious could have been tallied twice or even three times on the same day. Horace Mann reached the sobering conclusions that 'a sadly formidable portion of the English people are habitual neglecters of the public ordinances of religion' and that many of the working class were especially 'estranged from our religious institutions'.[43] There was the further problem of belief, since the census of 1851 could not divine whether attendance at a religious service represented sincere piety or the mere performance of a social obligation. So, was it all for show? Had the census revealed a formerly devout nation to be languishing in spiritual apathy?

Perhaps understandably, even though *The Times* accepted the numbers as 'substantially accurate and trustworthy', several religious leaders simply refused to accept the results.[44] Samuel Wilberforce, the Bishop of Oxford, had opposed the survey in the first place and now protested that 'to spread inaccurate statements on such a subject would be a fertile source of many evils'. Another bishop in the House of Lords agreed that Mann had overplayed the demise of the Church of England; he also alleged without evidence that, through malpractice which amounted to a 'departure from fair dealing', Dissenting ministers had made 'excessive returns' by inflating the number of children who attended their Sunday schools.[45] Anonymous pamphleteers joined them in battle, railing against the 'false and mistaken views' that the 'erroneous and misleading statistics' in Mann's report had promulgated to a trusting and gullible readership.[46]

They had a point. Although the figures appeared to suggest that millions of people in England and Wales had given up on religious worship, there were several reasons to mitigate any pessimism. For one thing, historians have deduced that attendance was often relatively high in towns that had been 'scarcely touched by manufacture' and much lower in the industrial heartlands where demographic change had been the most explosive.[47] This implied that, rather than people choosing not to worship, churches had simply failed to make room for them. For another, as George Eliot would evoke in *Silas Marner*, rural communities could be far from 'severely regular in their church-going' because of the practical difficulty in getting to services, but remain deeply Christian, with residents merely expected to 'take the sacrament at one of the great festivals'.[48] Accordingly, there were 'many pious Christians who did not attend church every Sunday', which suggested that the snapshot of 1851 was 'not necessarily ... typical'.[49]

Notwithstanding the pioneering research that Mann and his colleagues conducted, it remains a challenge to generalise about British religious practices in the 1850s. Differences abounded between Anglicans and Dissenters, and between rural and urban communities, and there were imperfections in the data. There were also conflicting, qualitative aspects to British religion that mere numbers could not analyse. For instance, even if millions of Britons appeared to have lapsed in their churchgoing, a young academic such as J. A. Froude could ruin his reputation by authoring blasphemy; in a similar vein, although personal tragedy and innovative theology were combining to winnow his faith, Charles Darwin was still reluctant to speak or write publicly about his heretical views on species. In these ways, Britain remained a Christian nation where at least the appearance of piety was essential to that most Victorian quality, respectability; indeed, it was because of this very religiosity that the implications of geology, the discovery of dinosaurs, and literature such as the *Vestiges* had caused such hysterical reactions in conservative quarters. Even so, as the events of 1851 would soon prove, there was no contradiction between this position and Britain's self-proclamation as the leading scientific power in the world.

If Mary Anning has somewhat faded from this narrative, it is simply because she made few major discoveries after the excavation of the *Dimorphodon* in 1829: 'there had been but little found', she complained.[50] Still, her life had not been uneventful. In 1833, she had cheated death not once but twice. First, Anning escaped burial under a landslide but lost her 'old faithful dog', Tray, to the same accident: 'The Cliff fell upon him,' she wrote to Charlotte Murchison, 'and killed him in a moment'. Soon afterwards, while walking to the beach-front at Lyme, a runaway cart-wheel pinned her to a wall. 'Fortunately,' noted one of Anning's friends, 'the cart [itself] was stopped in time to allow her being extricated from her most perilous position'.[51] In the wake of these traumatic events, it appears that Anning took solace in Christianity, reproaching her friend Anna Maria Pinney for 'speaking of religion as if ashamed of it'. Writing in her diary, Pinney noted that 'The Word of God is becoming precious to her after her late accident'.[52]

Without fresh finds in the lias, in the rocks that geologists now called 'Jurassic' after the mountain range in eastern France and Switzerland, there was less about Lyme to impress the great men of British geology. In 1832, Gideon Mantell went 'in quest of . . . the geological Lioness' but was not impressed: 'We found her in a little dirty shop', he wrote in his diary, 'with hundreds of specimens piled around her in the greatest disorder. She, the presiding Deity, [was] prim, pedantic, vinegar looking.'[53] Several years later, Richard Owen wrote scornfully of 'tak[ing] a run down to make love to Mary Anning at Lyme', and it was clear that London's palaeontologists regarded Anning more and more as an exploitable resource, not an equal partner.[54] Even 'in her own neighbourhood', at least according to a biography in Charles Dickens's *All the Year Round*, 'Miss Anning was far from being a prophetess'. Having been derided in the early days of fossiling for the absurdity of spending long hours picking through rocks on the beach, Anning was by middle age 'laughed at . . . as an unassuming person, who had made one good chance hit'.[55]

Visitors from further afield were kinder in their assessments. One German naturalist wrote of 'a strong, energetic spinster . . . tanned and masculine in expression', and, every morning, he saw this 'Princess of Palaeontology . . . walking and clambering about on the slopes of the Lias to see whether fossils have been brought to light by falls of rock'.[56] There was another German visitor in 1844, when Frederick Augustus II, the king of Saxony, toured Britain incognito. Without any of the tiresome ceremony that would have attended a formal trip, he went to the universities, explored London, and journeyed to Lyme Regis to procure fossils for his collection at Dresden. In the account of his doctor and companion Carl Gustav Carus, who was a friend of Goethe and a student of the painter Caspar David Friedrich, the two men had disembarked from their carriage and were walking into Lyme on foot, admiring the 'reddish-black coast walls' that enclosed 'the splendid azure sea', when they 'fell in with a shop in which the most remarkable petrifactions and fossil remains . . . were exhibited in the window'. Struck by the woman who kept the shop, who 'had devoted herself to this scientific pursuit', Carus asked for her name. With 'a firm hand' she wrote 'Mary Anning' in his pocketbook and advised him: 'I am well known throughout the whole of Europe.' Suitably impressed, the Saxon king paid £15 'for a perfect ichthyosaur'.[57]

Loss and pain marked Anning's last years. She had lived her whole life with her mother and, when Molly died in 1842, Anning copied a poem by Henry Kirke White into her notebook: 'It is that I am all alone', she wrote, 'I weep that I am all alone'.[58] Worse was to come in 1845 when Anning developed breast cancer, 'flying to strong drinks and opium to ease the pain', but none of it dulled her mind.[59] Indeed, in one of her few recorded comments on the prehistoric world, Anning showed remarkable prescience, writing that 'from what little [she had] seen of the fossil World and Natural History', she thought the connection 'between the Creatures of the former and present World excepting as to size [to be] much greater than is generally supposed'.[60]

When Mary Anning died in 1847, Henry De la Beche used his presidential address to the Geological Society to pay tribute. 'I cannot close this notice of our losses by death,' he declared, 'without adverting to that of one, who though not placed among even the easier classes of society, but who had to earn her daily bread by her labour, yet contributed by her talents and untiring researches in no small degree to our knowledge of the great Enalio-saurians'. Without Anning, he continued, the *Ichthyosaurus* and *Plesiosaurus* 'would never have been presented to the comparative anatomists in the uninjured form so desirable for their examination'.[61] It was Anning's work, her bravery under the rocks, and her expertise in preserving the fossils she found that had enabled others – at Oxford and London – to garner praise and high honour. She was the first woman to be accorded such a eulogy and it was a fitting testament, but rhymes remember her too: myths and legend have it that the lady who 'sells seashells by the seashore' is none other than Mary Anning.

PART THREE

The March of Progress

DRAMATIS PERSONAE

William Buckland, clergyman and geologist, Oxford and London
Charles Darwin, naturalist, Kent
George Holyoake, journalist, London
Joseph Dalton Hooker, botanist, London and overseas
Thomas Huxley, naturalist, London and at sea
Gideon Mantell, surgeon and geologist, London
F. D. Maurice, theologian, London
Richard Owen, anatomist, London
Alfred Russel Wallace, naturalist, Brazil and Malaya

12
The Palace

> London. Michaelmas term lately over, and the Lord Chancellor sitting in Lincoln's Inn Hall. Implacable November weather. As much mud in the streets as if the waters had but newly retired from the face of the earth, and it would not be wonderful to meet a Megalosaurus, forty feet long or so, waddling like an elephantine lizard up Holborn Hill.[1]
>
> Charles Dickens, *Bleak House* (1853)

The walls of glass and iron rose for fifty yards from earth to sky, the barrel-vaulted roof of the transept touching the highest leaves of the elm trees that were left in place. On the ground, the same walls stretched for a third of a mile into Hyde Park. Inside, twenty-six acres of red carpet and wooden slats swept under fountains, statues, and gates of bronze. The Royal Commission for the Exhibition of 1851, led by Prince Albert, had charged the architect Joseph Paxton with making a home for 'the Works of Industry of all Nations' and this, which *Punch* had christened a 'palace of very crystal', was the temple that they built.[2] For Benjamin Disraeli, it was an 'enchanted pile which the sagacious taste and the prescient philanthropy of an accomplished and enlightened Prince have raised for the glory of England'.[3] The purpose of the Great Exhibition was truly to glory in the miracles of machinery, to inspire and astound, and most of all to demonstrate British supremacy. When the doors of the Crystal Palace opened in May 1851, its avenues and chambers played host to Mancunian power-looms, Sheffield steel, ancient Celtic jewellery, the Koh-i-Noor diamond that British forces had just procured from the Punjab, and one of the world's largest telescopes; upstairs were watches and clocks, artefacts of naval architecture, and 'philosophical instruments'.[4] Less prominent billets gave space to the genius of other nations: there was Sèvres porcelain, Samuel Colt's firearms, and – though it came late – Russian armour.

Over the course of that summer, the Exhibition drew praise from the highest ranks of society. King Leopold of Belgium rejoiced that the 'truly colossal task' had met with 'unexampled success'; Queen Victoria's eldest daughter gave tours to Friedrich of Prussia, her future husband; and the queen herself visited some thirty times. Charles and Emma Darwin stationed their family in Mayfair so they could make the several visits they needed to see everything.[5] Yet with ticket prices falling gradually, eventually to a shilling for a day, the Exhibition also brought the marvels of the new technology to the masses, and Britain's booming railway network ferried families from the shires on discount fares, many of them on trips arranged by Thomas Cook's new travel agency. The satirist Henry Mayhew put these visitors into the pages of his fiction, where they 'wore a holiday aspect, the work-people with clean and smiling faces, and decked out in all the bright collars of their Sunday attire'; setting off from home 'shortly after daybreak' for their local stations, they were 'soon seen streaming along the road . . . towards the Crystal Palace, focus of the World'.[6] By the time this great jubilee of science and industry closed its doors in October 1851, more than six million people had paid to enter. Among them was a young man, recently returned from Australia, who despite his humble origins – and perhaps because of them – would transform the image and practice of science in Britain.

Thomas Henry Huxley was twenty-six, raven-haired, and possessed of his mother's 'piercing black eyes'.[7] And like so many others, Huxley had waited for the price of admission to the Crystal Palace to fall: he was the *enfant terrible* of London science, a rising star in the firmament, but he could not afford £3 for a season ticket. Yet as the people flocked to 'the great Temple of England' in Hyde Park over the course of 1851, just 'as the Jews came to Jerusalem at the time of the Jubilee', Huxley sensed a change in the times. While leaders of the Church were in rancour debating 'prevenient Grace' – the idea that the Lord would sooner or later reveal Himself to everyone – the people were taking communion in Paxton's church of glass and steel, under a ceiling that was loftier than 'the vaults of even our noblest cathedrals'. Huxley wrote that 'if Satan can laugh, I think he must do so at the perplexities of the English bishops'.[8]

He had come from Ealing, then a village ten miles to the west of Charing Cross. His father was a teacher, a master of mathematics, but he was no man of means: Thomas was born above a butcher's shop, and the failure of his father's once-prestigious Great Ealing School meant he would spend just two years in the classroom: 'I can hardly find any trace of my father in myself', he would reflect.[9] Still, the Huxleys were freethinkers, and when the family decamped to Coventry, rebuilding life among the weavers and watchmakers of the West Midlands, Thomas taught himself. He read Hutton's *Theory of the Earth*, Carlyle on history, even Johannes Peter Müller's *Physiologie des Menschen* in the original German.[10] He took to the last of these, to medicine. Despite the revulsion brought on by his first human dissection, which left him in a month-long pallor, he apprenticed at the age of fifteen to a surgeon in the East End of London. He saw 'strange things' there: 'tall houses full of squalid drunken men', women and children starving on a paltry diet of 'bread and bad tea', and an atmosphere 'filthy and brutal beyond imagination'.[11]

Huxley left behind these horrors, moving to Euston and lodging with his sister Lizzie. Now he crammed, taking courses on medicine, chemistry, and forensics at Sydenham College, and borrowing to pay the fees. This was not a reputable nursery. It was a last resort for unmoneyed men in a 'dingy purlieu', and the kind of place where dangerous ideas – Reform, Lamarck, atheism – could slip past the censors.[12] Huxley won prizes, then a scholarship, answering an exam paper from 11 a.m. to 8 p.m., his hands in cramp and his mind exhausted. There followed three years at Charing Cross Hospital, the wards a home for the casualties of West End construction; in the beds were men with crushed limbs and broken skulls, children in bandages, their bodies 'half-consumed by fire'. It was endless, draining, and Huxley lived with the owls, 'nocturnal natured'.[13] But there were more prizes to come and, by 1846, he had taken a gold medal in anatomy from Charing Cross. He was a published scholar too: in the

Medical Gazette, he described a new membrane, a 'hitherto undescribed structure' in human hair.[14] At twenty, however, he was too young to practise. So how, having borrowed £2 a week as a student, could he clear his debt?

The answer lay at sea, in the Navy, where assistant surgeons earned a daily rate of 7*s* 6*d*. A lengthy voyage, such as the *Beagle* had taken, would clean his slate and provide ample material for study besides. Huxley interviewed with Sir William Burnett, the physician-general of the service, and produced his bona fides: a certificate from the Royal College of Surgeons, a vicar's testimony to his 'good moral character', and naturally some proof that he was versed in Latin and Greek.[15] After several months in the naval hospital at Gosport, where wounded sailors disembarked directly into its beds, he found his berth. HMS *Rattlesnake* was an ageing troopship which had seen action during the first Opium War and its captain, Owen Stanley, had orders for the far side of the world: they would survey the straits between Australia and New Guinea, map the coral reefs that were snagging too many ships, and drop off gold – some £65,000 – at the Cape Colony and Mauritius. There beckoned several years at sea, several years of decent pay, and the chance to explore: 'New Guinea', Huxley wrote to his sister, 'is a place almost unknown, and our object is to bring back a full account of its Geography, Geology, and Natural History'.[16]

They cast off in December 1846, with Huxley of a mind to return as an expert on something, anything, and perhaps on jellyfish. Stopping first at Madeira, where he thought the chanting at Christmas Mass 'of a vile description', they caught the trade winds in the New Year. Flying fish decorated the decks of the ship; sharks and dolphins swam beside it. In the South Atlantic, Huxley fished Portuguese men-of-war from the deep and dissected them under a microscope in the chartroom; it was an urgent business, for the specimens putrefied quickly, 'semifluid & stinking' within a day of capture.[17] From Rio de Janeiro, where the stain of slavery appalled him, they sailed for the Cape. The crew risked superstitious ruin by firing on the albatrosses which circled above the ship, but Huxley kept keen on his jellyfish. This would be his domain, he resolved, and he fished daily for more evidence that even Richard Owen, the glowering doyen of British zoology, had described these creatures wrongly. Impressed with his surgeon's work, Captain Stanley dispatched Huxley's paper on the man-of-war to his father, who was the president of the Linnean Society and a potentially powerful sponsor.[18]

A month took the ship from Mauritius to Tasmania, and another eight days to Sydney. Fifteen years earlier, Darwin had marvelled at this outpost of British civilisation; now, Huxley had months to himself while a new ship was fitted for exploring the coral reefs, the *Rattlesnake* too deep in its draught for the coastal waters. There was loneliness, frustration, and hours peering through the microscope during these long days on shore, but there were also fishing trips and picnics in the bush; there were dinners, balls, and parties where he tripped the 'light fantastic' in dress uniform. At one of these he met a girl with hair of 'Australian silk', Miss Henrietta Heathorn. She took to him too, his eyes 'burning', his manner 'fascinating'. As the crew readied their new ship, the *Bramble*, Huxley embarked on a courtship. He warded off rival suitors with 'magic words', he surprised Miss Heathorn at lunch, and claimed her red camellia as a token of esteem. By the time of their fifth meeting, Huxley and 'Nettie' were engaged: 'You have tied your fate', he told her, 'to that of a young, poor . . . man, rich in nothing but his love for you'.[19]

Yet to marry he needed money, and that meant finishing the expedition and winning promotion. There now came months of sailing back and forth along the coast, charting reefs and inlets, rivers and islands. Some good news broke the tedium: the man-of-war paper had found favour in London and the Linnean Society would hear it read to them. Other news arrived at sea. In January 1848, as Huxley's sailship drifted in calm waters, a steamer powered past, the new leaving the old in its wake. These spells away from Sydney, without Nettie, were a torment. 'Here I am in this atrocious berth,' he wrote, 'without a soul to whom I can speak an open friendly word'. In the 'fearful, damp, depressing heat' of northern Australia, Dante's *Inferno* was a brave thing to read, but he had escaped another hell on land.[20] Led by the explorer Edmund Kennedy, a team of naturalists had set out to plot a path of 600 miles to Cape York in the north of present-day Queensland. In May 1848, Huxley's ship had escorted them to their point of departure, Rockingham Bay, but Kennedy would miss each rendezvous along the coast; it was only ten months later that three survivors, 'reduced almost to a skeleton', fell from a schooner onto the wharves of Sydney.[21] Of the others, rescue parties found only clothes and skulls.

The *Rattlesnake* departed on a further mission in the spring of 1849. Within a month, the ship was darting between the islands of the New Guinea archipelago, mapping sounds and shallow bays, bartering guns and metals with the 'man-eaters and sorcerers' who dwelled on the nameless, numberless islands. Yet for all that the uncharted beauty of these places could stir a man's soul, the ship's crew had been at sea for three years now, and they were failing in strength and spirit. Fever and 'nastiness' stalked the decks; the ship's carpenter took ill and died; even the captain's dog had suffered enough, walking out into the sea, not to return. Huxley too was in misery, 'sweated and stewed & bedeviled under the sun'. True, he had conducted years of valuable fieldwork, dissecting and collecting new fauna, and he had published in the learned journals of the metropolis, but where had this got him, really? As Adrian Desmond relates, he now pursued a 'deranged dialogue' with himself. 'It's all very well for young fools to talk of the nobility of knowledge', complained one of these Huxleys; 'You get no thanks for that,' answered the other, 'and the pay [is] just the same'. It was a sense of injustice that took root within him.[22]

The months dragged on, the heat weighing down the sailors. But having rescued a young Scottish woman who had been shipwrecked in the straits some years before, and having dined on turtles for Christmas, the crew of the *Rattlesnake* returned to Sydney in January 1850. Captain Stanley would not leave port again. 'Prematurely old' after a life spent at sea, he suffered an epileptic fit in the spring; Huxley found his captain sprawled on the floor of his cabin and cradled him as he died. While Sydney mourned Stanley, and as orders arrived for the *Rattlesnake* to return at once to London, Huxley reflected on his own circumstances. He had Nettie, her love the fire that had sustained him in the waters of New Guinea, but he still could not afford to marry, complaining to his mother that, if the couple were 'terribly prudent', they would only 'get married about 1870'. As the *Rattlesnake* left Sydney for the last time, therefore, Huxley strained his eyes for a last glimpse of his fiancée, who bade him farewell from her balcony on the coast. 'We part [as] friends, O land of gum trees', he reflected in his journal: 'I have much, much to thank you for.'[23]

They went home by the south Pacific, rounding Cape Horn in the deep mid-winter and calling upon a frozen Falklands in July. There was nothing to do in this 'Ultima Thule' but sleep and eat: beefsteak and geese for both breakfast and dinner.[24] For the Irish soldiers who barracked in nothing better than barns, Huxley felt a solemn pity. Another month took the *Rattlesnake* to Montevideo and warmer climes and, by late October 1850, hearts gladdened at the sight of smoke rising from the chimneys of Plymouth. A few days more and there were white cliffs and solid ground at Sheerness. Huxley now embraced with family and friends, the sunburned mariner regaling them with tales of adventure and romance, but there was a less welcome greeting from his bank manager: despite remitting cheques from Australia at every opportunity, Huxley remained £100 – roughly £10,000 in today's money – in debt.[25]

Discharged from the *Rattlesnake* in November, Huxley now sought the same patronage that in the 1830s had propelled Darwin into the first rank of science. The Australian research had prepared part of the way and, in the winter of 1850–51, he dined with Roderick Murchison, Charles Lyell, and the other genteel governors of geology. Perhaps surprisingly, given later events, he also made friendly contact with a man at the very zenith of his powers. Now in his mid-forties, Richard Owen was reigning supreme over natural history in London. He was professor and conservator at the Royal College of Surgeons, the author of more than 150 learned papers, and the honoured recipient of a £200 pension from the government. With his allies hailing him as 'the English Cuvier', Owen appeared to the world as a titan engaged in 'a continued series of labours for the promotion of scientific truth and its practical application to the well-being of mankind'.[26]

Owen would help Huxley with the Admiralty, whom the younger man begged for time ashore on half-pay so that he could write up his notes from the *Rattlesnake*. With Owen providing a letter of recommendation, and with the geologist Edward Forbes declaring that 'more complete zoological researches had never been conducted during any voyage', they acceded: Huxley became an 'additional assistant surgeon' on HMS *Fisgard*, a frigate which served as the flagship at Woolwich.[27] It was not much – and certainly not enough to pay for Nettie's passage from Sydney – but it was a start, and the young man grew in confidence. Fellowship of the Royal Society arrived soon afterwards and, though pipped to the Queen's Medal in 1851, he won it the next year. Yet for all these baubles, Huxley remained an outsider in a scientific community where rank and wealth could still prevail over acumen. Why were there so few jobs in science for self-made men? How could Britain erect a crystal monument to its science, but admit monied dilettantes to its halls before extending any welcome to working naturalists? These questions gnawed at Huxley in the early 1850s; they would never leave him.

In the autumn of 1851, as Huxley wondered whether science could ever throw off the shackles of class and religion, another radical thinker struck a blow for intellectual independence. It had been eight years since George Holyoake bade farewell to Gloucester gaol, where he had survived for six months on rice that had 'a blue cast, a saline taste, and a slimy look'.[28] And if Holyoake had ever flirted before with outright atheism, the man who left prison in early 1843 embraced it as a creed. He returned to the 'warpath of opinion' and, in his first article for the *Oracle of Reason*, he assailed 'christianity' – he refused to capitalise the word – as either 'a great evil or a great cipher', presenting the recent spate of prosecutions for blasphemy as merely the latest chapter in 'the history of despotism'. Why should Protestants have shaken off the iron dominion of Rome, he asked, only to confine the new dissenters 'to the Procrustean bed of *their* confined notions'?[29]

Holyoake re-joined his family at Birmingham and returned to the lecture circuit, imploring his allies to 'come and hear the Liberated Blasphemer', but there was no immediate upturn in his fortunes. A poor man before prison, he earned only pennies from the lectures and, at Alcott House in Richmond, where he and his fellow utopians communed, there was no more than raw cabbage for breakfast. But things would improve. First, he found work teaching in the freethinking schools of the capital and, as the *Oracle* moved into offices on Booksellers' Row off the Strand, with shop-window illustrations of the Old Testament attracting hundreds of daily visitors, he assumed the leadership of the Anti-Persecution Union. In its circular, the *Movement*, he declared that, 'after a severe struggle with law and religion, materialism ha[d] been rescued from the philosopher's closet' and that all religion was 'a broad, blazing, refulgent, meridian fraud'.[30] Despite this belligerence, the *Movement* was not a missionary magazine, for instead of converting the credulous to its cause, it sought only to unite existing unbelievers. With the establishment of *The Reasoner* in 1846, however, Holyoake embarked on a more ambitious enterprise.[31]

Here, in a magazine which enjoyed a circulation of between 1,500 and 5,000, Holyoake enlisted respectable radicals such as Harriet Martineau to broaden his readership. More pointedly, 'having discovered the insufficiency of theology for the guidance of man', he now wanted 'to ascertain what rules human reason may supply for the independent conduct of life'.[32] He also aimed to rally Britain's radicals, atheists, and freethinkers under one banner. First came the adjective. In November 1851, in the article 'Truths to Teach', Holyoake enjoined his allies 'to teach men to see that the sum of all knowledge and duty is secular – that it pertains to this world alone'.[33] Then came the self-description: 'Giving an account of ourselves in the whole extent of opinion,' Holyoake wrote two weeks later, 'we should certainly use the word Secularist as best indicating that province of human duty which belongs to this life'.[34] And last, in the new year, Holyoake published a letter from the Chartist journalist Charles Frederick Nicholls under the heading 'The Future of Secularism'. For Nicholls, 'secularism' was a means not only of thinking rationally about the world, but of practising an ethics that religious dogma could not pollute. 'If Rationalists have any mission,' he wrote, 'it is to secularise the world, that they may moralise it. A strong moral impulse', he averred, 'must be given to the secular army'.[35]

The secularists might yet have been few, but this was their call to arms. Over the next eighteen months, dozens of radical communities across Britain converted themselves into local chapters of the Secular Society, while the Chartist veteran James Watson took up the presidency of the parent London society in 1853. And where atheism once ran the risk of prosecution for blasphemy, secularism began to flourish. As Martineau explained to the American abolitionist William

Lloyd Garrison, by rejecting the label of 'atheism' and 'by the adoption of a new term, a vast amount of impediment from prejudice is got rid of, [and] the use of the name Secularism is found advantageous'.[36] For Britons who wished to conduct their lives – whether moral, political, or scientific – without the suffocating supervision of religion, Holyoake and the secularists had given them a home and, perhaps more importantly, a vocabulary.[37]

A former leading man of science was drifting from the stage. Though Gideon Mantell had published several well-received guides to geology during the 1830s, his medical practice and his museum at Brighton had failed; and with the royal court returning to London upon the death of William IV, Mantell had sold up. The £4,000 fee that the British Museum paid for his collection of fossils might have been more than £450,000 in today's money, but Mantell was unimpressed: 'And so passes away', he lamented, 'the labor of 25 years'.[38] Moving to London in the spring of 1838, he had taken over a surgery on Clapham Common, not far from the church which served as the spiritual home of British abolitionism. He would live there without his wife, Mary Ann, for a coldness had long grown between them. Separations had punctuated their marriage since leaving Lewes for Brighton and, in 1839, she walked out for good. Mantell's obsession with geology and his consequent neglect of family were the likely cause: 'There was a time when my poor wife felt deep interest in my pursuits,' he wrote, 'but in later years that feeling has passed away and she was annoyed rather than gratified by my devotion to science!'[39] Finding no fault in himself, he sulked: 'I am . . . downright savage in mind', he wrote, 'from the conduct of my wife'.[40]

1840 had brought no relief. His son Walter departed for New Zealand, where he rose through the colonial government, and then, in the second week of March, his 'sweet girl', his daughter Hannah, 'suddenly expired from haemorrhage'. She had been Mantell's favourite child, perhaps the only member of his family to whom he had been close, and now he had lost her 'sweetness of disposition and affectionate heart'. He marked the event in his journal with lines of sorrow from Byron: 'Before the Chastener humbly let me bow / O'er hearts divided and o'er hearts destroyed!' Mantell's mourning lasted through the year. That April, he languished 'in a state of depression almost unbearable', almost 'sunk under the accumulated evils' pressing upon him. 'Another year has rolled away,' he reflected in December, 'and I am alone!'[41]

Obscuring his grief through work, he published new papers on the *Iguanodon* and the *Hylaeosaurus*, then on the canine brain and human prostate in the newly founded *Lancet* magazine.[42] Engaging with London society, he showed specimens to Prince Albert, walked with bishops and dukes through the Royal Horticultural Gardens at Chiswick, and with no sense of irony dined with Edward Bulwer-Lytton on a 'wet and stormy' night. But high friends could not shield Mantell from calamity. Having purchased a splendid new carriage from 'the Pantechnicon', he was driving over Clapham Common in the autumn of 1842 when his horse became entangled in the reins. Jumping out to assist the animal, Mantell fell 'with great violence' to the ground. Though the wheels of his carriage only grazed his head, the greater mischief of the accident became clear when Mantell soon 'fell ill with symptoms of paralysis, arising from spinal disease'. With 'numbness of feeling extend[ing] to the feet', he could 'scarcely' walk. As the decade progressed, Mantell's debility prevented him from properly debating with Richard Owen over the constitution of the *Iguanodon*, and he became bitter and more critical of his colleagues. He thought one lecture on the *Megatherium* 'very poor', while a 'stupid' meeting of the Clapham Athenaeum was an utter 'waste of time'. His back was getting worse, too: the word 'neuralgia' appears more frequently in his diaries, and he was 'frantic with pain'.[43]

Despite these travails, Mantell maintained his surgery for the sake of solvency; he even summoned the resolve to keep writing. In new editions of the brief and accessible *Thoughts on a Pebble*, he distilled his philosophy of science, proclaiming the duty of a geologist 'to discover order and intelligence in scenes of apparent wildness and confusion'. The question that geologists should seek to answer was equally evocative: 'Can Science, like the fabled wand of the magician, call forth from the stone and from the rock their hidden lore, and reveal the secrets they have so long enshrined?'[44] Another success was the two-volume *Medals of Creation*, which sought 'to present such an epitome of PALAEONTOLOGY . . . as shall enable the intelligent

Reader to comprehend the nature of the principal discoveries in modern Geology'.[45] Written for the expert as well as 'the Tourist', and thus for the audience that Lyell had captured with the *Principles* a dozen years before, the *Medals* met with acclaim: the *Sussex Advertiser* prayed that it would 'speedily find its way into every public and good private library in the kingdom', while the poet Thomas Hood paid homage by imagining a 'fine morning [in the year] Anno Domini 2000' on which a troupe of geologists discovered even more prehistoric remains in Tilgate Forest.[46]

If this sustained his intellectual reputation, at least where Richard Owen had not undermined it, Mantell's body continued to deteriorate. He turned sixty in 1850 and, by the autumn of 1852, was suffering badly. Struggling to sleep and taking ever more laudanum, he appeared 'unusually ill' before the world, telling a friend that his very existence was like having 'a universal tooth-ache'. On his last Monday, he attended the Clapham Athenaeum to deliver a long-planned lecture on 'Petrifactions' and, by its end, he complained 'of extreme exhaustion' and went home 'overpowered with weakness'. On the Tuesday, he neither ate nor drank but attempted to visit patients. That evening, he went to bed at eight, hauling himself up the stairs by placing his weight on the banister, and requiring the assistance of his servant to undress. By the early morning it was clear that Mantell, who had taken an especially strong draught of opium, was 'drawing near his end'. On the Wednesday afternoon, his 'emaciated frame yielded' to eternity.[47]

Mantell had long planned to leave his body to science, convinced that the deformity of his spine could assist with medical inquiries, but he refused to let Richard Owen conduct the post-mortem; instead, his will set aside £10 to secure the services of the pathologist Thomas Hodgkin, who had discovered the eponymous lymphoma. Owen, however, could not resist levelling a final insult at his departed rival. In the *Literary Gazette*, he soaked an obituary in venom. Apparently, Mantell 'lacked exact scientific (and especially anatomical) knowledge', and Owen disdained the departed as 'peculiarly sensitive to any supposed oversight of [his work's] importance'. Even more cruelly, he sought to strip Mantell of credit for the *Iguanodon*. In Owen's telling, it was 'to Cuvier [that] we owe the first recognition of its reptilian character, to Clift [Owen's father-in-law] the first perception of the resemblance of its teeth to those of the iguana, [and] to Conybeare its name'. Most spitefully of all, Owen attributed the recognition of the *Iguanodon*'s 'true affinities among reptiles, and the correction of errors respecting its bulk and alleged horn' to himself.[48]

Others remembered Mantell more kindly. The Yale chemist Benjamin Silliman recalled an affectionate, sociable man whose 'powers . . . poured forth treasures of knowledge . . . with a natural eloquence and finished grace'.[49] Wrapping its obituary around suggested designs for handkerchiefs, the *Lady's Newspaper and Pictorial Times* paid tribute to Mantell's 'indefatigable industry . . . for diffusing knowledge amongst the people'.[50] The most striking feature of these memorials, however, was the sincerity of Mantell's Christianity. A councillor at the Clapham Athenaeum recalled Mantell's belief that 'he could never study too much the *works* of God, and was firmly convinced that they were all in accordance with His holy *word*'.[51] It was a faith that informed the 1849 edition of *Thoughts on a Pebble*, which invoked the spirit of 'He who formed the universe' and ascribed changing sea levels to 'laws which the Divine Author of the Universe has impressed on matter'.[52] Despite unearthing copious evidence that appeared to undermine the accuracy of Genesis, and despite suffering severely from bad fortune and vicious rivals, Mantell was unshakeable in his faith. But as Britain lurched into a decade of industrial change, intellectual turmoil, and the cataclysm of the Crimean War, that could not be said for everyone.

13
Death and the Dinosaurs

> Well, I sees these people. Their lives is pretty much open to me. They're real folk. They don't believe i' the Bible, – not they. They may say they do, for form's sake; but, Lord, sir, d'ye think their first cry i' th' morning is, 'What shall I do to get hold on eternal life?' or 'What shall I do to fill my purse this blessed day?'[1]
>
> Mrs Gaskell, *North and South* (1854)

In the early 1850s, Great Britain appeared to be peerless. Scientific marvels were filling the Crystal Palace, the City of London was the centre of the financial world, and British merchants were dominating global markets in the great commodities of the day: cotton, coal, metals, and chemicals. There was an urgent, unceasing sense of progress. Engineers were laying submarine telegraph cables from Britain to Ireland, and to France, the Netherlands, and Denmark; intrepid colonists were striking gold in New South Wales and Victoria; and the Royal Navy, larger than all other European fleets combined, was harnessing steam-power to enforce the Pax Britannica on the high seas. Parliament had even established the world's first compulsory vaccination programme, which from October 1853 inoculated all British children against smallpox.[2] Though a disappointing harvest had seen 'the Price of Provisions . . . enhanced', the Queen's Speech to Parliament in January 1854 observed that a 'Spirit of Contentment' nonetheless prevailed.[3]

This was why the carnage of the Crimean War and the apparent impotence of the British military would cause such a visceral shock.[4]

It was the first major conflict for forty years and, even if the French and Russians made noises about protecting religious minorities in the Middle East, it was really about the age-old Eastern Question: what should the great powers do with the sick man of Europe, the Ottoman Empire? Russia wanted to seize Turkish territory on its western borders, especially the Orthodox-majority principalities of Moldavia and Wallachia; Napoleon III wanted war and glory to restore French prestige; but the British would not tolerate the aggrandisement of any belligerent if it upset the continental balance of power. By the time that Lord Aberdeen's ministry entered the war in March 1854, a generation of Britons had lived and died since Napoleon's Hundred Days, and at last this was a new chance for heroics. Morale was high, for this was industrial, scientific Britain fighting against backwards, agrarian Russia. What could be easier? 'Thousands who had an appetite for a little excitement or hard knocks rushed to the Standard', recalled one soldier. 'Our only cry was, "Let's get at them!"'[5]

These hopes of glory collided with the brutality of modern fighting. In September 1854, 2,000 men died at the Battle of Alma. In October, at Balaclava, the Sutherland Highlanders formed 'the thin red line' before Lord Raglan dispatched the Light Brigade into the valley of death. The next month, thousands more fell in the fog of Inkerman. All the while, the allies besieged Sevastopol on the shoreline of the Black Sea: during ten months of cannon-fire, in fetid trenches and ramshackle hospitals where sepsis and cholera preyed upon the wounded, 20,000 more men perished. With the Reuters telegraph line now stretching from London to Belgrade in the Balkans, 'the terrible details of warfare', often relayed by field correspondents of *The Times*, reached British readers quicker than ever before.[6] Despite initial enthusiasm for war against Russia and for delivering 'a hit on the head' to the overbearing Tsar, the public now baulked at frontline descriptions of 'the groans of the dying', of 'indescribable agony'.[7]

The allies would secure their victory, with the Russians withdrawing from the Danubian regions and ceding naval rights in the Black Sea, but two years of slaughter had exposed a raft of problems. In the search for solutions some would turn to science, and Florence Nightingale's application of sanitary principles to battlefield medicine is perhaps the best-known example. Others sought political retribution for British failings, and the Aberdeen ministry fell dramatically in 1855, the prime minister taking his guilt to an early grave. Many more, however, sought solace in religion. Between 1854 and 1856, there were several days of 'fast, humiliation, and prayer' which 'called the faithful to humble themselves, to receive pardon for their sins, and to implore God's blessing' on British soldiers.[8] In Edinburgh, one minister reminded his congregation that 'God doeth according to His will in the armies of heaven' and that each man should 'do his individual duty to his country in the spirit of Christian patriotism'.[9] By 1855–6, Catherine Marsh had sold 80,000 copies of her biography of Hedley Vicars, an evangelical preacher whom God had 'called . . . to eternal life in the army' before 'a welcome from the armies of the sky' – or rather, a Russian bullet – had interrupted his mission at Sevastopol.[10]

Beyond these explicitly religious reactions to the war, Christian conviction – or at least Christian imagery – still pervaded British culture. From the spring of 1854, in riposte to the secular penny weeklies, the Religious Tract Society began publishing *The Sunday at Home*, which in its first issue urged readers to a 'firmness of religious principle' by way of rousing stories about the South Seas and China.[11] In the fine arts, Christian themes suffused the work of the Pre-Raphaelites: though he courted accusations of blasphemy, John Everett Millais had triumphed with *Christ in the House of His Parents* (1849–50), while William Holman Hunt's *The Light of the World* (1851–4) captured the beauty of a Bethlehem sunrise as Jesus knocked on the door of the human soul. All this was evidence that, for all the scientific progress and industrial innovation that had coloured the last forty years, and for all that palaeontology had asked searching questions about the literal truth of the Bible, there was a strong, unyielding core of Christianity that ran through British society. And if John Ruskin feared the clink of geology's hammer, which was beating his faith into 'mere gold leaf . . . at the end of every cadence of the Bible verses', the fate of one academic would prove that even minor theological differences could yet result in scandal.[12]

When J. A. Froude courted controversy with *The Nemesis of Faith*, one of the few churchmen to offer support was Frederick Denison ('F. D.') Maurice. Approaching middle age in the grandeur of a chair at King's College London, which figured as a conservative counterweight to the godless of Gower Street at UCL, Maurice had lived with religious disputes all his life, even from childhood. Whereas his father was a Unitarian preacher, his mother was Calvinist, and two of his sisters defied both parents to become Anglican and Baptist, respectively: 'Great differences arose between them', he reflected, 'which had a serious effect upon my life'.[13] At Cambridge, Maurice studied law and joined the Apostles, the elite intellectual society where undergraduates debated in the safety of secrecy; his influence, predicted a fellow Apostle, will 'be felt, both directly and indirectly, in the age that is upon us'.[14] It would take time for Maurice to attain that eminence but, by the 1840s, having affirmed the Thirty-Nine Articles to study further at Oxford, he had ascended to prestigious chaplaincies and twin professorships in history and theology. In the wake of 1848, when the 'Springtime of Nations' threatened revolution across Europe, and as the flame of Chartism burned out, he became a proponent of Christian Socialism, a movement which sought to harness the 'true functions and energies' of the British constitution with a view to establishing 'God's order', all while doing the work of the Lord among the most needy. It was from this place of comfort and confidence that Maurice had defended J. A. Froude as no 'false witness'.[15]

Yet if Froude owed his ostracism to self-conscious iconoclasm, the storm which engulfed Maurice in the mid-1850s came from something more innocuous. On joining the new theological department at KCL, he had preached and written prolifically on Paul's letter to the Hebrews, the Lord's Prayer, and the philosophical relationship of other religions to Christianity, all standard fare for a theologian.[16] Then, in the summer of 1853, he published *Theological Essays*. As he explained in the dedication to the poet Tennyson, Maurice had imagined the *Essays* as an artefact from which future Christians might learn something of the 'meaning some of the former generation attached to [religious] words' and phrases.[17] The collection therefore addressed Maurice's views on fundamental elements of Christian theology such as redemption, the unity of the Church, and the Athanasian creed. Yet if the volume met an underwhelming critical response in the round, with reviewers complaining of the 'tormenting indistinctness' of its prose, there was something in Maurice's final essay which raised serious concerns 'respecting his orthodoxy'.[18]

Here, he had contemplated the word 'eternal' in the context of John 17:3: 'This is life eternal, that they should know thee, the one true God, and him who thou hast sent, even Jesus Christ.' His argument on the meaning of the word 'eternal', laid out in intricate detail, was that it referred exclusively to the condition that a person attained upon knowing the Lord and accepting Christ as the saviour; it therefore bore no relation to the ordinary human concept of time. 'I cannot', he explained elsewhere, 'apply the idea of time to the word eternal. I must see eternity as something altogether out of time.'[19] This reasoning was not controversial in itself, but when applied to the reward or punishment that awaited mankind after death, it became taboo: if Maurice was right that eternity had nothing to do with time, then the eternal punishment which awaited sinners could not be 'everlasting'. Maurice simply would not allow that 'eternal death' meant everlasting punishment. For conservative Christians who had founded their eschatology on the punishment of sin with endless retribution, the fear of which was 'necessary for [shepherding] the reprobates of the world', this was unspeakable profanity.[20]

Malcontents grumbled in the press, damning Maurice as 'one of the chief oppugners of the faith' and his sermons as 'a whole course of heresy'.[21] Even more serious was the outrage and criticism of Richard William Jelf, the principal of KCL and Maurice's superior. A severe man with a sharp face who had passed through Eton and Christ Church at Oxford, flirting with Tractarianism before settling as a High-Church Anglican, Jelf demanded an explanation of his professor of theology. By August, he declared himself 'not at all satisfied with [Maurice's] explanations', on account of which he 'drew up a statement of his reasons for thinking that [he] was not fit to be a teacher in the College'. In October, KCL's council pronounced that 'the doubts indicated in [the] *Essays* respecting future punishments and the . . . day of judgment [were] of dangerous tendency, and likely to unsettle the minds of the theological students'; the council held further that Maurice's 'continuance as Professor would be seriously detrimental to the interests of the College'.[22] Fighting back, Maurice queried which canon law he had broken and asked that the college should sack him or acquit him, but this invited an easy choice: they sacked him.

Maurice published details of the controversy in a pamphlet that raced through multiple editions and sold thousands of copies, but his career at King's College was over. He also resigned as chairman of the Queen's College, the school for governesses that he had helped to found despite contemporary concerns that teaching mathematics to girls would have pernicious consequences; he was simply unwilling to let his reputation infect its prospects. Still, Maurice would remain in education. By October 1854, even before his dismissal from King's, he and the Christian Socialists had established the Working Men's College in Camden to provide a liberal education to artisans. Years later, under more auspicious circumstances, Maurice returned to the ivory tower, taking up a Cambridge chair in philosophy. But there were years in the relative wilderness, years of stress and anxiety. It was a stern warning – to theologians, philosophers, and naturalists – that unorthodox thinking, even on esoteric subjects, could jeopardise a career.[23]

At the close of 1851, as the Great Exhibition wound down, one great question had vexed its governing committee: what to do with the Crystal Palace? In an emotional pamphlet, its architect Joseph Paxton suggested that it should remain in situ and supply London's 'two and a half millions of inhabitants [with] . . . a Winter Park and Garden under glass'.[24] Henry Cole, the civil servant who had worked closely with Prince Albert on the Exhibition, likewise wanted a permanent venue for 'riding and walking in all Weathers among Flowers, Fountains, and Sculptures', the kind of refined public space that the Spanish and Viennese enjoyed but that climate denied to Londoners.[25] *Punch*, *The Economist*, and the *Westminster Review* supported the same ideas, and 'a universal regret prevailed at the threatened loss of a structure which had accomplished so much for the improvement of the national taste'.[26] Taking the opposite view, one pamphleteer complained that keeping the Palace in place would amount to the expropriation of public land at Hyde Park and that the cost of its maintenance would fall upon the national purse; others feared that any strong storm would throw shards of glass across central London. By late 1851, Dickens was damning the whole question as 'a horrible nuisance', and it was only in April 1852 that the House of Commons voted 221 to 118 against 'the preservation of the Crystal Palace'.[27]

But the Palace would not fall for good, because a consortium of businessmen now entered the debate. Led by Paxton and several directors of the London to Brighton railway line, these investors proposed 'rescuing the edifice from destruction, and . . . rebuilding it on some appropriate spot'. Part of their interest was for the greater good: they truly did wish the Palace to 'rise again' as a monument to British greatness, as an 'accessible and inexpensive substitute for the injurious and debasing amusements of a crowded metropolis'.[28] Yet having seen the Exhibition make a profit of £213,000, much of which went towards the purchase of land in South Kensington, they had commercial ambitions too. The consortium purchased the Palace for £70,000, deconstructed it piece by piece, hauled the whole across the city, and rebuilt it 'greatly enhanced in grandeur and beauty' in 300 acres of parkland on Sydenham Hill. It was no coincidence that the new site of the Palace soon connected to the Brighton railway.

Inside this second Palace, historians and architects re-created the art and majesty of the great civilisations, allowing guests to wander through time from ancient Egypt to classical Athens, from a faux Pompeii to the Alhambra and Renaissance Florence. In the main hall, an orchestra of 4,000 musicians and singers sat around an organ of 4,500 pipes, and there was space for another 4,000 concertgoers. Outside, Joseph Paxton – whose heart lay always with landscapes – sculpted mazes, gardens, and more than 22,000 fountainheads. He also foregrounded British pre-eminence in a field that had been curiously absent from the original Exhibition: geology. Digging deep into the land of south-east London, he and his colleagues built models of coal formations, outcrops of limestone, and strata of iron, lead, and sandstone, 'all of which [had] helped so largely towards the prosperity of our commercial nation'.[29] Passing through these quarries and mines, emerging onto a plateau, then descending towards a lake, visitors to the Crystal Palace came before an even greater testament to British leadership in the exploration of the earth. Dotted around three islands in the lake, the waters of which separated the men and women of the 1850s from the antediluvian world, were life-sized, monstrous dinosaurs.[30]

First there was Buckland's *Megalosaurus*, the predator which had been trapped in the rocks of Oxfordshire; then came the *Iguanodon*, Richard Owen having bested Gideon Mantell to persuade his peers that the dinosaur's signature spike belonged on its forehead, not its thumb. Beyond them was the *Hylaeosaurus*, 'the great Spiny Lizard of the Wealden' that Mantell had unearthed in the 1830s.[31] After them were the beasts of the Dorset lias, Anning's *Ichthyosaurus* and *Plesiosaurus*. Other, stranger creatures were in attendance too. The *Labyrinthodon*, meaning 'maze tooth', was a squat, frog-like creature that belonged 'in scenes of midnight incantation' or 'the halls of the preadamite sultans'; H. G. Wells might have written of its 'inflexibly calm' gaze, but others shuddered.[32] Nearby was the *Teleosaurus*, a crocodilian animal with a hugely elongated snout that Cuvier and Geoffroy, when they were not fighting each other, had described in the 1820s. There were ancient mammals as well: the *Megatherium* of South America had 'a forefoot, terminating in an immense claw, which was more than a yard long and twelve inches wide'; and the tapir-like *Palaeotherium*, found widely across Europe, was 'so powerful that a lasso of sufficient strength to capture the bull [would be] snapped by its rush'.[33] It was an awesome, fearsome menagerie that would appal and astound in equal measure.

The inclusion of dinosaurs and other prehistoric creatures in the Crystal Palace Park was entirely fitting. Although the Dutch, French, Germans, and South Americans had found and described certain fossils, it was primarily in Britain that geology and palaeontology had coalesced into the systematic, revolutionary study of the prehistoric world. In other words, by erecting 'representations of those extant species which existed in our globe thousands of ages before the birth of man', and by allowing Britain's fossil-hunters for the first time to gaze upon the fruits of their industry, the directors of the Crystal Palace had engaged in a mission of intellectual nationalism. All but one of the featured dinosaurs had been found in Britain and, as one guide to the park put it, the nation now gloried in being 'the metropolis of the "dinosaurians"'.[34]

The honour of making the models had fallen to the sculptor

Benjamin Waterhouse Hawkins, who had served as a superintendent at the original Exhibition. A London native whose illustrations of the animal kingdom had earned him fellowship of the Linnean Society, Hawkins took up residence in a massive workshop in the grounds of the new Palace Park in September 1852. Starting with the 'models of the extinct animals that [he] might find most practicable', he studied 'the elaborate descriptions of Baron Cuvier' and worked closely with Owen in making 1:6- or 1:12-scale sketches of the skeletons.[35] To these he added 'the skin and adjacent soft parts' in the fashion of 'the nearest allied living animals'. When satisfied with the accuracy of his drawings, Hawkins made a miniature model in clay, which he 'rigorously tested in regard to all its proportions'.[36] He then collected extraordinary volumes of material for his creations: into the *Iguanodon*, for instance, whose frame he made with 100 feet of iron hooping, went 600 bricks, 650 drain-tiles, 900 plain tiles, thirty-eight casks of cement, and ninety casks of broken stone.[37]

In December 1853, elated with their progress, Hawkins and Owen decided to celebrate. Imprinting their invitations with the image of a *Pterodactyl*, they planned a dinner: on New Year's Eve, beginning at

The Iguanodon Dinner at the Crystal Palace,
depicted by the *Illustrated London News*

six o'clock, Owen, Hawkins, and twenty of Britain's leading naturalists would eat together *inside* the half-completed model of the *Iguanodon*. Owen, naturally, sat at the top of the table in the *Iguanodon*'s head; other diners occupied 'capacious premises in the rear of the monster'; and everyone sheltered under a vast tent, which had been erected not only to keep out the cold, but also 'to prevent the illustrious Plesiosaurus . . . from prying too closely into the nature' of the evening. The menu, provided by the landlord of a local tavern, was exceptional. Mock turtle, julienne, and hare soups preceded cod in oyster sauce and fillets of turbot. For entrees there were lamb cutlets, rabbit curries, and sauteed partridges. Main courses of turkey, ham, pigeon pie, and boiled chicken complemented pheasant, woodcock, and snipe. Afterwards came jellies, pastries, nougat, meringues, fruit, and nuts. Filling the party's glasses were sherry, madeira, port, moselle, and claret. It took more than seven hours for the diners to clear their plates and empty their glasses, and the merriment concluded with a chorus of song so fierce 'as almost to lead to the belief that a herd of iguanodons were bellowing from some of the numerous pitfalls' in the park: 'The jolly old beast is not deceased', they sang, 'There's life in him again! ROAR!'[38]

The model dinosaurs were not complete when the second Palace opened to the public in June 1854. 'The huge, gigantic restorations did little honour to the great science which had embodied them', complained the *Illustrated London News*.[39] Yet as Hawkins and his 'small but valiant staff' persevered into the next winter, with 'mud boots and woollens' their only protection against the 'wet and blast', the models underwent 'a daily process of transmigration'. Painted first with a layer of brilliant red, then with a milky white, they at last attained the green-grey 'regular tint which they [were] believed to have possessed when they moved upon the face of the earth'.[40] The auctioneer and antiquary Samuel Leigh Sotheby hailed Hawkins as an heir to the great neoclassical architects of the Enlightenment, and William Buckland's son thought him 'the modern Pygmalion', a sculptor capable of awesome beauty.[41] Harriet Martineau meanwhile reported that, of all the attractions at the Palace, 'the school-children, the artisans and trades-people' liked the dinosaurs the best. Of course, that did not necessarily mean that they understood them: Martineau heard one spectator informing another that the monsters were 'antediluvian' because 'they were too large to go into the ark; and so they were all drowned'.[42]

By the summer of 1855, a trip to see the dinosaurs had become 'a treat of the first order'. They were so popular that Hawkins and his business partners were producing dinosaur-themed merchandise by way of models, posters, and 'diagrams'.[43] Only a few months later, however, Hawkins had been sacked. There was no misconduct, or dissatisfaction with his work; the directors of the Crystal Palace had simply flinched at the expense of the project. Since he began designing the dinosaurs in 1852, Hawkins had spent £13,729 on labour and materials, more than £1.57 million in today's money. In an open letter to the directors, Sotheby, who was a leading shareholder in the Company, expressed dismay that starting work on 'some of the most important features' of the park, only then to cancel them, was 'a case of suicide'. Why, he asked, had the directors dismissed Hawkins, who even 'in the depth of winter, and in severest weather', was always at his post, reconstructing the beasts of 'bygone ages'?[44] The *Observer* too regretted that the sculptures, which it thought 'alone unique' among the exhibitions, would not be completed. The protestors did not avail: the mammoths, mastodons, and glyptodons were abandoned.[45]

As Hawkins carved his dinosaurs from stone and metal, William Buckland's health was failing. Now in his seventies, the grand old man of British geology had published little since his Bridgewater Treatise of the 1830s. Though he had continued to deliver occasional lectures at Oxford, and though he found sufficient 'Traces of Glaciers round Ben Nevis' to suggest that glaciation and not the Deluge had worked the more profound changes in the British landscape, he had for a decade devoted the greater part of his energies to a new appointment.[46] In 1845, Samuel Wilberforce, the son of the abolitionist, had resigned as the Dean of Westminster to become Bishop of Oxford; then, as now, 'great' men rode a carousel of preferment between the places. Because the deanery fell under the purview of the Crown and not the Church, the task of finding Wilberforce's replacement had fallen to the prime minister Robert Peel and, for Buckland, who was growing weary of college life, this was a stroke of luck. Peel had long been a patron to Buckland, sponsoring him for early positions and more recently inviting him to soirées at Drayton Manor, where the geologist mingled with pioneering railway engineers, industrialists, and chemists. His appointment to Westminster was a welcome boon in the autumn of his career; for Buckland's wife Mary, it also marked a change to timid English traditions. 'On the Continent,' she wrote, 'where there is far less religion than in England, a man who cultivates Natural History, who studies only the works of his Maker, is highly considered and raised by common consent to posts of honour, as were Cuvier, Humboldt, etc.'. Yet in England, she believed, 'a man who pursues science to a religious end (even who writes a Bridgewater Treatise) is looked upon with suspicion, and, by the greatest number of those who study only the works of man, with contempt'. She thought that giving the sinecure to her husband, who had tried for thirty-five years to reconcile the mysteries of the Earth with the revelations of the scripture, was an act of intellectual bravery. Peel quite agreed. 'I never advised an appointment of which I was more proud,' he reflected, 'or the result of which was in my opinion more satisfactory'.[47]

Leaving Oxford for Westminster in 1845, the Bucklands found that the deanery, vast and grave, was not 'a very lively abode'. There were countless rooms of sombre oak panelling, sixteen stone staircases connecting sparse landings to distant wings, and long, dark corridors where the portraits of former deans stared down at the new residents; at night, 'queer noises and gusts of wind' escaped from the chasm of the Abbey to spook the souls who slept poorly. Little had changed since Elizabeth Woodville sought sanctuary here following the death of her husband, Edward IV. Still, this was now their home and, having opened its doors to society, a 'continuous stream of visitors' replaced the deanery's stale and solemn airs with life and movement. At a typical lunch in the summer of 1849, Michael Faraday joined Buckland to administer chloroform to 'Beast, Bird, Reptile, and Fishes'.[48] Buckland found plenty else to occupy his time at Westminster. For one thing, he looked to restore the fortunes of the associated school, which had fallen into physical disrepair and academic disrepute. For another, ever looking underground, Buckland prescribed the extensive redevelopment of Westminster's sewerage system, emptying the cesspits and advocating the separation of waste and drinking waters. In this context, the second book of Kings provided obvious inspiration for his sermons, which he used to admonish slum landlords and the generally indigent: 'Wash and be cleansed!'[49]

The onset of illness towards the close of 1849 put an end to this reformation. First came apathy and depression, then 'terrible weakness, torpor, and a loss of flesh'. London's physicians came to Westminster to assess the patient, but 'the cause of the illness baffled the highest skill'; Buckland's son blamed an accident which befell his father en route to Berlin, the old man having fallen from the heights of a stagecoach to bruise and almost break his neck.[50] By 1850, Buckland had relinquished many of his duties and retired to an Oxfordshire rectory that came as a grace-and-favour residence. The peace of the countryside and the sight of the garden, where his family had planted roses and strawberries, helped him to rally, but only briefly. 'Exercise of the brain in thought' aggravated his symptoms dreadfully, and in these final years he read only the Bible and *The Leisure Hour*, a safe, unchallenging 'Family Journal of Instruction and Recreation'.[51] As Buckland failed, his wife reflected on a man who, for the first time in his life, was content only 'when left in perfect repose – a strange contrast to his former existence'. He died in August 1856 at the age of seventy-three, but one last work of excavation remained: with Buckland having chosen his burial plot in the churchyard at Islip, a spot from which he had watched 'beautiful sunsets lighting up' the Oxfordshire countryside, the diggers would need powerful explosives to hew Buckland's resting place from the very same limestone he had spent a lifetime exploring.[52]

14
The Branching Tree

> As Professor Owen or Professor Agassiz takes a fragment of a bone, and builds an enormous forgotten monster out of it, wallowing in primeval quagmires, tearing down leaves and branches of plants that flourished thousands of years ago, and perhaps may be coal by this time – so the novelist puts this and that together: from the footprint finds the foot; from the foot, the brute who trod on it; from the brute, the plant he browsed on, the marsh in which he swam – and thus . . . traces this slimy reptile through the mud, and describes his habits filthy and rapacious.[1]
>
> William Makepeace Thackeray, *The Newcomes* (1854–5)

Darwin was not alone in his dangerous thinking; as he convalesced at Down House and examined his barnacles, fretting over the social ruin that might attach to his ideas, others were catching up. Alfred Russel Wallace was a young man, only thirty-two, but he had lived a lifetime already.[2] His father's unwise investments had cost him a formal education but after schooling himself – and after reading Humboldt, Lyell, and the *Vestiges* – he had embarked on a tropical odyssey just like Darwin and Huxley before him. He went first to Brazil, now an empire in its own right, and to the rainforests and delta-lands around the northern city of Belém; there, in the company of the naturalist Henry Bates, he collected insects and animals for the museums of London. With palm trees and coffee plants growing wild about the city, with people of 'every shade of colour' walking the rough and narrow streets that the Portuguese had hewn from the jungle, there was nothing here to recall a grey childhood in Wales, or of terms teaching cartography in Leicester.[3]

A bespectacled man whose black hair swept across a pensive brow, clean-shaven in his youth save for mutton-chop whiskers, Wallace immersed himself in the rainforest. There were monkeys, marmosets, tapirs, and armadillos; jaguars and pumas prowled in the undergrowth. In the trees and the air – always warm, 'wonderfully uniform' – there were macaws, parrots, toucans, and hummingbirds. From the soil sprang oranges, bananas, pineapples, and watermelons, cashews, mangoes, and cassava. All of life was here: 'its wonderful variety . . . never palled', and Wallace revelled in it. He collected and categorised, dissected and debated; the specimens piled up in their thousands. The Rio Negro was another quarry. The longest blackwater river in the world, in which decaying vegetation had leached into the water, turning it dark and bitter, it swept down to the Amazon from highlands in the north-west, and Wallace wanted to know where. It would be 'a journey of unknown duration' and, with fevers and dysentery weakening him by the day, it would almost kill him. But the promise of exotic treasure drew him further into the jungle; there were rumours in those parts of the magnificent, orange-plumed bird called the cock-of-the-rock and of 'white umbrella birds' too. Those lessons in Leicester now proved their worth: the map that Wallace charted – up from the Amazon, into the territory of present-day Colombia and Venezuela – was the best that anyone had yet made of the region.[4]

By the summer of 1852, Wallace was ready to leave South America. The rainforests were grand and beautiful, stocked with the marvels of nature and still 'virgin' land from the perspective of British naturalism, but Wallace was tired, and he had already collected thousands of curios to sell back home. He loaded his latest trove onto the *Helen*, a cargo ship departing from Belém, and for the first twenty-five days of the voyage all went well. Wallace had fought off the fever that often broke out in the early days of sailing, and he was reading in comfort when the captain brought dire news: 'I'm afraid the ship's on fire', he reported. 'Come and see what you think of it'. Thick smoke was billowing from the forecastle where a cargo of precious resins had combusted, and the ship was burning. As the crew readied the small boats, Wallace grabbed what he could: 'a small tin box containing a few shirts . . . my drawings of fishes and palms . . . my watch and a purse with a few sovereigns'. Into the boats they went with water and biscuit, raw pork and ham, and a few tins of vegetables. As the *Helen* sank, 'rolling over . . . like a huge cauldron of fire', the castaways considered their doom. The ship's last bearing was 700 miles from Bermuda, 300 more from the Virgin Islands, and Wallace 'hardly thought it possible we should escape'. They drifted for days, the sun scorching their skin, the sea-spray showering them with salt. How long would the food and water last?[5]

Ten days after the fire, some 200 miles from the coast of Bermuda, the *Jordeson* hove into view. Although this ageing brig could manage no more than five knots, and though its food was 'of the very worst quality' that Wallace had 'imagined to exist', it was salvation. By October, almost three months after first leaving Brazil, he disembarked at Deal in Kent and went straight to dinner: 'Oh, beef-steaks and damson tart,' he rejoiced, 'a paradise for hungry sinners!' Of course, there was much to grieve. Wallace's specimens had gone down in the hold of the *Helen*, and he rued that 'almost all the reward of my four years of privation and danger [in Brazil] was lost', but a pay-out from his insurers would sustain him for a while. Working from memory and his surviving notes, he rattled out papers for London's scientific societies – including a notable exposition of Amazonian monkeys – and a full-length book on palm trees. With his name now 'well known to the authorities', he was a 'welcome visitor' at their meetings, and at the Zoological Society in 1852 he met Thomas Huxley, whose reputation was swelling. 'I was particularly struck', Wallace reflected, 'with [Huxley's] wonderful power of making a difficult and rather complex subject perfectly intelligible and extremely interesting to persons who . . . were absolutely ignorant'; it was a gift of communication that many others were beginning to notice.[6]

By late 1853, having spent the best part of a year in the city, Wallace's feet began to itch. More specifically, the idea had taken him that 'the very finest field for an exploring and collecting naturalist was to be found in the great Malayan Archipelago', by which he meant the many thousands of islands of present-day Indonesia.[7] But for a man without a royal commission or independent wealth, getting there was a problem. Accordingly, Wallace called upon the patronage of Roderick Murchison, the one-time companion of Lyell and the acquaintance of Anning, who had since assumed the presidency of the Geographical Society. Ever alive to the imperial prestige which attached to these expeditions – the Dutch had annexed the islands politically, but the British could seize them for science – Murchison secured Wallace a berth on the *Frolic*, the next ship leaving London for Singapore.[8] This route proved abortive: Wallace had looked forward to comparing the crew to the characters of Frederick Marryat's naval novels, but the Admiralty requisitioned the *Frolic* as British forces built up in the Crimea, orders diverting the ship and its rations to the Black Sea.

Wallace tried again. This time a steamer took him to Gibraltar then to Malta; on board, he kept company with Scottish clerks, Portuguese soldiers, and Dutch ladies bound for Batavia. At Alexandria, he struggled through 'a vast crowd of donkey's [*sic*] & their drivers', despairing of the 'profane Mahometans' and the 'Philistines' around him. Passing the pyramids, 'huge and solemn', he went by carriage to Suez, where he boarded 'a splendid vessel with large and comfortable cabins'.[9] The rest of his journey was a tour of the eastern Empire: a day at Aden, 'desolate and volcanic'; the gems and coconut trees of Galle on the Ceylon coast; the waterfalls and spice-trees of Penang; and finally through the Straits of Malacca to Singapore. The first letter that he wrote upon disembarking assured his mother that he was safe: 'I have not seen any tigers yet & . . . there has not been a man killed at the place for two years.'[10]

Now began 'the central and controlling incident' of Wallace's life – his epic, eight-year-long exploration of British Malaya, New Guinea, and the Dutch East Indies. Sailing from place to place and from island to island in pursuit of new varieties of insects, all with the blessing of the Dutch colonial government, Wallace and his teenage assistant Charles Allen prescribed themselves a demanding routine. They rose at half past five each morning, showered, and drank coffee; they mended their nets and equipment and took breakfast at eight; and by nine they were walking into the jungle, 'dripping with perspiration'. On these 'entomologizing excursions', Wallace would discover thousands of new species of wasps, beetles, butterflies, birds, and ants. There were orangutans too, and Wallace shot a good many of them, killing the animals as they scrambled through the trees. At the same time, he tried to raise an orphaned infant, letting it suck on his finger, finding it a 'young hare-lip monkey' for a playmate, and observing its qualities of patience and affection. Alas, the experiment would not last: in the absence of mother's milk, a diet of 'well-soaked biscuit, mixed with a little egg and sugar and sometimes sweet potato' was fatal to his 'little pet'.[11]

On this second great expedition, in this wilderness of infinite variety, Wallace was not just collecting; he was observing, analysing, and writing. By early 1855, by the time he reached Sarawak on the north-western coast of Borneo, he had developed an idea about 'the Law which has Regulated the Introduction of New Species'. It was the wet season on the island, the rain producing a 'gloomy and monotonous' effect, and Wallace had 'a continual struggle to get enough to eat'.[12] But he was thinking clearly and, by September that year, London's *Annals and Magazine of Natural History* had received, accepted, and published Wallace's paper. Here, having spent years among the creatures of the tropics, among thousands of animals which differed only slightly from each other, Wallace argued that 'every species has come into existence coincident both in space and time with a pre-existing closely allied species'. Referring to the useless limbs 'hidden beneath the skin in many of the snake-like lizards' and to 'abortive stamens [and] undeveloped carpels' in plants, he rubbished the idea that divine omniscience could have caused each variation: 'If each species has been created independently,' he asked, 'and without any necessary relations with pre-existing species, what do these rudiments, these apparent imperfections mean?' Paying tribute to the exertions of geologists, whose research had exposed the creation and extinction of species from the more ancient periods of the Earth's history, he proposed 'the analogy of a branching tree, as the best mode of representing the natural arrangement of species and their successive creation'.[13] Wallace might not have been first to the idea, but he was the first to publish. Had Charles Darwin been scooped?

When Alfred Russel Wallace first arrived in the Dutch East Indies, Darwin had been studying barnacles for eight years, and that was quite long enough. He had cut them open, pulled them apart, contrasted them with one another, and written hundreds of thousands of words in description. He had not wanted for materials because, despite finally exhausting his haul from the *Beagle*, friends from around the world had furnished him plentifully: the British Museum broke protocol by shipping its collection from Bloomsbury to Down House; a friendly conchologist turned over troves from the Philippines and Polynesia; and the celebrated admiral James Ross gave Darwin his finds from the Antarctic, promising more from the polar north. Nobody had ever paid such close attention to so many barnacles, to the minute differences between types. Darwin was rising early, working for an hour, breaking off to walk the garden, then returning to the microscope before lunch. It went like this day after day, week after week, and the manuscript kept growing: it now demanded several volumes, dozens of illustrations, and even more time. Darwin might have become the world's leading expert on barnacles but, 'working like a slave', he hated them 'as no man ever did before, not even a Sailor in a slow-sailing ship'.[14]

The botanist Joseph Dalton Hooker, perhaps Darwin's best friend and his intellectual confidant, kept him at the work. Indeed, Darwin now accepted that without a serious publication on animal life, even on such lowly creatures as barnacles, his more explosive manuscript on species, which had lain in abeyance for so long, would not be taken seriously. Despite being 'wearied', despite asking himself 'what is the good of spending a week or fortnight in ascertaining that certain just perceptible differences blend together and constitute varieties and not species', Darwin persisted with the 'miserable work'. He feared 'frittering away' his time but, fully twenty years since he had first seen a barnacle burrow through a Chilean conch shell, he was done.[15] Across more than 2,000 pages on valves, antennae, larvae, and eggs, he laid out all that he knew of 'the sub-class cirripedia', the 'lepadidae', and 'the fossil balanidae and verrucidae of Great Britain'; and by noting 'slight variations' among the species and suggesting 'special adaptation for a cold climate' in specimens from Greenland, Darwin hinted at bolder things to come.[16] In the meantime, the work on barnacles would bring him the Royal Society Medal and establish him as one of Britain's leading zoologists. It would also bring him closer to Thomas Huxley. Though the pair had corresponded before, the younger man seeking advice on several papers, Darwin had gifted a copy of the first barnacle volume to Huxley, who could not yet afford his own; he thought it one of the more 'beautiful and complete [studies] of its kind', the work of 'a philosopher highly distinguished in quite different branches of science'.[17]

Freed from the burden of barnacles and now secure in his reputation, Darwin turned his mind to more interesting matters. First was the distribution of plant species across oceans, a problem which had vexed him since those days in the Galapagos: Hooker had since reported that the same plants grew on Tasmania as on Tierra del Fuego but, in the absence of a sundered supercontinent, just how and why did these varieties appear in both places? Using his garden and the outhouses at Down for laboratories, Darwin sought to disprove the notion that seawater invariably killed seeds. Buying salt from the local chemist, he pooled dishes of brine about the house: would anything grow in them? It would. Having cancelled all social engagements lest he miss the germination, and enduring the 'horrid' smell of the stagnant saltwater, Darwin found that, even after six weeks of bathing, seeds of celery, lettuce, cress, and carrots sprouted life.[18] It was proof that ocean currents – and not the myth of a lost Atlantis – explained why the same plants could bloom thousands of miles apart.

Darwin also began to 'look over [his] old notes on species'.[19] Even

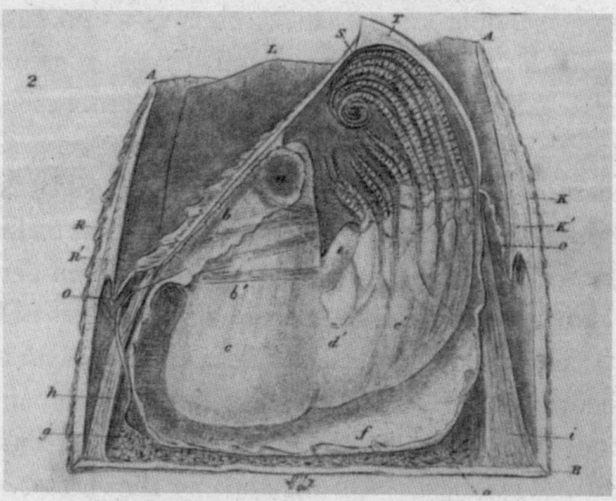

The focus of so much of Darwin's labour in the 1840s and early 1850s: barnacles

during the barnacle years, he had concluded that the most obvious example of his theory – that variation among species was attributable to the selection of preferred characteristics – was the domestication of certain animals. Accordingly, he took not to the exotic beasts of the zoo at Regent's Park, but to the farmyard. The homely, practical workings of this environment had long intrigued him. How did country squires breed their gundogs for hunting? How did farmers cultivate their herds? And how did fanciers breed vibrant colours, homing instincts, or plumed tails into their pigeons? In search of new specimens he wrote to Wallace in Malaya, asking the younger man to collect and dispatch 'Skins, Any domestic breed or race, of Poultry, Pigeons, Rabbits, Cats, & even dogs, if not too large, which has been bred for many generations in any little visited region'.[20]

The moment of these questions even wrested Darwin from his pastoral reclusion. He joined clubs across London to learn from the experts and, in grog shops and gin palaces, for the cost of their drinks, the 'odd little men' who knew pouters from runts furnished 'the Squire' with answers.[21] Of course, ever the practical naturalist, Darwin would conduct his own experiments. 'Get young pigeons', he instructed himself, and within a year Down House was home to hundreds of doves and ducklings. It also became a charnel house, a 'chamber of horrors', for it was never enough to breed the birds: Darwin had to dissect and 'skeletonise' them, and 'watch their insides' too. Cyanide gas would kill the chicks, then vats of potash and silver oxide stripped the flesh from the bones. It was a painful, horrible business. 'I have done the black deed', he wrote to a cousin in 1855, '& murdered an angelic little Fan-tail & Pouter at 10 days old'.[22]

The appearance of Wallace's article in the *Annals of Natural History* gave these enquiries a greater urgency. Writing to Darwin from Calcutta, the zoologist Edward Blyth – who had been developing his own ideas about variation – remarked that the jungles of Borneo had afforded Wallace 'capital data . . . to descant upon'. Blyth approved of Wallace's theory that 'the various domestic races of animals have been fairly developed into *species*', and begged to know of Darwin: 'What think you of [the] paper?'[23] On his copy of the *Annals*, the sage of Down House had noted that Wallace used a 'simile of [a] tree' to explain variation over generations, as he had done himself in his notebook so many years before, but thought the general theory unoriginal, dismissing it as 'nothing very new'.[24] But by the time he wrote again to Wallace, engaging him to procure further specimens, Darwin had revised his opinion. 'I can plainly see', he told the younger man, 'that we have thought much alike & to a certain extent have come to similar conclusions. In regard to the Paper in [the] Annals', he confessed, 'I agree to the truth of almost every word of your paper'.[25] As the events of coming years would prove, neither man knew the extent of that agreement.

While Darwin boiled pigeons, Thomas Huxley's luck had turned. The early 1850s might have been years adrift, with Huxley crying 'It is all a sham' as he lost appointments to candidates with better connections, but hard and endless work had repaired his fortunes. Cataloguing the marine collections at the British Museum by day and spending his nights with the *Rattlesnake* jellyfish, the Royal Medal, a 'scientific knighthood', had been his reward. He was coming into eminence and looked forward to his '*Meisterjahre*'.[26] He was making allies too, finding kin among the other young men who looked to secure a future in science on merit, and not through old conditions of class and creed. He bonded with the Irish physicist John Tyndall at the Ipswich meeting of the British Association in 1851, when the attendance of Prince Albert distracted all but twenty people from Huxley's lecture on men-of-war, and in 1853 he befriended Herbert Spencer.

Raised in the Midlands town of Derby, where Dissenting religion and industrial innovation forged a crucible of curiosity, Spencer learned from his father that 'the surrounding world gave no sign of any other thought than that of uniform natural law', and he buttressed this education 'by reading all kinds of books and hearing the conversations around me'.[27] There followed training as a civil engineer and work on the railways, but the lure of literary life took him to London: by 1848, still in his twenties, he had a post with *The Economist*, the magazine founded five years earlier to 'take part in a severe contest between intelligence, which presses forward, and an unworthy, timid ignorance obstructing our progress'.[28] That notion of 'progress', a paradigm of mid-Victorian thinking, was essential to Spencer's worldview. Living above *The Economist*'s offices on the Strand, he argued in *Social Statics* (1851) that 'man can become adapted to the social state' and that it was 'only by the process of adaptation' that society could ever achieve a sustainable balance.[29] The next year, articulating the 'development hypothesis' for the *Leader*, he went further: 'If a single cell may, when subjected to certain influences,' he supposed, 'become a man in the space of twenty years, there is nothing absurd in the hypothesis that under certain other influences, a cell may, in the course of millions of years, give origin to the human race'.[30] By slow and unsteady steps, British intellectuals were edging closer to an answer.

Spencer also introduced Huxley to the coterie of freethinkers who convened once a week in the offices of John Chapman, the publisher of the *Westminster Review*. On those Friday evenings, just across the road from *The Economist*, some of the great luminaries of the age gathered to debate and dissent. Among them were the secular journalist George Holyoake, the zoologist William Carpenter, and the philosopher-politician John Stuart Mill, who had called for the application of inductive logic to scientific inquiry: 'If two or more instances of the phenomenon under investigation have only one circumstance in common,' he had advised, 'the circumstance in which alone all the instances agree is the cause (or effect) of the given phenomenon'.[31] Joining them often were the Scottish publisher Robert Chambers, whom the congregants now recognised as the author of the *Vestiges*, and Chapman's assistant and mistress, Mary Ann Evans (that is, George Eliot), who was translating Spinoza's *Ethics* and Feuerbach's *Essence of Christianity* into English. In their discussions and in the pages of the *Westminster Review*, this gang had conceived of a 'Law of Progress' which held that 'the institutions of man, no less than the products of nature, are strong and durable in proportion as they are the results of gradual developments'.[32] They were good friends, like minds, allies in the cause.

Huxley's relationship with Richard Owen was a different matter. There was no question that Owen had helped Huxley upon his return from Australia, condescending to '*grant* . . . the *strongest and kindest testimonial* any man could possibly wish for', and some part of Huxley looked up to Owen, whom he knew was 'the superior of most' in their field. During the British Association meeting at Belfast in

1853, for instance, the two men walked together along the Giant's Causeway, polite in conversation beneath the basalt pillars. Yet there was always a distance between them, a tension between the radical upstart and the unrepentant Tory who had enlisted in the Honourable Artillery Company to defend London against the Chartists. There were intellectual differences too: Anglican and devout, Owen cleaved to the idea that all life stemmed from a set of divinely designed archetypes, while Huxley simply laughed that 'if Moses were ever really in possession of the laws of geology, they must have been written on the tablets which he broke and left behind on Sinai'. Around Owen, therefore, Huxley kept his counsel: 'I am as grateful as it is possible to be', he wrote, 'towards a man with whom I feel it necessary to be always on my guard'.[33] He knew the fate of Gideon Mantell; Owen could ruin a rival.

When he was not attending Chapman's brilliant soirées, Huxley now spent his Friday nights lecturing at the Royal Institution, with the *Morning Post* noticing his 8.30 p.m. lectures on 'Animal Individuality' under the heading of 'Fashionable Entertainments'.[34] He had begun writing for Chapman's *Review* as well, his early columns lambasting the popular craze for spiritualist seances and sub-standard works on philosophy. Never one to spare a friend from the lash, he tore into the latest edition of Chambers's *Vestiges* as mere 'waste paper' and a work of awesome stupidity: 'Time was when the brains were out,' he spat, 'the man would die'. Having seen too much of the chaos which abounded in Australian nature, Huxley simply could not abide the idea that organisms developed automatically and inexorably in accordance with 'a Logos intermediate between the Creator and his works', and he despaired that such a 'notorious work of fiction' had commanded such a readership.[35] It was the only review, he later regretted, that he 'had qualms of conscience about, on the grounds of needless savagery'; it was also a sign that even where two men agreed that, say, Richard Owen was wrong about the Lord's role in nature, there was no guarantee the same two men would agree on what was right.[36]

And at last, after years of relative penury, there came stability too, through lectureships in natural history and palaeontology at the School of Mines in Piccadilly. 'In the course of the week,' Huxley crowed, 'I have seen my paid income doubled . . . I am chief of my own department & my position is considered a very good one.' Secure in his station, he delighted in provoking the old order, with his course on natural history described as 'the most heretical Syllabus' that London had seen. It was a lot of work, often too much: 'What idiots we all are,' he complained to John Tyndall, 'to toil & slave at this pace'. But it paid well and, by late 1854, Huxley could reckon on an annual income of £500, some £40,000 in today's money. After four years 'full of dreams & nightmares', he was almost ready to clear the 'incubus' of medical debt and competent to marry. 'So, my darling pet,' he wrote to Nettie at Sydney, 'come [here] as soon as you will'.[37]

They would make their home in the city. 'I will *not* leave London,' Huxley had promised himself, 'I *will* make myself a name'. He found a three-storey terraced house to rent in St John's Wood and he waited for Nettie to 'teach [him] something of . . . gentleness and patience'. He expected her in May 1855, and it was fully five years since they had last seen each other: 'I am another man altogether,' he fretted, '& if my wife be as much altered, we shall need a new introduction'. Altered she was. Bankruptcy had scandalised her family in Sydney, incompetent doctors had made her sickness worse, and the voyage had drained her further. Still, two months of rest worked wonders and by July they had married: from the pews of All Saints' Church on the Finchley Road, Hooker and Tyndall saw their man into marriage and then off to south Wales on honeymoon. The years of strife were over and Huxley, in his own words, 'had reached the Promised Land'.[38] He was safe, solvent, and happy; and of course, he was ready to fight.

15
The Devil's Gospel

> It was presently clear that private tutors, natural history, science, and the modern languages were the appliances by which the defects of my organization were to be remedied. I was very stupid about machines, so I was to be greatly occupied with them; I had no memory for classification, so it was particularly necessary that I should study systematic zoology and botany; I was hungry for human deeds and human emotions, so I was to be plentifully crammed with the mechanical powers, the elementary bodies, and the phenomena of electricity and magnetism![1]
>
> George Eliot, *The Lifted Veil* (1859)

The idea came to him, fully formed, in a febrile dream. Three years had passed since Alfred Russel Wallace's first major missive from the jungles of the east – the article on the creation of new species – and, by early 1858, he had the substance of another. Wallace had been among the beetles, bees, and butterflies of Borneo; he had rambled past the ancient Hindu temples of Bali and Lombok, where he drew the enduring 'Wallace Line' that separates the ecosystems of Asia and Australasia. And now, having sought out birds of paradise on the Aru Islands towards New Guinea, he was on the 'earthquake-tortured' island of Ternate, renting a small, stuccoed bungalow in what, even for a well-travelled man, appeared as a 'wilderness of fruit-trees'. There was a well for cold, clean water, an 'educated' Dutchman for company, and a market manned by the Ternatean people which supplied Wallace with the rare luxuries of 'milk and fresh bread'.[2] He would spend three productive years in that house, walking from beach to mountain to jungle, but in January 1858 Wallace was 'suffering from a sharp attack of intermittent fever'.[3] Every day, for weeks on end, successive fits of heat and cold compelled him to bed, and it was during one of these bouts of delirium that the works of Thomas Malthus came to mind.

Malthus was an English clergyman and economist who, from the late 1790s, had articulated several influential ideas concerning demography. In his view, there was eternal tension between a population, the resources available to it, and external shocks such as war, disease, and famine; in this system, if a population ever grew in number, destructive forces would act as 'positive checks to increase' and return that population to its original level. (This was among the reasons that British authorities were so keen for 'surplus' people to emigrate, so as to prevent the necessity of a destructive 'check'.) It occurred to Wallace, as he sweated out his fever, that the same 'causes or their equivalents are continually acting in the case of animals also'. Was it the ability to survive famine and disease that explained not only 'how and why . . . species change, but [also] how and why . . . they change into new and well-defined species'? He turned the question over, contemplating 'the enormous and constant destruction which this implied'. Why, he asked himself, 'do some die and some live'? The answer was that 'the best fitted live'. The healthiest animals escaped from disease; the strongest escaped their enemies; the best hunters defied starvation. It 'flashed upon' him that such a 'self-acting process would necessarily *improve the race*', that the weakest in every generation would die out, that '*the fittest would survive*'.[4]

As Wallace recovered his health, he 'became convinced that [he] had at length found the long-sought-for law of nature that solved the problem of the origin of species'. He pondered the deficiencies of Lamarckian transmutation and the rigid developmental theory that Chambers had put forward in the *Vestiges* and, over two consecutive evenings, the weather cooler and his strength restored, he wrote down his theory 'carefully in order to send it to Darwin'.[5] Giving his paper the title 'On the Tendency of Varieties to Depart Indefinitely from the Original Type', Wallace wrote of the warfare that he now perceived in nature. There was a 'struggle for existence' and the 'abundance or rarity of a species' would depend on its capacity to adapt to conditions; and as useful variations would increase, the useless would diminish, so that the superior variety would replace the inferior, 'of which it would be a more perfectly developed and more highly organized form'.[6] Although the original paper and accompanying letter have been lost, we know that they arrived at Down House in June 1858. At once, in something of a flap, Darwin sent both to Charles Lyell. 'I never saw a more striking coincidence', he complained. 'If Wallace had my M.S. sketch written out in 1842 he could not have made a better short abstract [of my theory]! . . . So all my originality', he feared, 'whatever it may amount to, will be smashed'.[7]

Darwin had other worries besides Wallace. His daughter Etty was ill with diphtheria, while scarlet fever was epidemic in the village of Downe; and though Etty would recover, the fever would take the Darwins' youngest child, Charles junior. Only eighteen months of age, he had never been well, unable either to walk or talk. On the last Monday of the month, Darwin made a stark and desolate entry in his journal: 'Poor dear Baby died.'[8] On the next day, still 'prostrated', he wrote twice to J. D. Hooker. The first letter, in the afternoon, was an outpouring of grief: 'It was the most blessed relief', Darwin told his friend, 'to see [my son's] poor little innocent face resume its sweet expression in the sleep of death'. The second, in the evening, turned to Alfred Russel Wallace and species. Sharing the manuscript that Lyell had returned to him, Darwin begged to know Hooker's thoughts. Though he feared that 'all is too late', he proposed an intervention: 'I would make a similar, but shorter & more accurate sketch', he suggested.[9] Was it now the time, prompted by Wallace, to declare his own position?

At the start of the hot, fetid summer of 1858, when the Great Stink caused by sewage in the Thames poisoned the London air, Darwin and his allies laid their plans. Choosing the Linnean Society as their venue, where Hooker and Lyell were sufficiently influential to control the agenda, they determined that Wallace's paper and two summaries of Darwin's thinking would be read before a meeting on 1 July. Neither author would be present. Darwin stayed at home, sick and bereaved, while Wallace was still in the jungle; indeed, he did not even know what was happening, for even the fastest Peninsular & Oriental mail-ship took seventy-seven days to reach Ternate from London. But on a sultry Thursday evening, besides receiving news on 'the Portuguese territories in Western Africa' and a new American climbing plant, thirty or so gentlemen of the Linnean Society were the first to hear publicly of these ideas.[10] The meeting did not cause the expected waves; the Society's president would even complain, despite listening to the papers first-hand, that 1858 had not witnessed 'any of those striking discoveries which at once revolutionize ... science'.[11] It remained for Darwin to refine his theory and to prepare its grander exposition.

He took to the Isle of Wight, where he grieved for his late son and prepared a written paper for the *Linnean* journal. He also worried what Wallace might think of his conduct. In their correspondence, Darwin had confessed to Wallace that 'we have thought much alike & to a certain extent have come to similar conclusions', but had the wealthy squire now exploited the work of the penniless collector, a man who was suffering 'the excessive inconvenience & misery of being without cash in a strange country'?[12] He need not have worried. Though Darwin protested that he 'had absolutely nothing whatever to do' with the Linnean plan, Wallace was in fact delighted to stand in such company.[13] He boasted to his mother that appearing before the Linnean Society 'insures me the acquaintance and assistance of these eminent men on my return home', and he wrote directly to Hooker, thanking him and Lyell for 'the course you have pursued & the favourable opinions of my essay which you have so kindly expressed'. Wallace even praised Darwin for the 'excess of generosity [which] led him to make public my paper'.[14]

As Darwin worked on his manuscript, Lyell pitched the book to John Murray III, the son of the man who had published his *Principles of Geology* some thirty years before. Murray bought it, agreeing to a first run of 500 copies, and sharpened the title: Darwin's book would now be called *On the Origin of Species and Varieties by Means of Natural Selection*. By the autumn of 1859, and despite Hooker's momentary panic that his children had torn up and drawn upon the only copy of the manuscript, Darwin had received notes and encouragement from friends. Huxley suggested 'a flaw' in Darwin's analysis of domesticated crossbreeds, but took up the idea of 'the gradual modification of pre-existing species' in a paper before the Royal Institution.[15] Lyell fretted that Darwin's theory would impugn 'the dignity of man', but gave his friend 'very great *kudos*'.[16] And Hooker, who adopted 'the ingenious and original reasonings and theories of Mr. Darwin and Mr. Wallace' in a monograph on Tasmanian plants, described his own work as merely a 'ragged handkerchief beside [the] Royal Standard' that Darwin was about to unfurl.[17] With Murray stumping up £72 to pay for corrections and then doubling the print run, Darwin signed off the proofs on 1 October. It had been twenty years, a loss of faith, and countless nervous fits in the making, but now the *Origin* was ready.

Beginning with his time on the *Beagle* before treating of variation, natural selection, instinct, geology, and the distribution of species, all of Darwin's research and thinking now poured forth over 500 pages and 150,000 words. There were lengthy passages on pigeon-fanciers breeding specific characteristics into their progeny, and descriptions of the fauna of Tierra del Fuego and the Galapagos. The long-gestating analogy of the 'tree of life', which he had sketched in his notebook as early as 1837, found full expression: 'As buds give rise by growth to fresh buds, and these, if vigorous branch out and overlap on all sides many a feebler branch,' Darwin wrote, 'so by generation I believe it has been with the great Tree of Life, which fills with its dead and broke branches the crust of the earth, and covers the surface with its ever branching and beautiful ramifications'. Though the word 'evolution' did not appear in this first edition – Darwin would not use it for years to come – he was getting closer to it: 'Whilst this planet has gone cycling on according to the fixed law of gravity,' ran the last sentence of the book, 'from so simple a beginning endless forms most beautiful and wonderful have been, and are being, evolved'.[18]

The 'gravest objection' to his work that Darwin recognised was, perhaps surprisingly, the fossil record, for in 1859 he believed that 'geology assuredly does not reveal any . . . finely graduated organic chain' between varieties of animal, but the *Origin* now gave naturalists both the mechanism and the vocabulary with which to explain the emergence and divergence of species. Lamarck had long ago argued that animals developed and then directly bequeathed characteristics in response to their immediate needs; Chambers had imagined the progressive development of natural life in accordance with a plan where divine agency was absent; and Richard Owen had posited the divine creation of archetypes from which all life had spawned. Darwin and Wallace, however, had devised a theory of nature that was subtly but brilliantly different: here, the creatures who could adapt successfully to the demands of their environments would survive, the creatures who did not adapt would perish, and over time this process would see the rise to predominance of the characteristics that had been proven essential to sustaining life. If anyone could parse a work of such magnitude and complexity into a single sentence, it was Huxley, who summarised the Darwinian theory thus: 'Species originated by means of natural selection, or through the preservation of the favoured races in the struggle for life.'[19]

In late November, when the *Origin* went on sale, Darwin was nowhere near a bookshop. He was not even at Down House, for in the months before publication his health had been ghastly. 'I have been very bad lately', he wrote to Hooker, 'One leg swelled like elephantiasis – eyes almost closed up – covered with a rash & fiery Boils'; it was, he lamented, 'like living in Hell'.[20] He had therefore sought sanctuary at another hydropathy centre, this time at Wells House in Yorkshire, where the wealthy and infirm rested in luxury in a marbled mansion that boasted a dining room for one hundred guests, several drawing rooms, 'a coffee-room for general visitors', a billiards room, and even a bowling alley.[21] From the house, a pack of donkeys bore the patients in their gowns over rough-hewn paths to the baths which lay in the nearby spring. As Darwin took his daily plunges and dispatched copies of his 'child' to friends and the great men of science, he continued to worry: 'God knows', he confided in Wallace, 'what the public will think'.[22]

This was more than the typical apprehension which attacks an author during the days before publication. Darwin was suffering from a more acute anxiety because he understood the religious implications of the *Origin* perfectly well. As it happened, God was almost entirely absent from the book, confined to an epigraph from Francis Bacon and a scornful remark that to believe that each species had been created specifically and independently was to make 'the works of God a mere mockery'.[23] More pointedly, Darwin knew that, by positing natural selection as the mechanism for the development of species, he had removed the need for divine intervention in the natural world; and though he had restrained himself from commenting on the subject, he knew that his readers would apply this mechanism to the natural history of mankind itself. If natural selection could explain the creation and development of humankind without reference to the Almighty, what was the *Origin*, Darwin fretted to Huxley, but 'the Devil's gospel'?[24]

All this mattered because – as Froude, Holyoake, and Maurice had found – the penalties for sacrilege remained severe, and there was no sense (yet) of Christianity loosening its hold over British sensibilities. Just two years before the *Origin*'s publication, the Earl of Derby had sought to 'uphold the Christian character of a Christian country' by arguing against the admission of Jews to Parliament: 'Though among us', he told the House of Lords, 'they are not of us . . . they have interests wholly apart. Between them and us there is an impassable gulf.'[25] In the same year, the Lord Chief Justice of England and Wales strove to maintain the devout and even puritanical character of national life when he described pornographic literature as a 'poison more deadly than prussic acid, strychnine, or arsenic', an outburst which prompted the passage of the Obscene Publications Act.[26] And just as news of the Indian Rebellion was reaching Britain, the Presbyterian preacher 'Roaring' Hugh Hanna incited weeks of anti-Catholic rioting in Belfast by invoking 'faction's flame [and] hatred's gall' at massive open-air meetings: 'It is something too bad', noted one newspaper, 'to have a repetition in Ulster of the anarchy of Bengal'.[27]

These were also the halcyon years of 'muscular Christianity'. Heeding the call from W. J. Conybeare, the son of the man who described the *Plesiosaurus*, for the creation of a 'Broad Church' that could practise both 'Charity and Toleration', liberal theologians had articulated an ideal of Christian manliness which exalted the athletic and the heroic.[28] Eliding with the Victorian cult of self-improvement, which would receive its most persuasive expression in Samuel Smiles's *Self-Help* (1859) – a book that John Murray published in the same week as the *Origin* – these churchmen now preached a gospel of physical prowess, bravery, duty, and if necessary sacrifice.[29] As one historian has explained, their Christ was no longer an 'ethereal icon' but human and heroic, toiling and suffering and sharing in the struggles of his flock.[30] It was an explicitly masculine philosophy that, by marrying 'the almost unconscious instinct to do good' with 'every sort of athletic accomplishment', sought to bind young men closer to the Anglican Communion.[31]

The first of its great proponents was Charles Kingsley, the theologian who had supported J. A. Froude during his ostracism and who, in the 1850s, had emerged as a leading literary figure. His scholarship on Ptolemaic Egypt won him a professorship at Cambridge; friendship with Prince Albert, with whom Kingsley bonded over the notion of Anglo-Teutonic fraternity, meant he became chaplain to the queen and tutor to the Prince of Wales; and his novels were classic, popular works of historical fiction that he suffused with ideals of Christian manliness. In *Westward Ho!* (1855), Kingsley had narrated the derring-do of Amyas Leigh, who joined Francis Drake on his privateering missions and Walter Raleigh in his quest for El Dorado, fighting off the dastardly Spanish and resisting the Inquisition all the while.[32] It was the perfect, patriotic escape from the trauma of the Crimean War. Two years later, in *Two Years Ago* (1857), the protagonist Tom Thurnall figures as a hero who 'has been in every country, survived every species of disease, can perform any feat of strength, and knows every secret of practical success'; he even endures imprisonment in the Russian reaches of Circassia.[33] In reviewing the book for the *Saturday Review*, Dean Arthur Stanley coined the term: 'We all know by this time', he wrote, 'what is the task that Mr. Kingsley has made specially his own – it is that of spreading the knowledge and fostering the love of a muscular Christianity. His ideal is a man who fears God and can walk a thousand miles in a thousand hours.'[34]

It was the barrister and reformer Thomas Hughes, however, who wrote the canonical work of muscular Christianity, *Tom Brown's School Days* (1857).[35] Inspired by his own education under Thomas Arnold at Rugby School, Hughes puts Brown – a new boy, far from intellectual but honest and brave – through a gauntlet. The masters cane him, Flashman bullies him, and his best friend nearly dies. But Brown learns invaluable lessons from these trials and becomes the ideal Christian gentleman: a young man who prays solemnly at night, honours his elders, and masters what Hughes later calls the 'machinery of games' such as cricket.[36] In the sequel, Brown goes on to Oxford, where he resists the temptations of women, money, and unbelief. And here, while his hero marches to a 'wild beast show . . . in a part of the suburbs little known to gownsmen', Hughes preaches to the reader that 'a man's body is given him to be trained and brought into subjection, and then used for the protection of the weak, the advancement of all righteous causes, and the subduing of the earth'.[37] It was a persuasive philosophy of religion that informed 'the games-playing "cult" which [was sweeping] across Victorian Britain'.[38]

So how would Christian Britons receive the *Origin*?[39] The earliest news that reached Darwin, still recuperating at Wells House, was of commercial sensation. Booksellers had snapped up all 1,250 copies of the first print run, John Murray was planning for 3,000 more in January, and even the commuters at Waterloo station were buying the *Origin* to read on their journeys home. There were private letters of praise. Charles Kingsley wrote 'in awe' of the *Origin*'s 'heap of facts' and 'clear intuition' and confessed that it would transform his understanding of the world: 'I must give up much that I have believed & written.'[40] Writing from her retirement in the Lake District, Harriet Martineau declared to George Holyoake that the *Origin* was 'overthrowing . . . revealed Religion on the one hand, & Natural (as far as Final Causes & Design are concerned) on the other'.[41] Still in the East Indies, Wallace would by September 1860 have read the *Origin* 'through 5 or 6 times, each time with increasing admiration'. He described the book as 'the "*Principia*" of Natural History' and as 'a study that yields to none in grandeur & immensity'.[42] It was no small source of pride that his 1858 paper had helped to bring this great work to light.

Darwin's allies ensured a warm reception in parts of the press. Over three-and-a-half columns in the Boxing Day edition of *The Times*, Huxley proclaimed an 'immensity in the speculations of science to which no human thing or thought at this day is comparable'. Placing Darwin among 'the heirs of Bacon and the acquitters of Galileo', he exalted the idea of natural selection as 'perfectly simple and comprehensible, and irresistibly deducible from very familiar but well nigh forgotten facts'. It was 'a most ingenious hypothesis' and, with Darwin proving 'as greedy of cases and precedents as any constitutional lawyer', Huxley proclaimed that the *Origin* would 'carry us safely over many a chasm in our knowledge'.[43] On New Year's Eve, Hooker joined Huxley in praise. Writing in the *Gardeners' Chronicle*, a prestigious botanical magazine, he celebrated Darwin as an author of 'first-rate standing' and the *Origin* as a book that was 'so full of conscientious care, so fair in argument, and so considerate in tone'. Succeeding where Lamarck and the *Vestiges* had failed, and suggesting 'a plausible method according to which Nature might have ... produc[ed] suitable varieties', Darwin had in Hooker's view authored a work that was almost peerless 'in the whole range of the literature of science'.[44]

Parts of Darwin's Anglican readership, however, as he had anticipated fearfully, rose in outrage at the implications of natural selection. Adam Sedgwick, having read the *Origin* 'with more pain than pleasure', went as far as accusing Darwin of 'damage that might brutalize ... and sink the human race'.[45] Writing to the *Spectator*, the Archbishop of Dublin, Richard Whately, laughed that Darwin had shown 'a wonderful credulity'. Disparaging the idea that species could develop into others, he derided Darwin for seeming 'to believe that a white bear, by being confined to the slops floating in the Polar basin, might in time be turned into a whale; that a lemur might easily be turned into a bat; that a three-toed tapir might be the great grandfather of a horse'. More strikingly, Whately thought that natural selection was insufficient to explain the existence of dinosaurs. 'And how came the Dinosaurs', he asked, 'to disappear from the face of Nature, and leave no descendants like themselves, or of a corresponding nobility? By what process of *natural selection* did they disappear? Did they tire of the land, and become Whales, casting off their hind-legs? And, after they had lasted millions of years as whales, did they tire of the water, and leap out again as Pachyderms?'[46] Of course, these questions were rhetorical.

Richard Owen's reaction was entirely in character. He passed polite compliments in private, telling Darwin that he would welcome the *Origin* with a 'close & continuous perusal' and that he was deeply grateful 'for the application of [Darwin's] rare gifts' to the question of species, with Darwin taking the compliment that he had produced the 'best [explanation] ever published of [the] manner of [the] formation of species'.[47] But the April 1860 issue of the *Edinburgh Review* gave Owen scope – and the cloak of anonymity – to let loose upon an intellectual enemy. In a broad and sweeping article which covered works by Wallace and Hooker besides, Owen sneered that the 'real gems' of the *Origin* were 'few indeed and far apart', that the book left 'the determination of the origin of species very nearly where the author found it', and that Darwin was suffering from not only a 'confusion of idea as to the fact and the nature of the law, but an ignorance or indifference to the matured thoughts and expressions of . . . eminent authorities on this supreme question'. He concluded that the *Origin* was naught but the same 'abuse of science to which a neighbouring country [that is, France], some seventy years since, owed its temporary degradation'. Instead of recommending the *Origin* to the readers of the *Review*, therefore, Owen anticipated the art of academic peer-review by referring them to his own work.[48] It was a scathing, vituperative attack on the man as much as the work, and it shocked Darwin into sleepless nights. 'You [will] never read anything more envious & spiteful', he complained to Wallace.[49]

Owen would not confine his resentment to the pages of the *Edinburgh Review*. Only months before the publication of the *Origin*, he had spoken as president of the British Association about the 'continuous operation of Creative power'.[50] He would not concede the field to Darwin or to an increasingly belligerent Huxley, who was taking direct aim at Anglican orthodoxy through his Friday night lectures at the Royal Institution. And though Owen would seek cynically to capitalise on the sensation of the *Origin*, citing it before a parliamentary committee as evidence that the natural history collections of the British Museum now deserved a new and grander home, he would never accept its core principle: that competition within nature, not divine agency, had provided the mechanism for the variation of species. These were the battle lines behind which Owen, Darwin, Huxley, and all their acolytes would wage a culture war – a war for the soul of the British nation, over humanity itself – in the decades to come.

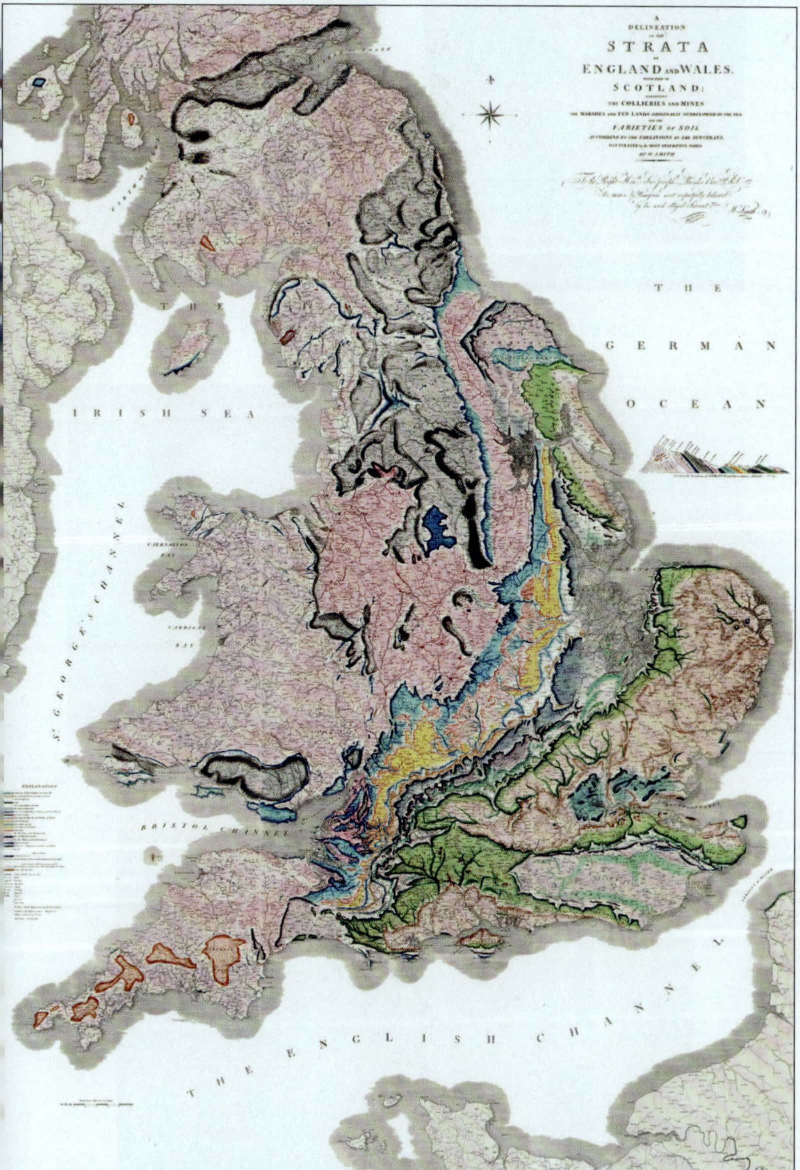

William 'Strata' Smith's geological map of Britain

James Ussher

Mary Anning

Georges Cuvier

Gideon Mantell

William Buckland lecturing at the University of Oxford

Henry De la Beche's *Awful Changes* (*above*) and *Duria Antiquior* (*below*), the first representation of a prehistoric landscape

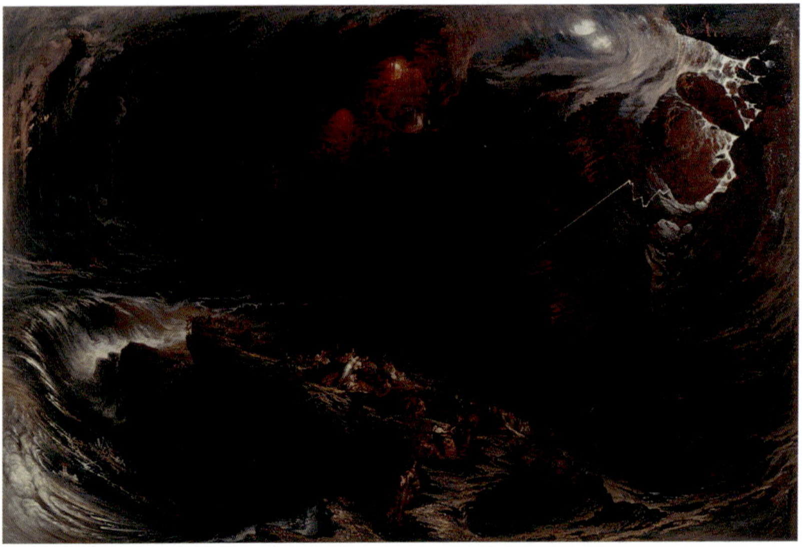

John Martin's *The Deluge* (*above*) and his *Country of the Iguanodon* (*below*), which he made for Gideon Mantell's *The Wonders of Geology*

Charles Lyell

Young Charles Darwin

The orangutans of the Regent's Park zoo whose human-like behaviour stunned London society – and Darwin too

A young Richard Owen at the Royal College of Surgeons

Thomas Huxley in his prime, lecturing on the gorilla

The Crystal Palace dinosaurs

Alfred Russel Wallace

The theologian, novelist, and historian Charles Kingsley, who coined the phrase 'impossible monsters'

Vanity Fair's depiction of Samuel Wilberforce and Thomas Huxley

John Tyndall

Charles Bradlaugh being arrested in Parliament

Othniel Charles Marsh (*back row, centre*) and the first Yale expedition to the American Badlands

Marsh's great rival, Edward Drinker Cope, depicted at his desk

Charles Knight's *Leaping Laelaps*

The Carnegie *Diplodocus* is unveiled at the Natural History Museum, London, in 1905

The two towers of London science: Owen's Natural History Museum (*left*) and Huxley's Royal College of Science (*right*)

PART FOUR

The Gorilla War

DRAMATIS PERSONAE

Charles Bradlaugh, journalist, London
John William Colenso, bishop, Natal
Charles Darwin, naturalist, Kent
George Holyoake, journalist, London
Thomas Huxley, naturalist, London
Charles Kingsley, theologian, Cambridge
Charles Lyell, geologist, London
Richard Owen, naturalist, London
J. R. Seeley, academic, London
John Tyndall, physicist, London
Alfred Russel Wallace, zoologist, Malaya and London
Samuel Wilberforce, bishop, Oxford

16

Monkeyana

> Nature repairs her ravages – but not all. The uptorn trees are not rooted again; the parted hills are left scarred: if there is a new growth, the trees are not the same as the old, and the hills underneath their green vesture bear the marks of the past rending. To the eyes that have dwelt on the past, there is no thorough repair.[1]
>
> George Eliot, *The Mill on the Floss* (1860)

According to one historian the 'shot reverberated through England, and indeed through other countries'.[2] In the words of another, more than a hundred years after the event, 'no battle of the nineteenth century, save Waterloo, is better known'.[3] There are few moments in the history of science that have attracted more attention, or that are more symbolic of the conflict between science and religion over evolution. Yet historians have known for decades that there is no comprehensive and objective account of what transpired on that Saturday afternoon in June 1860 at the annual meeting of the British Association for the Advancement of Science. So, what really happened when Thomas Huxley debated with Samuel Wilberforce? Has the tale merely grown in the telling?[4]

The British Association's meeting had been a movable feast since its inauguration in the early 1830s and, in keeping with its mission to evangelise across the British Isles, this 'parliament of science' had not confined itself to the south-east of England. There had been meetings in Aberdeen, Cork, Swansea, Hull, and Newcastle. In 1860, however, the Association had returned to Oxford, where the section on zoology and botany had scheduled a lecture from John William Draper, an English-born chemist who was now a professor at New York University. For his subject, Draper took 'the Intellectual Development of Europe' and the 'Progression of Organisms [as] determined by Law'. In doing so, he would address the arguments of the *Origin* from a platform in the Museum of Natural History, a magnificent new building on Parks Road which appeared to symbolise 'Oxford's concession to the claims of physical science'; it was only in the last few years, after all, that the university had established a degree in the natural sciences.[5] With a thousand people crowded onto the benches that carpenters had made specially for the occasion, Draper spoke from twelve o'clock on how progress 'in civilization does not occur accidentally or in a fortuitous manner, but is determined by immutable law'.[6] His audience listened 'with the profoundest attention ... no one stirred'. Perhaps the quietude stemmed from the heat of high summer; perhaps it was boredom, for Joseph Hooker complained about the droning of a 'yankee donkey'.[7] Or maybe it was nervous anticipation, since nobody was there for Draper's lecture, but for the debate that was expected to follow.

Joining Draper on the platform was John Stevens Henslow, the botanist who had exerted a formative influence over Darwin at Cambridge, and from one o'clock he moderated a discussion on the merits of developmental theory. He took interventions from a fellow botanist, a surgeon, and a Mr Dingle, whom the audience shouted down abruptly. Then he turned to Samuel Wilberforce, the Bishop of Oxford, whose powerful speeches in the House of Lords had earned him a reputation as one of the leading orators in the country; in less flattering terms, he was also known as 'Soapy Sam', a nickname he owed to Disraeli's barb that he was 'unctuous, oleaginous, [and] saponaceous'.[8] In the weeks before the Oxford meeting, Wilberforce had reviewed *On the Origin of Species* for John Murray's *Quarterly*, and in those pages the bishop had found 'much and grave fault'. Though he praised the *Origin* as 'most readable' and sparkling 'with the colours of fancy and the lights of imagination', Wilberforce decried its 'dishonouring view of nature' and lamented that Darwin 'should have wandered from [the] broad highway of nature's works into the jungle of fanciful assumption'.[9] Now, at Oxford, his audience awaited the eruption of an 'open clash between Science and the Church'.[10] What would the bishop say?

In the most detailed report, Wilberforce sought to undercut Darwin by insisting, like Cuvier, on the fixity of species. 'The permanence of specific forms', he said, 'was a fact confirmed by all observation. The remains of animals, plants, and man found in . . . the Egyptian catacombs, all spoke of their identity with existing forms of . . . an unalterable character.' In light of this evidence, Wilberforce could conclude only that there was no Lamarckian transmutation, and no adaptation by means of natural selection: 'The line between man and the lower animals was distinct. There was no tendency on the part of the lower animals to become the self-conscious intelligent being, man; or in man to degenerate and lose the high characteristics of his mind and intelligence.'[11] So far, so predictable, but then Wilberforce turned to the other occupants of the platform and to Thomas Huxley, of whom he begged to know: 'Was it through his grandfather or his grandmother that he claimed his descent from a monkey?'[12] The audience roared in laughter, the clergymen cheered their hero, and ladies waved their handkerchiefs in acclamation.

Huxley had not wanted to attend Draper's lecture, or to engage the bishop. He had spoken the day before on pyrosomes, the luminescent plankton that he had seen 'shining like white-hot cylinders in the water' around the *Rattlesnake*. And in the expectation that Wilberforce would merely 'appeal to prejudice' in attacking Darwin, he had planned to leave Oxford on the Saturday morning. It was only a chance meeting with Robert Chambers, the author of the *Vestiges* begging Huxley 'not to desert them', that gave him pause to reconsider.[13] And Huxley did hesitate. The previous year, he had vowed to abstain 'from petty personal controversies', but he had also resolved '[t]o smite all humbugs, however big'.[14] Thus was he in the museum, listening to Wilberforce's question about the monkeys, when he decided that the bishop's insolence deserved 'the severest retort [he] could devise'. Turning to the chemist Benjamin Brodie, who was sitting beside him, he whispered: 'The Lord hath delivered him into my hands.'[15]

The crowd now waited, the museum in silence, as Huxley rose to speak. 'If the question is put to me,' he declared, 'would I rather have a miserable ape for a grandfather or a man highly endowed by nature and possessed of great means and influence, and yet who employs these faculties for the mere purpose of introducing ridicule into a grave scientific discussion – I unhesitatingly affirm my preference for the ape.'[16] There is some disagreement over the precise words that were used, and there has even been the suggestion that it was Hooker, not Huxley, who delivered the most withering response to Wilberforce, but the sensation was real. As the bishop and the bulldog sparred on the stage, they engrossed – and scandalised – their audience: 'The battle waxed hot', Darwin learned, and 'Lady Brewster fainted'.[17]

On the next day the churchman Frederick Temple, who had succeeded Thomas Arnold as headmaster of Rugby School, preached a sermon before the university on 'the present relations of science to religion'. In the wake of Saturday's fireworks, it was in some respects a declaration of victory. 'There was a time', he said, 'when . . . if there were ever an appearance of collision [between science and religion], science was required to give place. That time ceased with Galileo, and can never return.' The religious man, moreover, no longer had any right 'to refuse to accept what he finds in the one than what he finds in the other' and so, 'whenever . . . there is a collision between them, the dispute becomes simply a question of evidence'. Turning to specific examples, Temple took up geology, which had 'already altered our conception of a great part of the Book of Genesis'. And though he denied that religion and science were foes, arguing for 'the harmony between [the Bible and nature] in character', Temple suggested that it was men such as Huxley, not Wilberforce, who might take the future with them: 'The student of science', he declared, 'knows that the most unlettered peasant can penetrate to the true reality of all things as surely as the wisest philosopher'.[18] The son of a penniless teacher, born above a butcher's shop, had bested a bishop, the son of a national hero; the weight of authority was shifting.

The Oxford showdown and the completion of the *Origin of Species* had occurred during a wider, bitter debate over humankind's relationship to the animal kingdom. It probably began in the London gutter press, far from the panelled halls of polite science, with the *London*

Investigator. Founded as a 'monthly journal of secularism' by George Holyoake's colleague Robert Cooper, the *Investigator* commenced with proclamations that science was 'the *only* providence of life' and that the Bible's claims 'to divine origin are an assumption'. In its first issue, the *Investigator* had inaugurated a series of articles on 'The Origin of Man: Science versus Theology' and, by the summer of 1854, it had concluded that humans were nothing special. 'Comparative anatomy', the journal stated, 'enforces the opinion that, instead of our organism emerging into existence impromptu, and distinct from the rest of the animal world ... [it] is only a link in the grand chain of animal forms', even down to the 'lowest of animated creatures' such as slugs, worms, and leeches.[19] It was a radical, provocative position which, besides disparaging the notion that the Lord had created humankind specially in His own image, had potentially revolutionary implications for society: if there was nothing to separate humankind from lower orders of creatures anatomically, what was there to separate classes of people socially, economically, or politically?[20]

It was for many naturalists a troubling thought. Even Charles Lyell, whose theories of geology had done so much to undermine biblical history, had worried while reading early drafts of the *Origin* that humankind could have descended 'from an Ourang'; accordingly, upon its publication, he hurt Darwin by failing to endorse the book's arguments fully.[21] For someone like Richard Owen – a staunch Tory, a grim captain of orthodoxy – the idea was intolerable. Though he had argued previously that apes and humans were anatomically separate, he entered the debate more formally in 1857 with a strident paper before the Linnean Society. Here, speaking across two nights in February and April, Owen contended that, even if zoologists could draw comparisons between the teeth and limbs of humans and other creatures, there was one enduring difference which marked a definite, unyielding distinction: the brain. It was the brain which not only 'distinguishes the Mammalia from all the inferior classes of VERTEBRATA', but which 'in Man ... presents an ascensive step in development, higher and more strongly marked than that by which the preceding subclass was distinguished from the one below it'. Owen held that the 'posterior development' of the human brain was so advanced that it had a unique third lobe, the hippocampus minor. It was this part of the brain, more than anything, which gave to mankind its 'peculiar mental powers'.[22] Owen's paper was the defining mid-century statement on the special place that mankind occupied in nature; two years later, he would deliver another version of it when he became the first person to receive an honorary doctorate from Cambridge.

Of course, not everyone agreed. When he received his copy in the post, Charles Darwin would applaud Owen's audacity: 'What a capital number of the Linnean Journal!' he wrote to Hooker; 'Owen's is a grand Paper.' But he could not accept his conclusions: 'I cannot swallow Man', he noted in a postscript, 'making a division as distinct from a Chimpanzee, as an ornithorhynchus from a Horse. I wonder', he asked, 'what a Chimpanzee wd say to this?'[23]

It was Thomas Huxley, though, who would respond the more vigorously. When he disembarked from the *Rattlesnake* in 1850, he had noted 'the intense feeling of hatred [with which] Owen is regarded by the majority of his contemporaries, with [Gideon] Mantell as arch-hater', but he had chosen to treat Owen with a cautious respect; in turn, Owen had rewarded this guarded deference by extending his patronage.[24] By the mid-1850s, however, their relationship had soured utterly. It was not that Owen thought species were incapable of change; rather, he just could not stomach a mechanism for change that was bereft of heavenly agency. He had therefore persisted with his belief in the divine archetype of anatomy, preaching at Leeds to the British Association about 'the ordained becoming of living things'.[25] But even before Owen savaged Darwin in the *Edinburgh Review*, the latent tension between him and Huxley had boiled over into open hostility. First, during an 1858 lecture at the Royal Institution, with Owen chairing the evening in helpless silence, Huxley ridiculed the concept of the divine archetype 'as fundamentally opposed to the spirit of modern science'.[26] Soon afterwards, in early 1859, he broke cover completely: 'It is as respectable to be modified monkey', Huxley wrote to the surgeon and naturalist Frederick Dyster, 'as [to be] modified dirt'.[27] Such was the ferocity of his contempt that some historians have wondered whether, in the late 1850s, he believed sincerely in any developmental theory – after all, he had savaged the *Vestiges* in a review article – or whether he simply wanted an enemy.

After the Oxford debate, tragedy propelled Huxley into more

aggressive combat. Noel was his and Nettie's eldest child, born on New Year's Eve 1856 with 'large blue eyes[,] golden curls[,] clear fair skin & regular features'; he was, wrote Huxley, 'our delight and our joy'.[28] But one Thursday in September 1860, not long after father and son had enjoyed 'a great romp together', Noel became 'sick & feverish'. Getting worse through the night, he was by Friday morning 'very ill'. A doctor thought it only a gastric complaint, but a rash on Noel's face soon erupted and 'the little fellow gradually became delirious'.[29] All things pointed towards scarlet fever and Huxley remarked that 'it was as if the boy had been inoculated with some septic poison'. On the Saturday, Noel's 'tangled golden hair [having] tossed all day upon his pillow', there came another violent attack; by nine o'clock, the Huxleys had 'a dead child in [their] arms'. Thomas picked up his son, carried him into the study, and 'laid [down] his cold still body'.[30] The mother and father went outside and walked up and down their garden in the rain.

Huxley reflected that the death of his son could, in other eras, have set him down a different path. He confessed to Charles Kingsley, who urged Huxley to recognise that the human soul was 'nearer to a God than to a Chimpanzee', that 'had I lived a couple of centuries earlier I could have fancied a devil scoffing at me ... and asking me what profit it was to have stripped myself of the hopes and consolations of the mass of mankind?'[31] But the Huxley of 1860 had 'convictions ... [that were] of long and slow growth and are firmly rooted'. No matter his grief, no matter the loss of 'our poor little son, our pet and hope', he was resolved that 'truth is better' than disingenuous comfort. There was nothing now that would throw Huxley upon the mercies of a priest or convince him to believe that Noel's soul could live on 'just as the humble bulb of an annual lives, while the glorious flowers it has put forth die away'. Rather, Huxley's loss had steeled him in his mission: 'My business', he told Kingsley, 'is to teach my aspirations to conform themselves to fact, not to try and make facts harmonise with my aspirations ... Sit down before fact as a little child, be prepared to give up every preconceived notion, [and] follow humbly wherever and to whatever abysses nature leads, or you shall learn nothing.'[32]

When Huxley returned to the fray, while Nettie was 'mend[ing] but slowly', it was Owen who would suffer his wrath.[33] In the relaunched *Natural History Review*, a failing Dublin journal that Huxley and his allies had bought and repurposed to their crusade, he polemicised on 'the zoological relations of man with the lower animals'. He wrote almost every paragraph, every line to assail that 'canting humbug' Owen, the avatar of the 'theologians and moralists' who 'have always tended to conceive of their kind as something apart, [as] separated by a great and impassable barrier from the rest of the natural world'.[34] Huxley, on the contrary, counted himself among 'the students of physical science' who had 'more distinctly admitted the closeness of the bond which united man with his humbler fellows'. And while he would write in this article on mankind's interactions with cats, dogs, and parrots – all of which could 'return love for our love, and hatred for hatred' – his real subject, once again, was the similarity of the human brain to those of primates. His conclusion? 'The Brains of the lower true apes and monkeys differ far more widely from the brain of the orang than the brain of the orang differs from that of man.'[35] It was the deliberate goading of Owen, a provocation that he would repeat each Friday from February to May during lectures for working men at the School of Mines in London. 'By next Friday evening', he boasted to Nettie in March 1861, 'they will all be convinced that they are monkeys'.[36]

Things escalated further with the arrival in London of the French explorer and zoologist Paul Du Chaillu, who had recently concluded a four-year exploration of equatorial Africa. Now touring the great cities of Europe and North America, Du Chaillu presented to the Royal Geographical Society in February 1861, and a young Francis Galton – the future eugenicist and Darwin's half-cousin – took notes on an evening when the Frenchman brought news and specimens of a creature that many Europeans had known only by rumour. Du Chaillu described a swathe of Africa 'characterised by mountains covered with forests, of tropical richness, and traversed by many rivers'. There were few wild animals here – no lions, no giraffes, no zebras – but there were 'cannibal races' and it was 'the chief abode of the gigantic ape', the gorilla.[37]

Owen, of course, was in the audience and, though he conceded 'the approximations . . . in this creature to the human frame', he seized upon Du Chaillu's discoveries as validation of his theory: 'As contra-distinguished from man,' he crowed, 'there was in the gorilla no trace of this posterior lobe [the hippocampus minor] beyond the cerebellum'.[38] Within a month, having examined the skulls that Du Chaillu bequeathed to the British Museum, Owen would lecture on 'The Gorilla and the Negro'. There was 'a gradation of cerebral development', he declared, 'from the lowest to the highest vertebrate species' but, crucially, there were interruptions in this sequence, and the gorilla's lack of a hippocampus minor was evidence of such an interruption between the most human-like and most advanced primate and even 'the Negro, [which Owen described as] the lowest variety of Human Race'.[39] While Owen's discourse might disturb the modern reader, he was in fact rejecting the scientific racism of anatomists such as Robert Knox and Anders Retzius, who had argued that the races of humankind constituted separate species; for Owen, the hippocampus was proof that mankind, and all of mankind, was special.

Would Huxley accept this? Absolutely not: he and Owen were enemies now, and when the former joined the council of the Zoological Society,

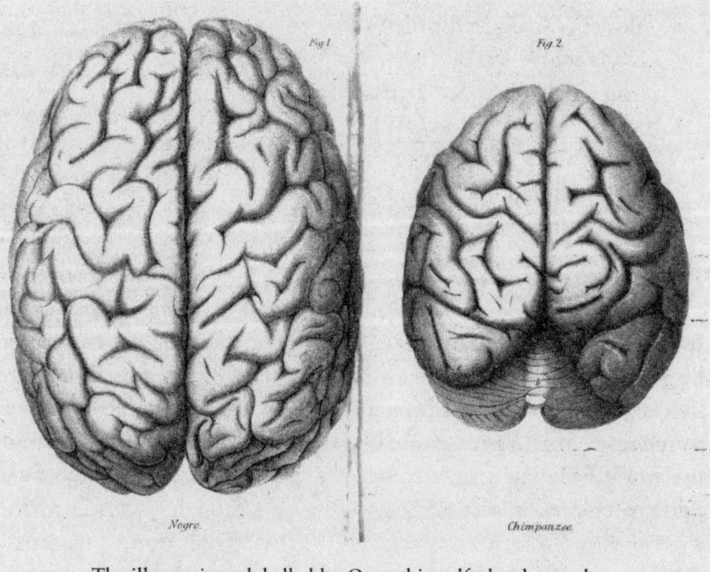

The illustrations, labelled by Owen himself, that he used to argue for the special nature of the human brain

the latter resigned in a strop. Over the next few months of 1861, the pair flayed strips off each other in lectures and the pages of the *Athenaeum*. When Huxley mocked the inaccuracy of Owen's illustrations and his confusion over 'structures which have no existence', Owen blamed 'the Artist' but maintained that he was right.[40] As Huxley pricked him again by complaining that 'life is too short to occupy oneself with slaying of the slain more than once', Darwin relished the contest.[41] He thought Huxley had been 'almost too civil' and professed himself 'more inclined to clap [him] on the back, than to cry hold hard!'[42] As 'the Great Hippocampus Question' became an intellectual blood sport for mid-Victorian scientists, the satirical magazine *Punch* offered its own analysis. In May 1861, under the heading 'Monkeyana' and in parody of the famous Wedgwood anti-slavery cameo, there stood a gorilla wearing a sign that read 'AM I A MAN AND A BROTHER?' In the same edition, the poet 'Gorilla' – really Philip Egerton – parsed the feud:

> Then DARWIN set forth,
> In a book of much worth,
> The importance of 'Nature's selection;'
> How the struggle for life
> Is a laudable strife,
> And results in 'specific distinction' . . .
> Then HUXLEY and OWEN,
> With rivalry glowing,
> With pen and ink rush to the scratch;
> 'Tis Brain *versus* Brain,
> Till one of them's slain;
> By Jove! It will be a good match![43]

It was a contest in which Huxley would not relent. 'What matters it', he asked during his lectures at the Royal Institution, 'whether a new link is or is not added to the mighty chain which indissolubly binds us to the rest of the universe? . . . Of what part of the glorious fabric of the world has man a right to be ashamed – that he is so desirous to disconnect himself with it?'[44]

While conservative churchmen licked the wounds that Huxley had inflicted in Oxford, they were also confronted by an enemy within.

For his 1860 volume *Essays and Reviews*, the publisher John William Parker had invited seven Anglican writers, six of them ministers, to 'reconcile intellectual persons to Christianity' by showing how new modes of thinking were compatible with sincere Christian belief.[45] The seven authors, writing 'without concert or comparison', fulfilled this mission expertly. First was an essay on 'the education of the world' by Frederick Temple – who, as we have seen, had preached on science and religion at Oxford – which argued that Christians, when interrogating the Bible, 'have no right to stop short of any limit but that which nature . . . has imposed upon us'. In this way, Temple asked his readers not to worry about censure from the pulpit, but to answer only to the demands of their own conscience and intelligence. Second came Rowland Williams, a professor of Hebrew, who deployed the work of German scholars to assail Anglican deference to Old Testament histories: 'The attitude of too many English scholars before the last Monster out of the Deep', he wrote, 'is that of the degenerate senators before Tiberius'. Next was Baden Powell, the professor of geometry at Oxford, who produced the syllogism that because the Lord operated by way of natural laws, and because biblical miracles *broke* those laws, it was foolish to believe in biblical miracles. More pointedly, Powell drew his readers' attention to 'the palpable contradictions disclosed by astronomical discovery with the letter of scripture', to 'the development of species', and to 'the rejection of the idea of "creation"'.[46]

The middle essay by the Oxford theologian Henry Bristow Wilson questioned the strictures of the Thirty-Nine Articles before Charles Wycliffe Goodwin, an antiquarian and Egyptologist, took aim at the scriptural literalists who were still seeking to reconcile Mosaic history with geological evidence: 'It would have been well', he sighed, 'if theologians had made up their minds to accept frankly the principle that those things for the discovery of which man has faculties specially provided are not fit objects of a divine revelation'. The penultimate essay by another Oxford cleric, Mark Pattison, offered an even-handed study of eighteenth-century theologians before Benjamin Jowett, Oxford's Regius professor of Greek, perorated on 'the interpretation of scripture'. Already regarded as a controversial figure for his work on St Paul's letters, Jowett now begged his readers to read the Bible like any other book. When it came to 'the meaning of words, the connexion of sentences, [and] the evidence of facts', he reasoned, 'the same rules [should] apply to the Old and New Testaments as to other books'; the Bible was a matter of study as much as a source of divine instruction.[47]

In the judgement of one historian, the *Essays* provoked 'the greatest religious crisis of the Victorian age'.[48] Within two years, the collection sold more than 22,000 copies – more than the *Origin* would sell in two decades – and provoked more than 400 replies in the press or in standalone publications. And while some of these correspondents were supportive, many more of them regarded the authors as heretics: the *Edinburgh Review* reflected that they could have been 'seven stars in a new constellation' but were more properly the '*septem contra Christum*', the Seven Against Christ.[49] The storm grew wilder when the archbishops of Canterbury and York allied with every consecrated bishop in the country to condemn the Seven in *The Times*. They grieved 'the pain it has given them that any clergyman of our Church should have published such opinions'; they wondered how any of the essays could be consonant with 'an honest subscription to the formularies of our Church'; and they asked whether the *Essays* were so vile as to merit the attention of the ecclesiastical courts or 'synodical condemnation'.[50] These complaints struck at the real core of the scandal: whereas clergymen such as Buckland and Conybeare had once sought to reconcile their science to the Bible, liberal Anglican ministers were now advising their readers that it was religion, not science, which might need to make way. Just as embarrassing was the fractious and public way in which the Church was dividing against itself.

Charles Darwin, for whom 'a bench of bishops [was] the devil's flower garden', attempted to defend the essayists, corralling several eminent signatures onto a letter which expressed belief that 'such enquiries must tend to elicit truth, and to foster a spirit of sound religion'. Without approving the *Essays* in explicit terms, he wrote to Frederick Temple to 'welcome these attempts to establish religious teaching on a firmer and broader foundation'.[51] But this remained a minority opinion. Much more influential was another intervention in the *Quarterly* by Samuel Wilberforce, who, only months after attacking Darwin in those pages, penned another acerbic review. 'This volume', he began, 'has met with a circulation, and excited a measure of remark, which appear to us to be far greater than it would naturally have obtained by its mere literary merits'. Scorning the *Essays* as 'well-suited . . . to the metaphysical mind of Germany', he represented the collection as a threat to the soul of the British nation: 'infidelity, if not Atheism', he warned, 'is the end to which this teaching inevitably tends'. There were truths at stake here, the bishop had decided, that were 'more precious [by] far than any which Plato could have endangered'.[52] John Murray paid a hundred guineas for Wilberforce's piece, almost £10,000 today.

There were further rounds of outrage as a Victorian Inquisition took shape. Charles Longley, who became Archbishop of Canterbury in 1862, declared that 'no graver matter since the Reformation or in the next 200 or 300 years could be imagined'.[53] So, what would be done with these turbulent priests? Two of them, Williams and Wilson, found themselves before the Church courts, charged with heresy for denying the divine inspiration of the scriptures. Williams's barrister was the son of an eminent abolitionist but he found no favour with the judge, the anti-slavery stalwart Stephen Lushington, who declared both men guilty. It was only when Williams and Wilson appealed to the Privy Council, which 'dismissed hell with costs', that they secured their acquittal.[54] Yet even this would not quench the controversy. Wilberforce was still in arms – and angry enough now to put his name to his convictions – and he would soon take a petition, signed by more than 11,000 churchmen, to the Convocation of Canterbury. By June 1864, in a move which attracted vocal support from the House of Lords, the Anglican synod had condemned the *Essays* as 'containing teaching contrary to the doctrine received by the United Church of England'.[55] Such dramatic censure seemed at the time, reflected one historian, 'to be the only possible course to allay the popular anxiety' among clergymen, to meet the growing concern that the new ideas of the Seven, and of Darwin and Huxley, would triumph over the certainties of old.[56]

17
Going the Whole Orang

> You must not say that this cannot be, or that that is contrary to nature. You do not know what Nature is, or what she can do; and nobody knows; not even Sir Roderick Murchison, or Professor Owen, or Professor Sedgwick, or Professor Huxley, or Mr. Darwin, or Professor Faraday, or Mr. Grove, or any other of the great men whom good boys are taught to respect. They are very wise men; and you must listen respectfully to all they say: but even if they should say, which I am sure they never would, 'That cannot exist. That is contrary to nature,' you must wait a little, and see; for perhaps even they may be wrong.[1]
>
> Charles Kingsley, *The Water-Babies* (1863)

In the early modern period, confecting a classical soubriquet was the fashion. The physician John Keys returned to England with continental pretensions and took the name 'Caius', which posterity remembers through a Cambridge college and a character in *The Merry Wives of Windsor*. At Leiden in the Dutch Republic, the humanist lawyer Hugo de Groot became 'Grotius'. And at Cremona in northern Italy, several hundred violins and cellos had the name 'Stradivarius' on their labels. Even ordinary folk looked to Latin and Greek for a dash of class. In the early seventeenth century in northern Germany, a musician called Neumann put his family name into Greek, giving 'Neander'. Fifty years later, Herr Neumann's grandson Joachim had become a charismatic, well-loved preacher at Düsseldorf. A skilled musician in his own right, whose compositions remain part of the German hymnal,

Neander was also an enthusiastic naturalist. He often led open-air services beneath the 'pretty, high limestone cliffs' which overlooked the Düssel valley and, according to legend, he lived there one summer in a cave, 'communing with nature and with God'.[2] In tribute to his popularity, the valley – in German, '*thal*' – in time took his name.

By the mid-1850s, the Neander Valley was part of the Prussian Rhineland, the heartland of Germany's Industrial Revolution. In seeking to exploit the region's rich resources – there was coal, lead, and limestone – a joint-stock company had sunk several quarries in the valley, but in the summer of 1856 its workmen had found something else. Deep within the mud of a cave now known as the Feldhofer Grotto, they made a discovery of an 'unusually savage aspect': there was a skull, some ribs, and several other bones.[3] The skull did not appear to be human. It was long, shallow, and a sturdy pair of ridges protruded above the eye-sockets. At first, the diggers thought it belonged to a bear; someone else suggested a 'rickety "Mongolian Cossack"' who had chased Napoleon across Europe with the Russians in 1814. But a local schoolteacher realised the bones belonged to something or someone else. In search of an expert, he took the remains to Bonn and to Hermann Schaaffhausen, a professor of natural history who had already written about 'the constancy and transformation of species'.[4] Working over the winter of 1856–7, the two men prepared a detailed study of the Neander specimens and, in spring 1857, Schaaffhausen presented their findings 'on the crania of the most ancient races of man' to a local scientific society.

Britain, of course, had its own share of caves to explore. In 1858, the discovery of Brixham Cave some nine miles south of Torquay had prompted the Geological Society to create a 'Cave Committee'. Within a year, under the leadership of Hugh Falconer, it had retrieved thousands of animal bones and dozens of man-made artefacts.[5] Yet when, in 1861, the zoologist George Busk translated the Schaaffhausen paper for Huxley's *Natural History Review*, it became clear that the Neander Valley discovery was vastly more important. It appeared, first, 'that the extraordinary form of the skull was due to a natural conformation hitherto not known to exist, even in the most barbarous races'; second, that 'these remarkable human remains belonged to a period antecedent to the time of the Celts and Germans, and were in all probability derived from one of the wild races of North-western Europe'; and third, that 'it was beyond doubt that these human relics were traceable to a period at which the latest animals of the diluvium still existed', which was to say that humankind – or this ancestor of it – could have shared the Earth with the creatures of pre-history. For Busk, who added commentary to the translation, it was not only proof of 'the geological antiquity of man' but a fearsome weapon that he and Huxley could deploy in their war with Richard Owen over mankind's

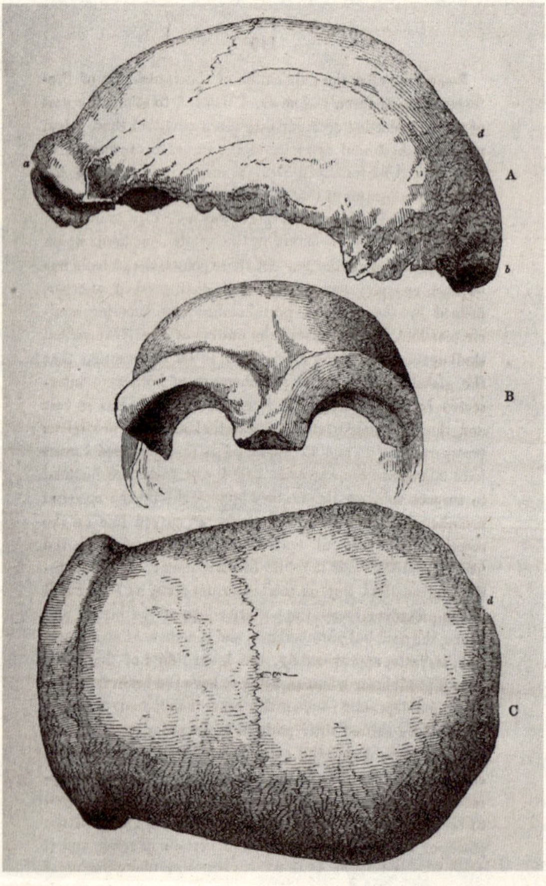

The Neanderthal skull from the Neander Valley in the Rhineland

relationship to primates: 'The conformation of the cranium', he wrote, 'is so remarkable, as justly to excite the utmost interest, approaching as it does in one respect that of some of the higher apes'. Had these German quarrymen found the missing link between humankind on the one hand, and 'the Gorilla and Chimpanzee' on the other?[6]

Speaking before the Royal Society in January 1862, the anatomist William Henry Flower surveyed the state of the Great Hippocampus Question. He reiterated that the presence or absence of the posterior lobe of the brain, the hippocampus minor, had become the decisive point of debate. While Owen insisted that only humans possessed this feature, Huxley had accused Owen of incompetence – or worse – in failing to account for 'the posterior lobes in *Ateles*, a monkey of the . . . New World group'. Though Flower declined to take a position on 'the transmutation of species, or origin of the human race', and though he attempted to defuse the controversy by pleading that the proper resolution of the question should depend only 'upon an appeal to nature, and not to authority', his conclusions were hostile to Owen: 'the posterior lobes', he declared, 'exist in all the Quadrumana', which was to say 'in all primates'.[7]

Of course, Owen would fight back, for he had too much to lose. He reigned at the Zoological Society, he had taken charge of the British Museum's collection of fossils, and he was negotiating with the government for the construction of a specialist museum for natural history. Prince Albert had even asked Owen to deliver a series of lectures at Buckingham Palace, where in the spring of 1860 he mocked up a theatre in one of the drawing rooms. With sofas and armchairs arranged 'in a little semi-circle', and with Owen fastening diagrams to 'a large green-baize curtain' which served as a blackboard, between thirty and forty members of the royal family and household attended talks on mammals, birds, reptiles, and fishes. 'They are all attentive', Owen reported to his sister, 'and seem often to be deeply interested'. They were indeed. When Owen gave an encore, the queen remarked in her journal that he had described birds 'most beautifully' and explained the behaviour of kangaroos 'in a most curious & interesting manner'.

Owen waited until the Cambridge meeting of the British Association in October 1862 to launch his latest salvo against Huxley and the very concept of natural selection. First, he presented the case of the aye-aye, the long-fingered lemur from Madagascar which feeds by gnawing through the branches of trees to find grubs. In Owen's view, the peculiarities of the aye-aye – and there were several 'perfect adaptations of particular mechanical instruments to particular functions' – were fatal to 'the Lamarckian and Darwinian Hypothesis of the Transmutation and Origin of Species'. It had strong feet to grasp branches, powerful incisors to gnaw through bark, and long, thin fingers for extracting its prey from trees. Owen insisted that 'all this must have a cause', but he refused to accept natural selection as the source of the aye-aye's peculiarities. Why should this lemur, over generations, have adapted itself in such ways to hunt grubs when birds, fruits, and insects – much more accessible sources of sustenance – 'abounded' in Madagascar? For Owen, it made more sense to regard the aye-aye, in all its strangeness, as an original creation of the Lord or as an example of degeneration from an original 'stock'.[8] Few in the audience were convinced. The 'Aye-Aye paper fell flat', reported Huxley to Darwin: 'All the people present who could judge saw that Owen was lying & shuffling: the other half . . . regarded him I think rather as an innocent old sheep, being worried by three particularly active young wolves.'[9]

Owen tried again with a paper on the 'Zoological Significance of the Cerebral . . . Characters of Man'. Here, he returned to the brain and, in what he thought would be a vivid demonstration of human exceptionalism, presented the casts of three skulls: one belonging to 'a male European', one to a 'Negro', and another to 'a full-grown male Gorilla'. Declaring that 'there had been no province of intellectual activity in which individuals of the pure Negro race had not distinguished themselves', Owen insisted once more upon 'the importance and significance of the much greater difference between the highest ape and lowest man, than [the differences which] existed between any two genera' of apes. If his audience wanted further proof of the superiority of even 'the lowest of the human races', Owen directed them towards the feet: gorillas had four thumbs, humans only two, and so 'the contrast between the foot-structures of the Gorilla and Negro was [just] as great' as that between their brains.[10]

Late in the week, as the delegates considered the latest skirmish in the 'gorilla war', copies of a brief but bitter satire made their way around the university. Printed privately, with no hint of the authorship, it was titled 'The Speech of Lord Dundreary ... on the Great Hippocampus Question'. Though largely forgotten today, Dundreary was one of the touchstone characters of nineteenth-century theatre. First appearing in the Tom Taylor farce *Our American Cousin* (1858) – the last play, of course, that Abraham Lincoln would see – Dundreary was a witless, gullible aristocrat who sported enormously bushy sideburns and spouted nonsensical aphorisms such as 'Birds of a feather gather no moss'. Imagining someone as Dundreary was no compliment, but who did this satirist have in mind? It might have been Huxley, who had 'drunk all our healths ... [in a] very handsome way' and 'said [the hippocampus] was in [the ape's] tail'. It might have been Owen, for if Huxley could see the hippocampus minor in humans, 'why can't [he]?' Or more likely it was the passive spectator of their feud, the casual naturalist who believed that 'we had all hippopotamuses in our brains' and that 'monkeys are men, only they won't work for fear of being made to talk'.[11] If the last of these, the Dundreary skit was symptomatic of frustration with 'the annual passage of barbed words between Professor Owen and Professor Huxley', a now-routine hatred which the *British Medical Journal* would describe as 'a hindrance and an injury to science ... and a scandal to the scientific world'.[12] How much longer, commentators were asking, would the decorous progress of knowledge remain hostage to the egos of two men, neither of whom would yield to the other? As Owen and Huxley sustained their vendettas into the autumn of 1862, some naturalists even asked whether the combatants were debasing themselves to the level of the quarry, 'reducing man to a monkey, or elevating the monkey to the man'.[13] It would take three telling blows, all landed within a few weeks of each other in the spring of 1863, to swing the question decisively in one direction.

The earliest of these interventions came from Charles Lyell. In the years since writing *Principles of Geology*, Lyell had issued a further eight editions of that great work as well as five editions of a popular survey of the science, *Elements of Geology* (1838). Having married and set up home on Harley Street, he and his wife had then travelled in the United States and Canada, with Lyell twice giving the prestigious Lowell lectures in Boston. The two travelogues that he published on returning home are fascinating examples of British ambivalence towards antebellum America: as much as Lyell delighted in describing the Great Lakes, the valleys of upstate New York, and the coalfields of Appalachia, and though he remarked upon the 'retarding' effect of slavery, he would describe enslaved Virginians, whom he saw playing quoits, as being as 'well fed and as cheerful as possible'.[14] But now, having played a key role in publicising Wallace's theory and encouraging Darwin to finish the *Origin*, Lyell grappled with the implications of natural selection.

This was an emotional as much as an intellectual struggle. Though his own theory of gradual change in the structure of the Earth had imploded biblical chronologies, and though he had little trouble in applying natural selection to flowers or barnacles, it was another leap altogether to accept that 'the case of Man and his Races and of other animals . . . is one and the same'.[15] Indeed, Lyell confessed to an American colleague that he was 'rather surprised at the popularity of the doctrine of a chain of beings leading up to man'.[16] Fearing for 'the dignity of man', he had winced at Huxley's belligerence in conducting his campaign against Owen. And so, even if he praised the *Origin* as 'a splendid case of close reasoning and long sustained argument', and even if he anticipated the discovery in the Neander Valley by predicting that 'we shall in time discover extinct fossil varieties of Men', Lyell still hesitated to 'go the whole orang'.[17]

Instead, he approached the question of human descent obliquely, consulting in the early 1860s with archaeologists, European geologists, and scholars of dead languages before conducting expeditions in France and the Low Countries. The result was *Geological Evidences of the Antiquity of Man* (1863), a book which swept through the extensive proofs, gathered from around the world, that humankind was truly ancient. Lyell had already learned from Leonard Horner, a Scottish geologist, that shards of Egyptian pottery dated back to the twelfth century BC, and now he addressed 'articles of human workmanship' from the peat bogs of the Danish islands, ancient mounds in the Ohio Valley, and human remains in Belgian caves. He then took his readers through the soil of the Somme, the silt at the bottom of the

Seine, and then to 'the Fossil Man of Denise', which comprised 'the remains of more than one skeleton, found in a volcanic breccia near the town of Le Puy'. Finally, after visiting American sites where mastodon fossils had emerged from melting glaciers, Lyell deliberated on the true age of humankind.[18]

After accounting for 'the Roman, the bronze, and the stone periods', and referring some geological phenomena to 'an interval of several hundred thousand years', Lyell accepted 'the assignment of 4,000 and 7,000 years before our time as the *lowest* antiquity which can be ascribed to certain events and monuments'. In this way, he could maintain his hostility to biblical histories of the Earth, but was that enough time for natural selection to apply to humans? Not quite, for Lyell doubted that men had walked with mastodons 'more than a thousand centuries ago'. Moreover, in the chapters that he devoted to theories of transmutation and natural selection, and despite affirming his approval of natural selection as applied to plants and animals, Lyell continued to resist.[19] Noting this reticence, Darwin complained: 'I have been greatly disappointed that you have not given judgment & spoken fairly out [on] what you think about the derivation of Species.'[20] Writing to Hooker, Lyell justified his position by declaring that he had 'spoken out to the full extent of [his] present convictions, and even beyond [his] state of *feeling* as to man's unbroken descent from the brutes'. It was simply too much for Lyell to accept that he could 'have come from tadpoles'.[21] Even so, the *Antiquity of Man* had served a purpose, for it was the work of a conservative, Christian knight of the realm which acknowledged not only that humankind was much, much older than commonly thought, but also that certain species – even if Lyell did not count humans among them – could have adapted by means of natural selection.

The *Antiquity* teed up a much more aggressive, much more daring account of the deep history of humankind that arrived in bookshops only weeks later: Thomas Huxley's *Evidence as to Man's Place in Nature* (1863). A revised collection of his recent lectures and articles on human descent, it begins as dramatically as any book of the nineteenth century. On the frontispiece, opposite the title page, Huxley confronts his reader with five ascending illustrations by Benjamin Waterhouse Hawkins (who, of course, had built the dinosaurs for the

Crystal Palace). The first is the skeleton of a gibbon, the second of an orangutan; third comes a chimpanzee and fourth a gorilla. With each new illustration, the specimens grow taller, the pelvis rises higher, the arms shorten, and the skull transforms. And then, at last, comes 'man'. It was a brilliant, deliberately incendiary depiction of what Darwinian theory looked like when applied to humankind, and it anticipated Rudolph Zallinger's *The March of Progress* – the famous but scientifically dubious illustration of primates 'progressing' into modern man – by more than a century.

After devoting his first chapter to 'the natural history of man-like apes' such as the mandrills of Guinea and, more briefly, to 'African cannibalism in the sixteenth century', Huxley discoursed on 'the relations of man to the lower animals'. Reprimanding those who would shrink from the difficult question of 'whence our race has come', he proceeded on the basis that 'every living creature commences its existence under a form different from, and simpler than, that which it eventually attains'; just as the oak was more complex than the acorn, so was mankind more complex than its ancestors. But who were those ancestors? He found them, of course, among the apes, and so he asked his readers: 'Is Man so different from any of these Apes that he must form an order by himself? Or does he differ less from them than they differ from one another, and hence must [man] take his place in the same order with them?' Naturally, Huxley preferred the latter answer. Although there were differences – often great differences – between man and the higher apes when it came to skeletal proportions, cranial capacity, and dentition, it remained the case that 'the structural differences which separate Man from the Gorilla and the Chimpanzee are not so great as those which separate the Gorilla from the lower apes'. What this meant, especially in the wake of Darwin's 'hypothesis regarding the origin of species', was that naturalists and zoologists no longer had any 'rational ground for doubting that man might have originated . . . by the gradual modification of a man-like ape'. In Huxley's view there was a 'complete and crushing argument against the intervention of any but what are termed secondary causes in the production of all the phenomena of the universe'; in other words, science had provided both theory and evidence that the separate, divinely ordained creation of humankind was nonsense.[22]

It proved to be one of Huxley's most successful polemics. 'Hurrah', cried Darwin, 'the monkey book has come'; Hooker praised 'the magnificence of Huxley's language'; and Friedrich Engels recommended the book to Karl Marx as 'very good'.[23] Early reviews echoed this acclaim. The *Leeds Times* praised Huxley as a 'clever' man who was working 'on topics the most interesting of the day', and lambasted the 'piteous appeals [of his enemies] to the orthodoxy of the people'.[24] The *Hereford Times* meanwhile reported that doctors as far away as Australia were 'admitting in point of anatomy the complete analogies of the human and the monkey brain'.[25] Of course, Huxley's enemies 'opened fire' in time. In the *Edinburgh Review*, the young palaeontologist Charles Carter Blake, 'a jackal of [Richard] Owen's', spat fury at Huxley's 'semifabulous anecdotes' and 'illogical fallacy'. For the most part, however, Huxley marvelled at 'how little abuse the book has met'.[26]

Huxley had now put the insurgent radicalism of his lectures into print, but the most influential intervention of 1863 probably came from Charles Kingsley, whose observation to Darwin that God could have 'created a few original forms capable of self-development into other and needful forms' had appeared in the second edition of the *Origin*.[27] *The Water-Babies* (1863), however, was Kingsley's most

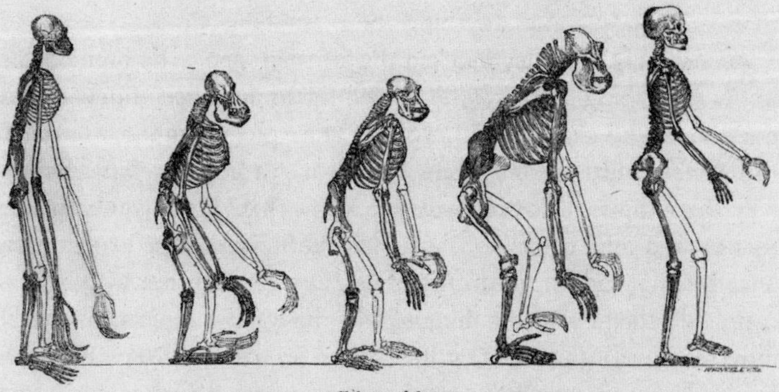

The celebrated frontispiece to *Evidence as to Man's Place in Nature* (1863)

significant contribution to the debate of the 1860s. Originally serialising it in *Macmillan's Magazine*, he had conceived of the book as a gift for his youngest child: 'Rose, Maurice, and Mary have their book', Kingsley's wife reminded him, 'and baby must have his'.[28] On one level, therefore, *The Water-Babies* is a simple, fantastical adventure for children that tells of a young chimneysweep called Tom who falls into a river, transforms into a tiny, gill-breathing creature known as a 'water-baby', and embarks on a redemptive quest before resuming human form. But more than 'a fairy tale for a land baby', and more than an enduring classic of children's literature, Kingsley's novel was a satire not only of the Victorian exploitation of child labour, but of the conservative Anglicans who had refused to accept the implications of geological evidence, palaeontological discoveries, and Darwinian theory. There was no question that Kingsley was sympathetic to natural selection. Writing to Darwin just before the publication of the *Origin*, he confessed that he had 'gradually learnt to see that it is just as noble a conception of Deity to believe that [H]e created primal forms capable of self improvement into all forms needful . . . as to believe that He required a fresh act of intervention to supply the lacunas' with new species.[29] Put more simply, Kingsley had no problem with reconciling his Christian faith with a natural mechanism which obviated any need for what he later called 'an interfering God, a master-magician'. He had tried 'in all sorts of queer ways', he explained to F. D. Maurice, 'to make children and grown folks understand that . . . nobody knows anything about anything'.[30]

Accordingly, Kingsley mocked the 'learned men' who pretended to know everything already, who denied the importance of new discoveries, and who shunned the evolving state of human knowledge. To those men who asserted 'there are no such things as water-babies', Kingsley simply asked: 'How do you know that? Have you been there to see?' If a man could not find a fox in the woods, he explained by analogy, that did not mean there were not foxes in the woods. Such men had called the 'flying dragon', like the water-baby, an 'impossible monster' until Mary Anning had found its fossilised remains in the Dorset limestone; the same men had then invented the name of 'pterodactyl', meaning 'winged fingers', because they were ashamed to admit that flying dragons really did exist. It was a brilliant, biting satire of the limited and conventional approach to natural history that men like Owen and Wilberforce had defended so fiercely. Yet in the very same passage, and even more enticingly, Kingsley referred to another palaeontological breakthrough that would soon revolutionise human understanding of nature: 'And has not a German', he asked, 'only lately discovered, what is most monstrous of all, that some of these flying dragons, lizards though they are, had *feathers*?'[31]

18
X

> I am trying to recall the peculiar instincts of the monsters of the preadamite world . . . The world then belonged to reptiles. Those monsters held the mastery in the seas of the secondary period. They possessed a perfect organization, gigantic proportions, prodigious strength. The saurians of our day, the alligators and crocodiles are but feeble reproductions of their forefathers of primitive ages.[1]
>
> Jules Verne, *Journey to the Centre of the Earth* (1864)

For Darwin the *Origin* was only the beginning, only a prelude to the 'big book' on natural selection and the variation of species; after all, he had written it rapidly, compelled to action by Alfred Russel Wallace's paper from Malaya.[2] In the early 1860s, therefore, as he watched Huxley and Owen trade blows over the dreadful paternity of humankind, Darwin had worked assiduously on the greater exposition of his theory. His occupation now was plant life. As a younger man he had watched the bees at work in the gardens of Shrewsbury; in the summer of 1861, during 'eight weeks and a day' on holiday at Torquay, he looked again.[3] Spending 'hours on his hands and knees', he observed how the petals of plants guided bees with their pollen to all the right places, and he mined 'capital information' from the new *Journal of Horticulture*.[4] He also tapped his network of naturalists for specimens and samples: 'I am got profoundly interested in Orchids', he confessed to Joseph Hooker.[5] By the time the Darwins returned to Kent he was quite ready to quit boiling pigeons, and the flowerbeds of his garden were a welcome relief from the fumes and viscera of the dissecting parlour. They also had real utility since, if natural selection were a universal mechanism for explaining variation, it would inhere in plants as well as in animals.

Writing to one botanist, he noted that the early marsh-orchid was much less successful in attracting insects than the heath spotted-orchid, so was the latter more likely to thrive? Huxley had once laughed at the idea 'of finding an utilitarian purpose in the forms and colours of flowers', but the orchids had convinced Darwin.[6] They convinced John Murray too. Collecting orchids was a popular pastime for wealthy Victorians, so there was a market for a book: 'I think this little volume will do good to the *Origin*,' Darwin wrote to his publisher, 'as it will show that I have worked hard at the details'.[7] The result was *The Fertilisation of Orchids* (1862), which argued in large print and straightforward language that the 'beautiful adaptations' of the orchid 'have for their main object the fertilisation of each flower by the pollen of another flower'.[8] It represented the prosaic application of his theory to a field that did not excite the same indignation as the descent of humankind: 'I have found the study of Orchids eminently useful', he told Hooker, 'in showing me how nearly all parts of the flower are co-adapted for fertilisation by insects, and are therefore the results of natural selection, even the most trifling details of structure'.[9]

After the orchids, Darwin stayed with plants. Workmen built a hothouse in the garden and Hooker sent a cart from Kew with seeds to stock it.[10] Darwin's health, however, remained frail. On a rare venture into London, speaking for only a few minutes at the Linnean Society had provoked weeks of retching, while 'devilish headaches' and 'a good severe fit of Eczema' confined him to the sofa in the summer of 1863; at one point, he recorded vomiting on twenty-seven consecutive nights.[11] Yet even in this state of infirmity, even when he struggled with the hundred-yard walk to the hothouse, he could study his new subjects. Darwin now focused on tendrils, the spiralling appendages that certain plants use to twine around structures for support. Why, he asked, should clematis climb? Why should a grapevine twirl around its stake? 'It has often been vaguely asserted', he wrote, 'that plants are distinguished from animals by not having the power of movement. It should rather be said that plants acquire and display this power only when it is of some advantage to them.'[12] In other words, just as orchids had developed their flowers to attract bees, other plants had climbed towards the light, towards the fresh air and rain that sustained their existence; their tendrils gave them the base, the support, for this mission of survival.

By 1865, then, Darwin had articulated the general theory of variation by natural selection and applied it to pigeons, orchids, and climbing plants; Wallace had observed the same phenomenon among the wildlife of the East Indies; and in the wake of the Neander Valley discovery, Huxley had placed humankind on the same biological plane as the animals. The evidential case for adaptation grew stronger with news from Gibraltar, the British territory on the southern tip of Iberia. There, and as long ago as 1848, a lieutenant in the Royal Navy had found a skull in one of the fissures which burrow into the great Rock, itself made of Jurassic limestone. But with servicemen thinking it merely a human skull, the fossil had lain in a library cupboard for more than a decade.[13] It was not until the zoologist George Busk visited Gibraltar in 1862 that its significance became clear. Upon examining the skull within a larger collection of human fossils, Busk noticed its striking 'lowness'; and when he reported to the British Association, comparing the skull to 'those of the *intertropical negro*, *Australian*, *and Tasmanian* races', British authorities were sufficiently intrigued to requisition the whole collection, with nearly 400 specimens being shipped promptly to London.[14] Upon taking charge of the 'monstrous' specimens that had 'excited the astonishment of all . . . who have beheld them', Busk declared that the Gibraltar discoveries added 'immensely to the scientific value of the Neanderthal specimen'. Why? Because they suggested that the Neanderthal skull was not an anomaly, and that there had been a separate 'race extending from the Rhine to the Pillars of Hercules'.[15] There was a growing body of proof, it seemed, that there were ancestors of humankind within the earth.

Yet if Darwinism appeared to be ascendant, that was not enough for Hooker and Huxley, who wished to place their movement on a more secure footing. 'I feel the want of association with my brother Naturalists', Hooker had complained: 'We never meet except by pure accident ... Without some recognised place of resort that will fulfil the conditions of being a rendezvous for ourselves ... we shall be always ignorant of another's whereabouts and writings.'[16] To this end, these two great friends and allies invited six like-minded men of science to dinner on 3 November 1864 at the St George's Hotel in Mayfair. There came Busk, the expert on the Neanderthal and Gibraltar specimens; the chemist Edward Frankland, who would soon discover helium; the banker and anthropologist John Lubbock, who was a neighbour to Darwin in Kent; and the mathematician Thomas Archer Hirst. Joining them all was Herbert Spencer, the philosopher and long-time acquaintance of Huxley's who, earlier in 1864 in *The Principles of Biology*, had compressed Darwinian theory and his own thinking on the mechanical competitions of nature into the phrase for which posterity remembers him: 'the survival of the fittest'.[17]

The last of Huxley's diners was the physicist John Tyndall.[18] A native of the Irish county Carlow, Tyndall had worked for the Ordnance Survey and several railway companies, often gambling on their fortunes on the stock market, before teaching mathematics at a Quaker school: 'The desire to grow intellectually did not forsake me,' he reflected, 'and, when railway work slackened, I accepted ... a post as master in Queenwood College'.[19] Here, upon befriending Edward Frankland, who had come to the same school to teach chemistry, Tyndall instructed himself in advanced science; within a year, the pair had removed themselves to Germany to study under Robert Bunsen, the inventor of the eponymous burner. Returning in the 1850s to London, where he impressed Michael Faraday with his experiments on magnetism, Tyndall found his lectures at the Royal Institution, where he became professor of natural philosophy, attracting crowds from all ranks of society: 'By voice and pen,' one historian has reflected, 'he spread [science's] discoveries and points of view amongst the great mass of the more intelligent public'.[20]

At their initial meeting on that night in November, the eight men agreed to reconvene on the first Thursday of every month from October through June. (The mathematician and publisher William Spottiswoode became the ninth and last member of the club.) Dinner would be at six, giving the men time to eat and talk before attending the Royal Society, and the club would have no rules except to have no rules. Huxley and Hooker were founders of sorts, but each man in turn would serve as treasurer and secretary, the latter forbidden from keeping anything other than informal minutes. As for the name of their new club, nothing seemed to suit. The 'Thorough' Club? No. The 'Blastodermic'? No, they would not name a club after a layer of embryonic cells. Instead, they decided on the 'X' Club, the letter committing the men to nothing other than 'devotion to science, pure and free, untrammelled by religious dogmas'.[21] In the words of one historian, the members of the X Club were 'nine men who wanted to change the world' and, over the course of the next twenty years, they would go some way towards achieving that goal.[22]

In fact, their efforts bore fruit within weeks. The Copley Medal was the Royal Society's highest honour: Benjamin Franklin had won it for his work on electricity, Humphry Davy for his discoveries in chemistry, and William Buckland – as we have seen – for his descriptions of Kirkdale Cave. It was supposed to represent the cutting edge of scientific inquiry. In the late 1850s and early 1860s, however, the Society had inclined to favour more conservative candidates, among them Richard Owen and Charles Lyell. The most recent recipient, in 1863, was Adam Sedgwick, the Cambridge geologist who had carped to Darwin that parts of the *Origin* were 'utterly false & grievously mischievous'; Sedgwick had even described the mechanism of natural selection as 'machinery as wild I think as Bishop Wilkins's locomotive that was to sail with us to the Moon'.[23] (In the 1630s and 1640s, the future Bishop of Chester had speculated about mechanical flight to, and the inhabitation of, the Moon.) How better to correct the Royal Society's course, therefore, and how better to assert the authority of the 'X', than to give the Copley to Darwin?

They began manoeuvres. All members of the 'X' bar Spencer were fellows of the Society and, that year, Busk and Hooker were on its council. Busk nominated Darwin for the medal and the geologist Hugh Falconer, who by no means approved of natural selection, acted as a 'neutral' seconder: 'This great essay on genetic biology', he wrote to the Council's president, was 'a strong additional claim on behalf of Mr Darwin'.[24] But this was no coronation. There followed 'furious politicking' with the Council's 'Cambridge men', the ancient guardians of Anglican science putting up stout resistance. 'Some of the older members of the Council', reported Lyell, 'were afraid of crowning anything so unorthodox'.[25] Still, enough of them relented for Darwin's nomination to pass, ten votes for and eight against. This early victory surprised even Hooker, who thought Darwin was much too far 'ahead of [his] day to be appreciated', and much too far ahead of 'the ruck of candidates who the Council bring forward for medals'.[26] Inevitably, Darwin was too sick to attend the ceremony, but the award delighted him. 'How kind you have been about this medal', he told Hooker: 'I am blessed with many good friends, & . . . I often wonder that so old a worn-out dog as I am is not quite forgotten.'[27]

It was a signal moment. When Mantell and Buckland were finding megalosaurs and hylaeosaurs in the rocks of southern England, there had been relatively little tension between 'science' and 'religion', at least for those two pioneers. They could regard both their discoveries and their faith as parts of a universal whole; when they went digging, they could practise Paley's natural theology, where 'science' could reveal the Lord's workings in the natural world. By the 1860s, things had changed dramatically. There was a welter of geological and palaeontological evidence that even the most literally-minded Christian would struggle to reconcile to the Mosaic record; there was a powerful impulse, coming most directly from Huxley and his 'X' clubmates, to segregate scientific inquiry from issues of religion and belief; and there was sophisticated biblical criticism which in some quarters had rendered unthinking belief in the scriptures a matter of derision. Now, the Royal Society had honoured Charles Darwin, who had put forward an explanation for the variation of species from which divine power was absent. In doing so, they had honoured a man who no longer went to church, whose wife despaired for his soul, and who regretted 'the constant inculcation in a belief in God on the minds of children'; indeed, he now thought it would be as difficult for a child 'to throw off their belief in God, as for a monkey to throw off its instructive fear and hatred of a snake'.[28] How, therefore, should a religious person now approach scientific inquiry? How, conversely, should a scientist approach religious belief? And how could a man who read the Bible, openly and in detail, accept the implausible and even the immoral within its books? These were questions that were troubling men at home and around the world, even on the new frontiers of the British Empire.

The early colonial history of Natal was confused and chaotic. In December 1838, during their Great Trek to the east that followed British conquest of the Western Cape, the Boer colonists of southern Africa had won a crushing victory over the Zulu at the Battle of Blood Bridge. A few weeks later, in defence of their promised land, they expelled British forces from the stronghold at Durban on the coast. The result, in the spring of 1839, was the proclamation of the Dutch-speaking Natalia Republic. The British, however, were not content to abandon these south-eastern stretches of Africa. Although the Peel ministry in London believed from 1841 that 'the establishment of a colony there would be attended with little prospect of advantage', the governor of the Cape Colony – who was making decisions some 6,000 miles from Westminster – had other plans.[29] In early 1842, on the pretence of preventing bloodshed between the Boers and the Zulu, he ordered an expeditionary force to seize 'the Natal'. This was not a straightforward war, and the redcoats' first encounter with the Boers was a disaster. In one night of fighting outside the village of Congella, now part of the city of Durban, the British forces lost fifty men and all their heavy guns, and soon faced starvation under siege in the heat and dust of the African summer. It was only when local resident Dick King rode 600 miles to summon reinforcements that the British prevailed and the Boers surrendered. Natal now joined the British Empire and within a decade it was flourishing: there was coal in the ground and the prospect of cotton in the fields; thousands of emigrants had arrived from the motherland; and the harbour at Durban was a way-station on the oceanic route to India. What the colony needed now, of course, was an Anglican bishop.

The man whom the Church authorities dispatched to Natal in 1853 was John William Colenso. Born to a Cornish family of modest means, Colenso had studied mathematics at Cambridge, become a fellow of St John's College, left the university to teach at Harrow, and faced down financial ruin, all by his late twenties. Having run a boarding house to supplement his salary, Colenso was trapped in debt when the house burned down: he found redemption by selling the copyright to textbooks he had written on algebra and arithmetic and, by the 1840s, he was content with a clerical living in Norfolk. From there, Colenso developed a friendly correspondence with F. D. Maurice and began to edit the journal of the Society for the Propagation of the Gospel, the principal missionary vehicle of the Anglican Church. Now, having loosed his bonds of debt 'like Peter in the prison', he concluded that there was 'a great missionary work to be set on foot' in Africa.[30] He accepted the offer of the bishopric of Natal and, after being consecrated on St Andrew's Day, embarked for the far end of the world with his wife and four young children.

In his early years at Pietermaritzburg, Colenso played the part of the earnest proselyte, ministering to British settlers, bringing Zulu converts within the Anglican Communion, and encouraging the growth of the colony. In *Ten Weeks in Natal* (1855), his first public report to the motherland, he promoted vistas of 'romantic, round-topped heights [that were] green and luxuriant to the very water's edge', gentle breezes along the shoreline, and 'bosky dells and hollows which would have been the glory of some gentleman's estate in England'.[31] By 1859, he had better news: the Zulu chief Ketchwayo, 'a powerful young prince' who was handsome, tall, and 'stout-limbed', had gifted the diocese some land for a missionary station in 'the Zulu country'.[32] A short time later, Colenso had produced an Anglo–Zulu dictionary, an essential aid 'for the missionary, traveller, and trader, but also for the statesman, and the ordinary colonist of these parts'.[33] He had even established a school for Zulu boys, which he called a 'Kaffir Harrow'.[34] Despite occasional friction with white settlers, who once protested at the imposition of established religion by burning Colenso in effigy, the mission redounded to the Empire's glory.

But not all was well. Colenso had been translating the Bible into Zulu, an undertaking which had obliged him to study and deconstruct the English, Hebrew, and Greek versions of the scripture in minute detail. In doing so, he began to notice oddities and contradictions that, if they had ever registered with him before, he had been able to ignore.[35] 'I have been brought again face to face with questions', he explained, 'which caused me some uneasiness in former days . . . [when] it was easy to draw from [the Old Testament] practical lessons of daily life, without examining closely into the historical truth of the narrative'.[36] Now, with the time and space to enquire more seriously into the integrity of the Bible, Colenso began to doubt the truth of what he read. It was showing, too: in December 1860, when the bishops of the southern hemisphere gathered in Cape Town, his brethren noticed Colenso's 'changed manner' and 'gloomy reserve and half-restraint', and how he 'broke forth with opinions which grieved them much'.[37]

The first articulation of these doubts came with his *Epistle to the Romans* (1861), where Colenso, in descent from F. D. Maurice's commentary of the previous decade, questioned whether divine punishment could last for ever: 'the word "endless"', he wrote, 'is not a proper representative of the word "Eternal"'.[38] Colenso continued to read the 'higher' German criticism and, as incredulous Zulus asked him searching questions about inconsistencies within the holy book, his doubts were amplified. 'How is it possible', he asked himself, 'to teach the Zulus to cast off their superstitious belief in witchcraft, if they are required to believe that all the stories of sorcery and demonology which they find in the Bible – the witch of Endor, the appearance of Satan in the court of heaven – are infallibly and divinely true?'[39]

These concerns found fuller expression in Colenso's epic study of the Pentateuch and Joshua, the first six books of the Bible. Writing in Pietermaritzburg or when travelling within his diocese, dispatching his manuscript to Durban for carriage on the ships which sailed to London, and waiting months for the opinion of his publisher, it was not until 1862 that Colenso released the first volume of *The Pentateuch and Book of Joshua Critically Examined*. Here, pained by the questions of his Zulu congregation – 'Is all that true? Do you really believe that all this happened?' – he had asked himself another biblical question, this one from Zechariah: 'Shall a man speak lies in the Name of the LORD?' He found that he could not. Instead, and well aware of the 'very general demand' that the Church was already making on the authors of *Essays and Reviews*, he dissected the early history of the Israelites and found that it was 'untrue, as a matter-of-fact, historical, narrative'.[40] Some of Colenso's conclusions were shocking, even to the man himself. Beyond the accepted criticisms of the Creation and the Deluge, he found that almost none of the Pentateuch was written by Moses or during Moses' lifetime; that the person who led the Exodus from Egypt was less likely to have been Moses than 'a personage as shadowy and as unhistorical as Aeneas . . . or . . . King Arthur'; that the Levitical law was not handed down in the desert, but composed during captivity in Babylon; and that Joshua was an entirely mythical character.[41] How, moreover, could news of the impending Passover have reached all two million Israelites in Egypt? And how could they have found and killed enough lambs to daub so many lintels with blood?

When Colenso's book arrived in the rectories and vicarages of England, conservative tempers flared. George Eliot might have translated critical German works into English, and J. A. Froude might have given voice to doubt through fiction, but this? This was new. As Colenso himself had recognised, 'the phenomena in the Pentateuch . . . which show so decisively its unhistorical character, have not yet . . . been set forth, in this form, before the eyes of English readers'.[42] Even worse, this appalling impiety was not the work of a scurrilous radical in the penny press, but the considered opinion of an Anglican bishop. 'Had he been *Mr.* Colenso still', reflected Edward Pusey, 'his book would have been stillborn. Now it is read by tens of thousands because he is a Bishop. It is his office of Bishop which propagates infidelity.'[43] For Colenso's critics, there was an obvious connection between the bishop's wickedness and the work of Darwin, Huxley, and the 'X'. One furious pamphleteer considered that 'The use that is made of science at present in certain quarters is utterly unlike the use that was made of it by Newton, the prince of science . . . Now-a-days the study of many scientific men seems to be to discover discord, and not harmony, and to find in science, not the handmaid of revelation, but the antagonist.'[44]

There was consternation in the colonies too. Robert Gray, the Bishop of Cape Town, was the African metropolitan – the senior Anglican clergyman on the continent – and he had been jousting with Colenso for years. They had disagreed over matters as diverse as polygamy 'among the converted heathen' and the voting rights of Church councils and so, even before *The Pentateuch*, Gray had been spoiling for a fight: 'I am very anxious about [the Bishop of] Natal', he wrote as early as 1861. Yet when 300 copies of the *Pentateuch* sold quickly among the freethinking Dutch colonists of Cape Town, Gray decided that something should be done, for he would not stand for 'the sight of a Bishop pulling the Bible to pieces'.[45] In December 1863, therefore, having secured the blessing of the Anglican bench, he convened an ecclesiastical court at Cape Town. Sitting with the bishops of Grahamstown and the Free State, he charged Colenso with 'holding and promulgating opinions which contravene and subvert the [true] Faith, as defined and expressed in the Thirty-nine Articles, and the formularies of the Book of Common Prayer'.[46] Colenso would not attend his trial, nor would he even recognise the jurisdiction of the court, but that would not matter: Gray and his fellow inquisitors found Colenso guilty, deposed him from the bench, and excommunicated him from the Anglican Church.

When Colenso returned to London to defend his honour, he became one of the *causes célèbres* of the 1860s. The public had already purchased 10,000 copies of the *Pentateuch*, and now the defrocked bishop found champions among the most prominent captains of the culture war. Hooker subscribed to the bishop's 'defence fund on principle', while Huxley took him to the Athenaeum Club, where Colenso's attendance outraged the ruddy-cheeked Baron Overstone into resignation; the bishop returned the favour by offering to write for Huxley's provocative *Reader*.[47] When the appeal advanced, Natal prevailed over Cape Town: the judicial committee of the Privy Council, which even today remains the supreme court for many Commonwealth countries, found that Gray had no 'coercive legal jurisdiction' over Colenso's diocese, and so his judgement was 'null and void in law'.[48] This would not, however, conclude the affair.

In 1867 more than seventy bishops from across the Anglican world convened at the first Lambeth Conference to resolve 'the strong divergence of opinion upon the legal aspect of . . . Colenso's deposition and excommunication'. Though the Archbishop of Canterbury sought to preserve the dignity of this first gathering of the Communion by prohibiting 'any distinct resolution of condemnation', the bishops nonetheless resolved, by forty-nine votes to ten, that the whole of Anglicanism had been 'deeply injured by the present condition of the Church in Natal' and that they should consider urgently how it 'may be delivered from the continuance of this scandal'.[49] It was a stinging rebuke, and Colenso's treatment by his colleagues was another warning of the fate which could await apostasy, but the farrago would not prove fatal to his career. Fifty years beforehand, men like William Hone and Richard Carlile had been imprisoned for printing the literature of infidels; as late as the 1850s, accusations of blasphemy had derailed the careers of Froude and Maurice; but now Colenso returned to Africa and, with funds raised by his supporters in England, devoted the remainder of his life to the Natal mission, producing six more volumes on biblical history. This was not to say that the Anglican attack dogs had lost their bite, or that scholars were now free to assail the Holy Bible, but it was a sign, perhaps, that the tenor of debate was changing.

19
Feathers

> And God said, Let the waters bring forth abundantly the moving creatures that hath life, and fowl that may fly above the earth in the open firmament of heaven.
>
> And God created great whales, and every living creature that moveth, which waters brought forth abundantly, after their kind, and every winged fowl after his kind: and God saw that it was good.
>
> Genesis 1:20–21

They had been digging again in Germany, and they had found more wonders in the earth. In the forested land around the town of Solnhofen, roughly halfway between Munich and Nuremberg, they had gone deep into the limestone of the Altmühl Valley and, by 1861, the quarries had produced two slabs of special interest. The first man to inspect them, or at least to describe them, was Hermann von Meyer. An esteemed geologist, now in his sixties, Meyer was no stranger to dinosaurs and their ancient brethren. In the 1830s he had described and named the *Plateosaurus* (meaning 'broad lizard'), a two-metre-tall herbivore whose leg bones had come to light in Bavaria. By the 1840s, he had proved the existence of a new species of pterosaur, the *Rhamphorynchus*. And by 1858, in recognition of this work and the taxonomy of fossilised reptiles that he offered in *Palaeologica* (1832), Meyer had received the Geological Society's great prize, the Wollaston Medal. But in the early 1860s, Meyer beheld a truly extraordinary fossil. It was a feather.

At first, as Peter Wellnhofer relates, Meyer did not believe it. There was a giddy market for fake fossils and the Solnhofen feather – two inches long, half an inch across, 'transformed into a black substance' – looked too good to be genuine. But with a perfect mirror of the feather appearing in the opposite slab of limestone, Meyer accepted the fossil's provenance: 'It [cannot] be imagined', he concluded, 'that a feather was compressed between two stone slabs and transformed by some treatment into an artificial fossil'.[1] It was real. But what to make of it? Upon consulting with colleagues in mineralogy, several astonishing possibilities presented themselves. Were dinosaurs actually feathered and, if so, why had palaeontologists not found any feathers until now? Or were birds more ancient than any ornithologist had imagined? Either way, it seemed clear that the fossilised feather came from a new species of animal. The name that Meyer gave to the fossil meant 'old wing' in Greek, *Archaeopteryx*.

Only two miles to the west, and just as Meyer was writing up the feather for Germany's journals, quarrymen had found something even more spectacular. It was the skeleton, almost complete, of 'a New Fossil Reptile Supposed to be Furnished with Feathers'. Again, the precise circumstances of the discovery are obscure, but the workmen appear to have presented this fossil as payment in kind to Karl Haberlein, a country doctor who conducted his rounds by carriage in summer, by sleigh in winter. They paid him well. Though the skeleton of the *Archaeopteryx* was incomplete, missing its neck and its skull, it provoked a bidding war, with Haberlein remarking that the 'wonders of creation . . . have more value than gold'. Local experts begged the Bavarian state to intervene, but one professor at Munich advised against it: in 'absolute incredulity', he protested that 'a feathered creature' such as the *Archaeopteryx* had to be a bird, but that it was impossible for Jurassic rocks, in which only reptiles had been found, to have contained the remains of a bird.[2] With news of the discovery spreading rapidly across Europe, the grandees of London science took a different approach. Urged on by Richard Owen, the British Museum stumped up £700 for the fossil, a fee that proved a handsome dowry for Haberlein's daughter, and by 1862 this 'London specimen' of the *Archaeopteryx* was in Owen's custody.[3]

Despite coming off the worse from the great 'gorilla war' with Huxley, Owen remained a commanding figure in British science. He

The fossilised *Archaeopteryx* feather from Bavaria

gave lectures the length and breadth of the country; he was friends with John Ruskin and Charles Dickens, though he chided the latter for 'the scanty beard he has now grown'; and the new king of Italy, Victor Emmanuel II, ennobled him as a Chevalier of the Order of St Maurice and St Lazarus.[4] It was Owen's urging that had secured the purchase of the *Archaeopteryx*, and announcing the fossil to the English-speaking world would only burnish his reputation. But someone beat him to it. In *The Times* of 12 November, appearing under the heading 'A Feathered Enigma', the pseudonymous correspondent 'Y' related the news that even 'in an age of marvels . . . a discovery has recently been made which has thrown the geological world into convulsions'. Recounting the roles of Haberlein and Meyer in its excavation, and describing the *Archaeopteryx* in perfect detail, 'Y' informed the paper's readers of a feathered animal that was 'intermediate

The almost complete *Archaeopteryx* skeleton known as the 'London specimen'

between birds and reptiles, wholly different from any creature previously known'. In reproach of Owen's stolid resistance to natural selection, 'Y' predicted that 'the followers of Darwin will not be slow to avail themselves of this new discovery, and adduce it in support of the transitional hypothesis respecting the origin of animals'.[5]

The identity of 'Y' remains a mystery, even if his choice of pseudonym suggested allegiance to the X Club, but Owen would present his own findings to the Royal Society. Devoting the whole of his paper to the analysis of 'an animal with impressions of feathers radiating fanwise from each anterior limb', Owen gave detailed descriptions of the bones, comparisons of their structure to other birds and vertebrates, and magnificent illustrations in the plates. His speech was 'verbose and minute' and though he conceded that the *Archaeopteryx* had emerged from older, ancient strata of limestone, he rejected the notion that it was a transitional fossil which could strengthen the Darwinists in their convictions. 'This animal is not an intermediate between the reptiles and the birds,' he is reported to have stated, 'but in fact simply the earliest known example of a fully-formed bird, not dissimilar to several types of modern raptors and other birds. Furthermore', he told the assembled fellows, 'I can with confidence say that this bird would in life have been a powerful bird of flight'.[6] Perhaps this conclusion was inevitable. Owen had deplored Lamarckian transmutation, sniped at Darwin, and railed against the notion of human fraternity with the ape. So, what would it profit him now to describe the *Archaeopteryx* as a transitional fossil between prehistoric reptiles and the animals of the present? No, Owen insisted that, if the *Archaeopteryx* was related to anything, it was only to the pterosaurs that fossil-hunters had found in rocks of a similar age.[7]

Other observers made much more of this new discovery. In the fourth edition of the *Origin*, which appeared in 1866, Darwin remarked upon 'that strange bird, the Archeopteryx, with a long lizard-like tail, bearing a pair of feathers on each joint', commenting that 'hardly any recent discovery shows more forcibly than this, how little we as yet know of the former inhabitants of the world'. He even gave mischievous thanks to Owen, on whose authority Darwin and his colleagues now knew that 'a bird certainly lived during the deposition' of the Jurassic rocks.[8] In the press, the *Intellectual Observer* noted the creature's 'deceptive appearance' and suggested the discovery of something truly novel by warning against 'Cuvier's theory that a similarity of particular parts indicated a similarity of other parts, or of the whole'.[9] British provincial newspapers remarked upon the *Archaeopteryx* as well. The *Loughborough Monitor* reported on the discovery of a 'feathered fish' and urged 'naturalists to be cautious in rejecting negative evidence; for as this instance shows, negative evidence may at any time be confirmed by positive facts'.[10] The *Cardiff Times* meanwhile predicted that 'We may . . . shortly expect to hear [of] as fierce a dispute between the Palaeontologists and the Ornithologists as that which has taken place on the Gorilla question'.[11] It was prescient enough.

Although Thomas Huxley had long accepted – indeed, propagated – the idea that humankind could have descended from primates, the potential relationship of dinosaurs to birds had occupied little of his time; when it did, he had focused on large, flightless creatures such as the ostrich and the emu. In the meantime, he and the X Club had married Darwinian theory to Bishop Colenso's scepticism by way of a manifesto, which Huxley used almost to declare war on religion: 'Science exhibits no immediate intention', he announced, 'of signing a treaty of peace with her old opponent, nor of being content with anything short of absolute victory and uncontrolled realm of intellect'.[12] Still, for all that Huxley and Tyndall relished this belligerence, they would fight each other over the Morant Bay Rebellion of 1865 in Jamaica, where Governor Edward Eyre's violent reprisals against the queen's black subjects had split informed opinion. Thomas Hughes, Herbert Spencer, Darwin, and Huxley – the last two having already shunned the new Anthropological Society over its pro-slavery racism – joined the Jamaica Committee that sought to prosecute Eyre.[13] Tyndall and Charles Kingsley, however, had subscribed to the governor's defence fund, and it was the only time during Huxley's friendship with Tyndall when each man 'would have been capable of sending [the] other to the block'.[14] There was other trouble for Huxley, besides. The death of his brother George in 1863 was grievous, and George's debts brought further pain: Huxley sold his Royal Medal for £50 to pay them off. There was also the endless fear that haunted all parents of young children in Victorian England, and scarlet fever, which had carried off young Noel in 1860, now threatened the life of Huxley's third daughter. It took the prescription of ammonium chloride to save his 'little black eyed girl Rachel'.[15]

It was not until late 1867, therefore, that Thomas Huxley paid serious attention to the *Archaeopteryx*. The eureka moment came during a trip to Oxford to inspect the precious relics within the late William Buckland's collection of *Megalosaurus* specimens. There, in the university's Museum of Natural History, where he had once duelled with Samuel Wilberforce, one bone stood out before all else: the ilium, the uppermost part of the hip which juts out from the spine. The year before, as Adrian Desmond relates, Huxley had lectured that scaly-legged birds were an 'extremely modified and aberrant reptilian type', but now he saw the *Megalosaurus*'s ilium as so perfectly avian that it became plausible for all birds – not just emus and ostriches – to owe their paternity to dinosaurs.[16] In the wake of this revelation, the work of the young Prussian naturalist Ernst Haeckel made greater sense. Huxley had just finished reading Haeckel's *Generelle Morphologie der Organismen* (1866), a vast and dense but enormously learned work which had synthesised the German tradition of *Naturphilosophie*, Lamarckian theory, and natural selection into a system that the German zoologist called '*Darwinismus*'; it was a term that would soon feed into English as 'Darwinism'.[17] Could Huxley, who would soon anoint himself as 'Darwin's bulldog', now deploy the fossil record to prove his master right? He thought he could, but he would need more evidence than the *Archaeopteryx* alone could furnish. Other palaeontologists, on both sides of the Atlantic, were happy to provide it.

Almost ten years beforehand, in the summer of 1858, William Parker Foulke and his family had escaped from the heat of the city. Leaving Philadelphia and taking the new railway into rural New Jersey, they had been renting a house for the season near Haddonfield when Foulke, an inquisitive man in his forties, heard tell of a local legend. Twenty years ago, the story went, a marl pit on a nearby farm had yielded a smattering of strange-looking bones and teeth. Foulke was a lawyer by trade and a zealous campaigner for penal reform, but an amateur geologist too. His interest piqued by news of the local fossils, he sought out the farmer and enquired of the truth: had there been fossils on his land? The farmer confirmed the rumours but confessed that he had given away the bones 'to curious friends, or casual acquaintances', or simply lost them.[18] But this would not deter Foulke, who secured permission to dig and hired a crew of workmen.

Finding the site of the initial excavation proved 'no easy matter', for there had been twenty years of growth.[19] Yet at the bottom of a ravine some twenty feet below the surrounding farmland, where a brook meandered to the east, Foulke found the signs of the former digging. He and his men got to work and down they went, down through a yard of soil before their shovels struck something harder; here was the marl. After another six feet, an eternity in the mid-Atlantic summer, they found their treasure: enveloped in 'the tough, tenaceous bluish' rock were dozens of 'ebony black' fossils.[20] Some of the teeth and vertebrae recalled the bones that the farmer had found and then given away, but others were astounding. There was 'a thigh-bone ... 40 inches long' and a tibia not much shorter; close by lay a humerus and an ulna, both approaching two feet in length. Working with knives and trowels, drawing and measuring each specimen as it lay in the ground, Foulke and his men realised that they had found 'a Reptile of . . . huge proportions'.[21] Wrapping each bone in cloth and straw before hauling them up and out of the ravine, they dispatched the fossils to the Academy of Natural Sciences at Philadelphia.

Back in the city, Foulke worked through the specimens with Joseph Leidy, a professor of anatomy at the University of Pennsylvania who, according to one biographer, competed with the British polymath Thomas Young for the epithet 'the last man who knew everything'.[22] What lay before Foulke and Leidy was for two reasons unique in the annals of palaeontology, at least for the late 1850s. First, it was unquestionably a 'new', previously undiscovered species of dinosaur: while its teeth resembled those of the *Iguanodon*, therefore marking it as 'an immense herbivorous saurian', its 'huge hind legs and very small arms' distinguished the creature as something altogether different; in honour of its size, Leidy named it the *Hadrosaurus*, the 'bulky lizard'.[23] Second, there were eight teeth, sixteen vertebrae, and several bones from the arms and legs, meaning that Foulke had found more of a dinosaur's skeleton than anyone before; indeed, the remains of the *Hadrosaurus* were so complete that Leidy would help to transform human understanding of how dinosaurs stood and moved.

During the 1840s, Richard Owen and Gideon Mantell had sparred over the constitution of the *Iguanodon* and of dinosaurs generally. Owen, of course, had won that battle and it followed that when Benjamin Waterhouse Hawkins made the models for the Crystal Palace under Owen's supervision, they appeared on all fours, their limbs jutting out to the side. The discovery in New Jersey, however, suggested that the fore-limbs of the *Hadrosaurus* were so weak, and the rear-limbs so long and so powerful, that it could only have been bipedal, which is to say that it could only have walked on its hind legs. Though the outbreak of the American Civil War delayed the publication of Leidy's full conclusions, he was clear in his mind: 'The enormous disproportion between the fore and hind parts of the skeleton', he wrote, 'has led me to suspect that this great herbivorous Lizard sustained itself in a semi-erect position on the huge hinder extremities and tail while it browsed on plants growing upon the shores of the ocean'.[24] In time, when the Academy of Natural Sciences summoned Hawkins

Two representations of the *Hadrosaurus:* the earlier, four-legged reconstruction on the lines proposed by Richard Owen is wrong; the later, bipedal reconstruction is correct

to Philadelphia to make models of the *Hadrosaurus*, the beast would stand tall on two legs; it had the poise, in other words, of a bird.

Soon after this, the Bavarian geologist Johannes Andreas Wagner announced the discovery of the complete skeleton of a two-legged creature that was long-tailed, slight, and roughly the size of a chicken. Finding these bones in the same stretch of limestone that had given up the *Archaeopteryx*, Wagner called the new creature *Compsognathus* after the Greek meaning 'delicate jaw'. It would take the intervention of a young American, however, to connect the *Compsognathus* to birds. The scion of a family of wealthy Quakers, Edward Drinker Cope had escaped the carnage of the American Civil War by travelling and studying in Europe and, upon returning to Philadelphia, he turned down a chair in zoology in favour of fieldwork in the American West.

By 1867, having moved his family to New Jersey and to the very same fossil beds where Foulke had found the *Hadrosaurus*, Cope was ready to report on the fruits of the German quarries and 'the extinct reptiles which approached the birds'. Here, he compared 'the modified bird Archaeopteryx' with 'the ordinary Dinosauria' in which class, in contradiction of Wagner, he placed the *Compsognathus*. There were differences to be sure, but Cope remarked upon 'the union of the tibia and fibula [of the *Compsognathus*] with the first series of tarsal bones, a feature formerly supposed to belong to the class Aves [that is, birds] alone'. He also looked at 'the transverse direction of the pubes', the hip-bones, and again observed 'an approach to the birds'. After describing other 'bird-like features' such as the number and nature of its vertebrae, Cope suggested that the *Compsognathus* stood 'intermediate between the position in most reptiles and in birds'.[25]

Back in London, keeping abreast of both German and American discoveries, Thomas Huxley would spend the winter of 1867–8 preparing his own explanation of it all.[26] Now there were pressing questions to answer. Were the *Compsognathus*, the *Hadrosaurus*, and the *Archaeopteryx* truly related? And were they all ancestors of the modern bird? This would not be a polite paper before the Royal Society, even if Huxley and the X now commanded greater authority in its chambers. Instead, he would lecture to his natural constituency, to the working men of the Royal Institution. And so, on the first Friday night of February 1868, Huxley delivered one of the most outrageous and important speeches of nineteenth-century science. 'Those who hold the doctrine of Evolution, and I am one of them,' he began, 'conceive that there are grounds for believing that the world, with all that is in it and on it, did not come into existence in the condition in which we now see it, nor in anything approaching that condition'. There was boundless evidence for this – in the sky, in the water, in the flora and fauna of the Earth – and in praising Herbert Spencer's *System of Philosophy*, Huxley proclaimed that 'One who adopts the nebular hypothesis in astronomy, or is a uniformitarian in geology, or a Darwinian in biology, is . . . an adherent of the doctrine of evolution'. But what more could they learn, and what more could they prove, from the recent discoveries in Germany? Huxley accepted that 'to superficial observation, no two groups of beings can appear to be more entirely dissimilar than reptiles and birds', for he knew that the hummingbird did not look like the tortoise, nor the ostrich like the crocodile. But those 'who believe[d] in evolution' believed that, once, there were no such gaps between creatures, and that the fossil record could demonstrate the fraternity of reptiles with birds. He therefore guided his listeners through the anatomies of these natural cousins, to the bird's pinion and the reptile's claw, to the breastbones, vertebrae, and hips. There were similarities and there were differences, but two questions trumped everything else: '1. Are any fossil birds more reptilian than any of those now living? 2. Are any fossil reptiles more bird-like than living reptiles?' If the answer was 'Yes' it would prove their relationship, and he thought the recent bequests from Bavaria answered those questions entirely.[27]

Huxley described the *Archaeopteryx* first. It had feathers and so it was a bird, but its tail, claws, and metacarpal bones were quite reptilian; accordingly, 'the oldest known bird . . . exhibit[ed] a closer approximation to reptilian structures than any modern bird'. Then came the dinosaurs: the *Iguanodon*, the *Hadrosaurus*, the *Megalosaurus*, and the rest. Here, Huxley paid special attention to the size and structure of their spines, pelvises, legs, and feet, and he remarked upon the relative weakness of their fore-limbs compared to their hinds. There was only one conclusion that he, like Mantell and Leidy before him, could draw: 'the *Dinosauria* . . . may have supported themselves for a longer or shorter period upon their hind legs'. Gone was the notion of a dinosaur slinking along on all fours, its legs jutting out and its belly touching the ground; gone were Owen's dreadful reconstructions at the Crystal Palace. 'There can be no doubt that the hind quarters of the *Dinosauria*', he declared, 'wonderfully approached those of birds in their general structure, and therefore that these extinct reptiles were more closely allied to birds than any which now live'. Now, at last, Huxley deployed the *Compsognathus* – that small, lithe, chicken-like dinosaur from the Solnhofen quarry – 'to obliterate completely the gap . . . between reptiles and birds'. It clearly 'hopped or walked in an erect or semierect position, after the manner of a bird'; it was the best example yet of 'the "missing link"' between those classes of creature; and it was further, compelling evidence that birds had their ancestry 'in the Dinosaurian reptiles'. All this was proof, he concluded, that 'the facts of palaeontology ... are not opposed to the doctrine of evolution, but, on the contrary ... enable us to form a conception of the manner in which birds may have been evolved from reptiles'. The 'fowl that may fly above the earth', supposedly created by the Lord on the fifth day of the first week, had in fact evolved from the sixth day's creatures 'that creepeth upon the earth'.[28] The book of Genesis lay in ruins, the dinosaurs had triumphed, and even Richard Owen recognised the quality and force of the bulldog's performance.

Not everything that Huxley attempted in these years worked so well, not least his 1868 analysis of deep-sea detritus and 'Atlantic mud'. Inspired by Ernst Haeckel, who had suggested that an albuminous material that he called 'protoplasm' was 'the original active substratum of all vital phenomena', Huxley would use the British Association's meeting at Norwich to announce that he had discovered the source of all life.[29] He argued that deep beneath the surface of the water, binding single-celled organisms and calcified materials into a sticky mass, there was a 'transparent gelatinous substance', an ancient slime that he christened *Bathybius haeckelii*.[30] Was this, he asked, the missing link between inorganic and organic matter, the primordial soup from which all life had sprung? Before the Geographical Society he declared that such a 'living scum' extended for 'thousands upon thousands of square miles' from the Red Sea to the Pacific.[31] In another essay, he proclaimed this abyssal mud as the 'matter which is common to all living beings': from the brightly coloured lichen clinging to a rock, to the botanist who might study it, he claimed, bathybius connected them all.[32] Of course, Huxley was wrong: there is no primordial ooze uniting the sea-beds of the world, and the protoplasm that he 'observed' was merely calcium sulphate. With good reason has one naturalist described the episode as 'one of the most peculiar and fantastic errors ever committed in the name of science'.[33]

Other endeavours were more profitable. Together with the X Club, he threw his weight behind a new journal that the astronomer Norman Lockyer, who had collaborated with Edward Frankland in the discovery of helium, would name after a line from Wordsworth: 'To the solid ground of nature trusts the Mind that builds for aye'. In *Nature*, established in 1869 and which figures today as one of the two great scientific journals, Huxley found a new home for his polemics and proselytising, and it was the obvious place to announce the discovery of further 'dinosaurs'. In the very first issue, therefore, besides noting correspondence about 'the all-engrossing topic of . . . the Suez Canal' that French engineers were carving through the desert, *Nature* gathered intelligence of new dinosaurs from around the world. From the Permian rocks of the Urals came the *Deuterosaurus* and *Rhopalodon*; from Prince Edward Island off the coast of Canada, the *Bathygnathus*; and from the 'reptiliferous beds of Central India . . . the little *Ankistrodon*'. Later analysis would reveal that none of these creatures was properly a dinosaur – rather, they belonged to other classes of primeval animal known as the synapsids and therapsids – but it was yet more proof that dinosaurs and other treasures of the earth were driving science further and further from the safe ground that Buckland once occupied, and that Owen still sought to defend.[34]

Yet Huxley was now rivalling Owen for mastery of scientific society and, besides directing the X Club and its campaign for secular inquiry, he assumed the presidencies of three major institutions in just two years. The first was the Royal Geological Society, where in February 1869, in a speech republished as 'Geological Reform', he presented 'Evolutionism' as the pre-eminent theory of the day and swiped at the philistines who deferred to 'the Mosaic cosmogony'.[35] Next, and three weeks later, was the Ethnological Society, where he discoursed on the Celtic tribes which had peopled the British Isles at the time of Caesar's invasion.[36] Finally, and most importantly, came the British Association in 1870, when even Lord Stanley, the son of the Earl of Derby, dared not stand against Huxley in a vote for the office: 'the president', he conceded, 'should be the mouthpiece of English science'.[37] At that year's meeting at Liverpool, the self-made upstart would give the principal address at the parliament of science. It was another symbolic shift in power, a sign that Huxley and the X were wresting control of British science from the conservative clique, and it troubled *The Times*: 'He is the champion of views', the paper complained, 'to which large classes of people entertain very strong objections'. Would it not have been wiser, it asked, to elect 'a president whose name is not identified with one side of an unsettled question, and whose declared opinions are not calculated to provoke any kind of antagonism?'[38]

There had always been something of the preacher in Huxley: the clear-eyed vigour, the adamant belief of a godless Calvinist, the soaring oratory that evangelised for the evidence of the natural world. When he gathered his recent speeches and lectures for publication, there was an obvious title for them, *Lay Sermons* (1870), and Darwin gushed in admiration.[39] 'It [is] a much greater shame & injustice [than the unequal distribution of wealth]', he told Huxley, 'that any one man sh[oul]d have the power to write so many brilliant essays as you have lately done'.[40] In recognition of these missionary faculties, the *Spectator* had spitefully crowned him a pope, reproaching 'the more than Papal arrogance of his recent tone' and his pretensions to 'personal infallibility'.[41] In fairness, Huxley had now professed something of a creed, although as Adrian Desmond notes, he preferred to describe it as 'a method of inquiry'.[42] At a meeting of London's Metaphysical Society, where Victorian luminaries such as Tennyson, Cardinal Manning, and Walter Bagehot declared the attainment of a certain 'gnosis' – a certain knowledge – Huxley went the other way. It took time to define what 'agnostic' meant, but it was always a reaction to certainty. 'I took thought', Huxley recalled, 'and invented what I conceived to be the appropriate title . . . It came into my head as suggestively antithetic to the "gnostic" of Church history, who professed to know so much about the very things of which I was ignorant'.[43]

But what would the agnostic pope preach to the British Association at Liverpool in 1870? *The Times* need not have worried. Unusually, Huxley restrained himself: the Franco-Prussian conflict was raging at the time and that was war enough. He therefore did not call for a crusade to vanquish those enemies who remained; instead, he confined himself to the history of science, to the works of naturalists, chemists, and mathematicians from all Europe. But if Huxley had sought to pacify at Liverpool, he was building his own basilica in London. He had been professor at the School of Mines for years now. He had also become principal of the Working Men's College, which the Christian Socialists, led by F. D. Maurice, had founded at Blackfriars: 'The chess-board is the world', he explained during his first lecture there, 'the pieces are the phenomena of the universe, the rules of the game are what we call the laws of Nature'.[44] And now in South Kensington, a suburb that the late Prince Albert had imagined as a great campus for British science and culture, as 'Albertopolis', Huxley raised his own tower to rival the spires of Oxbridge.

Where once he had relied on Owen for a letter of recommendation, Huxley now sat on a Royal Commission for Scientific Instruction that could appropriate the necessary money. Joining him on the Commission was the new MP for Maidstone and the X clubman John Lubbock, who in *Pre-Historic Times* (1865) had become the first author to use the terms 'Palaeolithic' and 'Neolithic' to describe the early periods of human history that lay well beyond the scope of the Ussher chronology.[45] Their ally Norman Lockyer, the editor of *Nature*, served as the Commission's secretary. The result of their deliberation was £66,000 in funding for Huxley's project, more than £6 million in today's money, and the construction of the four-storey Science Schools Building on what became Exhibition Road.[46] On the top floor, Huxley laid out his teaching laboratory and, in June 1871, the first class of students arrived. Here, trained and taught by the great man and his disciples, they would learn to dissect, analyse, question, and teach. It was the first significant movement in the creation of a biological profession and, unlike Huxley in the 1840s, these students would not want for fees. Joseph Whitworth, the inventor and manufacturer, was a kindred spirit. A serious, practical man, and one of the new generation of Victorians who had risen to eminence through expertise, he donated the equivalent of £10 million to Huxley's new school. It would be a pioneer in the field of scientific education, and today it ranks with Cambridge, Oxford, and Harvard among the great universities of the world; in time, it would take the name of Imperial College London.

20
Tides of Faith

The sea of faith
Was once, too, at the full, and round earth's shore
Lay like the folds of a bright girdle furl'd;
But now I only hear
Its melancholy, long withdrawing roar,
Retreating to the breath
Of the night-wind, down the vast edges drear
And naked shingles of the world.[1]

Matthew Arnold, 'Dover Beach' (1867)

Was there a Victorian crisis of faith? The question has long vexed historians. But if Victorian faith ever *was* in crisis, it was probably during the 1860s. By now, Wallace and Darwin had unveiled their theories on the variation of species, removing divine agency from the natural history of the world; German academics and British churchmen were exposing the inaccuracies and inconsistencies of the Bible; and Huxley was chasing priestcraft from the laboratory's door. Parliament appeared to be losing faith, too. In the summer of 1868, despite the serial outrage of bishops such as Samuel Wilberforce, it passed William Gladstone's bill for the abolition of the taxes that funded Anglican ministries and parish churches.[2] Within a year, in the aftermath of the Fenian Rising, it disestablished the Episcopal Church in Ireland, prompting ferocious accusations that the Liberal government had 'underhand dealings with the papal authorities' in return for Catholic votes.[3] Another domino would fall when the Universities Tests Act removed the remaining religious restrictions upon the students and fellows of Oxbridge; one of Huxley's disciples promptly secured a prestigious post at Trinity College, Cambridge.

These developments had coincided with another moment of religious scepticism. In 1863, the French polymath Ernest Renan had published *The Life of Jesus*. A scholar of the Bible, eastern civilisations, philosophy, and language, Renan worked primarily from the gospel of John and the histories of Josephus to reconstruct a life of Christ that he rid of the supernatural and miraculous. This Jesus, entirely human, led a life that Renan's 'narrative pace and . . . literary style' presented almost as the subject of a novel.[4] There is 'courage' in the face of betrayal, calmness and dignity during the trial and the 'grotesque' abuse that the Roman soldiers dispensed, and the (questionable) construction of an enemy in 'the old Jewish party'. Renan then describes the Passion and the crucifixion 'in all their atrocity'. Christ refuses wine on the cross because 'he preferred to go out of life with his mind perfectly unclouded', while Renan has Jesus's mother, Mary, present at the scene as a mark of 'personal tenderness'; this is the death of a man from 'the rupture of a blood-vessel', not the expiration of a deity.[5] In France, as Robert Priest relates, Renan's readers 'either thrilled or dismayed cultural authorities by getting their hands on the book', and Pius IX placed *The Life of Jesus* on the Index of Prohibited Books.[6] In Britain, conservative critics fumed. Local newspapers carried reports from France which had denounced the book as 'rash, impious, scandalous, sacrilegious, blasphemous, erroneous, heretical, and smitten with anathema'.[7] The next year, when the English translation appeared, a Dunfermline newspaper despised Renan's 'attack upon the character of Christ'; a correspondent of the *Leeds Intelligencer* revolted against a 'subversive and dangerous book'; and the *Belfast Newsletter* spewed that Renan had 'sought to impeach the historic purity of the Gospel'.[8]

Where Renan led others followed, even Englishmen. In 1865, London's bookshops began stocking a 'survey of the life and work of Jesus Christ' titled *Ecce Homo*. It was anonymous because, according to its publisher, Alexander Macmillan, the author desired 'to see the effect on men's minds of the thought that had seemed important to himself unbiased by reputation, or lack of reputation'.[9] That 'thought', as with David Strauss in the 1840s and Renan more recently, concerned Christ as a man and moral philosopher, not the son of God. *Ecce Homo* therefore sought to reconstruct a world in which 'he whom we call Christ bore no such name, but was simply, as St Luke describes him, a young man of promise, popular with those who knew him and appearing to enjoy the Divine favour'. So, what did Jesus feel? What did he want and why? What was his character as a human being? *Ecce Homo*'s conclusions were not those 'which church doctors or even apostles have sealed with their authority, but [those] which the facts themselves, critically weighed, appear to warrant'. This Christ was calm and confident, 'the founder, legislator, and judge of a divine society' that the author called a 'Christian Commonwealth'.[10]

Ecce Homo was a phenomenon. By the spring of 1867, six editions had sold more than 16,000 copies. Booksellers ran rapidly out of stock and, when they had it, W. H. Smith's boys walked the length of railway platforms, shouting: '*Daily Telegraph*, *Standard*, *Morning Star*, *ECCE HOMO*!' The writer Matthew Arnold, whose poem 'Dover Beach' figures as one of the most elegant expressions of Victorian doubt, reported to his mother that 'Every one is beginning to talk of [this] new religious book'.[11] Carefully and dangerously, some readers praised the volume. The *North British Review* admired its 'power and truthfulness', while the *Guardian* remarked upon 'the surpassing ability' of the author and 'the deep tone of religious seriousness which pervades the work'; Macmillan thought William Gladstone was the reviewer. Others, however, were lost in apoplexy. The journal *John Bull* complained that the book was a Trojan Horse, 'an attempt to familiarize the English mind to the German sceptical habit of reasoning'. The evangelical Earl of Shaftesbury, who had previously condemned Bishop Colenso for his 'puerile and ignorant attack on the sacred and unassailable Word of God', told one London meeting that *Ecce Homo* was 'the most pestilential book ever vomited from the jaws of hell'.[12]

The anonymity of the author only fuelled the craze and, just as the authorship of the *Vestiges* had beguiled the previous generation, Victorians of the mid-1860s now obsessed over *Ecce Homo*. Was Dean Stanley the author? Had J. A. Froude resumed his pursuit of infamy? Was it even Napoleon III or George Eliot? Macmillan exploited the mystery perfectly. As Ian Hesketh relates, on one winter's evening he lured 'sixteen of London society's leading intellectual and literary figures' to a dinner party by promising the author's attendance.[13] But as the men ate, drank, and talked, Macmillan refused to reveal his client's identity; his guests would leave that night not knowing. Among them, however, had been John Ronald Seeley, the professor of Latin at University College London. Not yet thirty, another prodigy, he had come top of his year at Cambridge before taking an interest in the 'positivist principles' that were in vogue at UCL.[14] Some of that empiricism had worked its way into *Ecce Homo*, and while Seeley now figures in memory as a historian – who suggested that Britain had acquired its empire 'in a fit of absence of mind' – his first intervention in public life had brought the controversies of European theology even further into the drawing rooms of Victorian London.[15]

In less salubrious settings, Britain's freethinkers were pressing ahead with their own mission. Since coining the term 'secularism' and publishing the *Reasoner* in the 1850s, George Holyoake had won his war on the press-stamp, which was one of the 'taxes on knowledge' that British newspapers had carried since the early eighteenth century.[16] By advising and giving shelter to fellow dissidents, and by encouraging the revolutionary spirit in France, Poland, and Hungary, he also won the esteem of many of Europe's leading democrats, in particular of one 'handsome, fiery young Italian [who had] escaped from his Austrian dungeon', the exiled nationalist leader Giuseppe Mazzini; Holyoake then served as secretary of the British committee which recruited volunteers for Giuseppe Garibaldi's military campaigns to Sicily and southern Italy.[17] But twenty years of campaigning, lecturing, and fending off prosecutorial attention had exhausted Holyoake, and his 1858 pamphlet *Self-Help by the People*, which told of the Rochdale Pioneers and their establishment of co-operative shops, was his last great work of propaganda: 'Human nature was different in Rochdale to what it was elsewhere', he had exulted.[18]

By the mid-1860s, therefore, British secularists were looking elsewhere for leadership, and many of them had turned to Charles Bradlaugh. Now in his early thirties, Bradlaugh came from the East End of London, leaving school at eleven before working as an office-boy on five shillings a week. Yet rather than taking the omnibus across the city as he delivered messages for his employer, Bradlaugh would run, and the money that he saved went on second-hand books. This learning proved a dangerous thing. While he was studying for confirmation in the Church of England, Bradlaugh found that 'he was totally unable to reconcile' the Thirty-Nine Articles with the Gospel; his vicar promptly denounced him as 'atheistical' and suspended him from Sunday school. Instead of repenting, however, Bradlaugh was soon listening to Richard Carlile's free-speech disciples and their 'anti-theological discourses' in London's parks. By 1850, at the age of seventeen, he had joined their ranks, leaving his disapproving family in order to lodge with Carlile's widow, and delivering his own lectures on 'The Past, Present, and Future of Theology'.[19]

There followed spells as a (failed) merchant and a cavalryman in Ireland before settling into legal work as a solicitor's clerk. Adopting the pseudonym 'Iconoclast' to protect his employer's reputation, Bradlaugh by the late 1850s had begun a career in journalism. He penned portraits of freethinking heroes such as Spinoza, Shelley, and Descartes; he delivered critical examinations of the Bible, mocking the 'plagues of Egypt' by pointing out that some cattle must have been killed four times 'by murrain, hail, then plague, then drowning'; and he edited the *National Reformer*, a Sheffield-based journal which proclaimed itself 'Atheistic in theology, Republican in politics, and Malthusian in social economy'.[20] In 1866, having taken a leading role in the Reform League that agitated for the further expansion of the franchise, Bradlaugh made the *Reformer* the house journal of a new association, one that would seek to unify British secularism. As Edward Royle tells us, Bradlaugh took soundings from allies in June, resolved upon the structure of the new body in August, and by September had drafted a programme and statement of principles for the National Secular Society. Its purpose was to advance the idea 'that human improvement and happiness cannot be effectually promoted without civil and religious liberty; and that, therefore, it is the duty of every individual . . . to actively attack all barriers to equal freedom of thought and utterance for all, upon Political and Theological subjects'.[21] Appointing himself as president, an office that he would hold for most of the rest of his life, Bradlaugh promised a national conference and began recruiting hundreds of members, galvanised into action by provincial lecturers, into branches across the British Isles.

The headquarters of this new movement was the Hall of Science, a vast and long-demolished meeting-place on London's Old Street. In the description of one journalist, it was 'not an attractive place outside', while inside it was 'less of a hall and more of a barn'. But 'Sunday night after Sunday night' upwards of a thousand Londoners – mostly of the 'tradesman and artisan class' – came flooding through the doors to listen to Bradlaugh.[22] He knew how to put on a show. On 'the very minute' of the hour, the doors to the Hall of Science would close and Bradlaugh, 'a tall, commanding figure', strode from the back and through the crowd to the platform, all to the sound of 'a roar of cheering'.[23] Described by one biographer as 'a vigorous, powerful-looking man, with a clean-shaven face, a massive jaw, and a full head of hair brushed straight back from his high forehead', Bradlaugh was a masterful speaker. He used his voice 'like a trumpet', hitting notes of 'eloquence, fire, sarcasm, pathos, and passion' as he tore into Christian superstitions; by the end, the audience that had been 'hung silent, breathing soft' would burst into 'a hurricane of cheers'.[24] The irony of these occasions – the evangelical zeal, the intense emotion, the cathartic release – was immaterial: 'It is always one story', Bradlaugh reflected. 'Enormous audiences, great applause, few opponents, and fresh recruits for our ranks'.[25]

Within only a few years, Bradlaugh was circulating 6,000 weekly copies of the *National Reformer*, which his daughter described as 'my father's voice, my father's sword, my father's shield'.[26] It was also his weakness and, on 3 May 1868, it carried the following notice: 'The Commissioners of Her Majesty's Inland Revenue having commenced proceedings to suppress the *National Reformer*, a special fund is opened, to be entitled "The *National Reformer* Defence Fund", to which subscriptions are invited.'[27] The problem was the Newspaper Stamp Duties Act, the last of the Six Acts that Lord Liverpool's government had introduced in the 1810s to repress dissenting speech. According to its provisions, and although many of Britain's taxes on knowledge had already been repealed, publishers were still obliged to pay sureties to the government against 'the appearance of blasphemy or sedition' in their publications.[28] Naturally, this was impossible for

Bradlaugh, since publishing his newspaper *without* blasphemy would be like performing 'Hamlet at the Lyceum [without] the Prince of Denmark'. He would welcome the controversy, bidding defiance to the Revenue: 'You may kill the *National Reformer*', he promised, 'but it will not commit suicide. Before you destroy the paper, we shall have to fight the question.' With the great philosopher John Stuart Mill pleading the *Reformer*'s case in the House of Commons, and with 'hosts of small subscribers' donating pennies to the fund, Bradlaugh sustained the fight long enough to see the new Liberal government repeal the very laws that he had broken. It was the end at last of the Six Acts, the end of the taxes on knowledge, and Mill hailed Bradlaugh a hero: 'You have gained a very honourable success', he wrote from Avignon.[29] It was far from the last time, however, that Bradlaugh would fall foul of the law.

For all that Seeley scandalised and Bradlaugh rallied, and for all that many Victorians might have shared in the pessimism of 'Dover Beach', where the sea of faith ebbed away on a falling tide, Christian naturalists had not yet given up the fight; they had refused in particular to concede that humankind was just another species in the panoply of natural life. George Campbell, the Duke of Argyll, was a senior politician who served as Lord Privy Seal, Postmaster-General, and Secretary of State for India in both Liberal and Conservative cabinets. He was also an accomplished ornithologist and geologist who, in the 1850s, had announced the discovery of fossilised leaves on his land in the Scottish islands, something that would make Charles Lyell wonder whether 'the cones of the Hebrides . . . were [once] as high as Etna or the Peak of Tenerife'.[30] Yet Argyll, even if he thought the *Origin of Species* 'delightful', and even if he accepted that natural selection could affect 'the preservation and extinction' of species once they had been created, paled at the notion of a secular creation and the application of Darwinian theory to humankind.[31] In *The Reign of Law* (1867), where he railed against a functional, utilitarian approach to natural history, Argyll argued that 'material forces have been always used as the instruments of Will' and that nothing yet discovered by science had conflicted with 'the nature and working of Creative Power'.[32] Two years later, in *Primeval Man* (1869), he took on

Huxley by claiming that 'the most ancient of all known human skulls is so ample in its dimensions that it might have contained the brains of a philosopher', and that such evidence was 'conclusive . . . against any change whatever in the specific characters of Man since the oldest Human Being yet known was born'.[33] It is not, perhaps, surprising that the Neander and Gibraltar specimens do not appear in that volume.

Argyll was not alone in this resistance. St George Mivart, the son of a successful hotelier (the family business became Claridge's), was a zoologist and had been, in the early 1860s, 'a hearty and thorough-going disciple of Mr Darwin'.[34] Attending lectures at the School of Mines, where Huxley 'impressed with the lucidity of his thought and the admirable clearness with which he gave expression to it', Mivart had been convinced of natural selection, accepting it 'completely'.[35] Within a few short years, however, he had resiled from his former surety. As he wrote extensively on lemurs and other primates for London's learned journals, his devout Catholicism began to clash with the fundamental implications of Darwinian theory and, by 1869, Mivart had published anonymously on 'the difficulties of the theory of natural selection'. He now deplored how 'half-educated men and shallow thinkers' had rushed to adopt Darwinism simply because they understood the phrase 'the survival of the fittest'.[36] Quoting liberally from Argyll, he mocked the idea that American men had descended from American apes, or that European men had descended from the apes of Africa and Asia. And just like Argyll, he refused to accept that the human mind could have developed from anything other than the genius of the Lord.

The third member of this holy alliance was the physicist William Thomson. Born in Belfast and educated at Cambridge, where he thrived on the track and the river, Thomson was by the late 1840s the leading expert on thermodynamics; by the turn of the next decade, he was consulting on the construction of the first transatlantic cables. But Thomson had sustained the stern Calvinism of his youth and this led to conflict both with uniformitarian geology, which suggested that the Earth was many millions of years old, and Darwin, who advised his readers to put down the *Origin* if they could not 'admit how incomprehensibly vast have been the past periods of time'.[37] It was not that Thomson was loyal to the Ussher date of 4004 BC, or that he refused to countenance the idea of an older Earth; rather, and in view of 'the more recently discovered laws of thermodynamics', he just could not see how the Earth was old enough to allow either uniformitarian geology or natural selection to have had the necessary effect.[38] Thomson launched his first attack in February 1868 in Glasgow, where he declared that British geology had become untethered from basic scientific principles, and that it was now 'in direct opposition to the principles of natural philosophy'. Indeed, Thomson believed that, in playing fast and loose with the very concept of time, geologists were subjecting their science to 'dogmatic hypotheses'.[39]

As president of the Geological Society in 1869, Thomas Huxley was honour-bound to fight back and, in his anniversary speech, he sought to repel 'so high an authority' as Thomson. 'We have exercised a wise discrimination', he told the assembled geologists, marking his territory, 'in declining to meddle with our foundations at the bidding of the first passer-by who fancies our house is not so well built as it might be'.[40] But Thomson would not leave it there. In reply to Huxley, he argued that because the Earth had been molten-hot at the point of its creation, and because life was sustainable only beneath a certain temperature, there simply had not been time for the Earth to cool sufficiently to allow natural selection to shape nature as the 'Evolutionists' claimed. By suggesting that the Earth was no older than ninety-six million years, Thomson was wrong, and wrong by an order of magnitude, but he was the pre-eminent physicist of his day and a committed Christian who, in seeking 'to divorce biological evolution from the uniformitarian concept of time', had approached systemic questions of science from a strictly religious perspective.[41]

Reading habits in other fields reflected this underlying strength of Christianity. The best-selling book of any description in the 1860s appears to have been *Jessica's First Prayer* (1866), a children's novel that the evangelical journalist Sarah Smith, inspired by her experience in the slums of Manchester, wrote under the pseudonym 'Hesba Stretton'.[42] Here, a drunken actress abandons her young daughter, Jessica, to the streets of London, where she must overcome the condescension of the local Methodist congregation and the coldness of the coffee-seller who in time adopts her; and as Jessica learns of Christ from the coffee-seller, he in turn learns humility from her. It was a sickly, sentimental story about the redemptive power of Christianity which originated in the serialised pages of *Sunday at Home*, but it was incredibly successful. During Smith's lifetime, *Jessica's First Prayer* would sell more than two million copies in fifteen languages including Braille, and its renown was such that Tsar Alexander II ordered copies for every school in Russia. It was also the kind of publishing sensation which reminds us that intellectual dogfights over science and religion did not always spill over into everyday life and conversation, not even after William Forster's Education Act of 1870 created a whole new generation of readers by massively expanding the provision of elementary schooling. For those who did read of science, however, at least one new controversy would excite and enthral.[43]

In 1870, in the pages of the magazine *Scientific Opinion*, John Hampden of Swindon proposed a wager. Hampden was wealthy, fifty years of age, and though dropping out of Oxford had prevented him from taking holy orders, he had become a well-known conservative polemicist who desired the 'reform of the Church of England on strict Protestant lines'.[44] He was also a disciple of Samuel Rowbotham, a socialist inventor who had championed the system of 'zetetic astronomy'. Under the pseudonym of 'Parallax', Rowbotham had argued that 'the true figure and condition of the Earth . . . instead of its being a globe, and moving in space . . . is . . . A PLANE, without motion'. In other words, Rowbotham believed that the Earth was flat, that its centre lay at the North Pole, and that 'deep barriers of ice' bound the southern oceans.[45] Now, in defence of his teacher's honour, Hampden challenged the readers of *Scientific Opinion* to prove this theory wrong. 'He will acknowledge that he has forfeited his deposit', ran the terms of the wager, 'if his opponent can exhibit, to the satisfaction of any intelligent referee, a convex railway, river, canal, or lake'.[46] The stake was not inconsiderable: £500, some £40,000 in today's money.

Among the readers of *Scientific Opinion* was Alfred Russel Wallace, who had returned from the East Indies in 1862 with more than 30,000 specimens of birds, beetles, and butterflies. Delivering papers to the Linnean and Zoological Societies, writing for the journals, and converting his notebooks into the classic *The Malay Archipelago*

(1869), Wallace had established himself as a leading naturalist, although the obligations of family and unwise investments would preclude financial security. He also travelled on less common paths. A rare anti-colonial voice who recognised the debts he owed to his Malayan assistants, he wrote in Huxley's *Reader* that 'the white men in our colonies are too frequently the true savages'.[47] More curiously, he was devoted to the pseudo-science of phrenology and he obsessed over the supernatural, 'swallowing' the 'humbug' that spiritualists were serving out at seances.[48] Wallace even began to doubt the validity of natural selection, the very theory that he had helped to articulate: 'Let us not shut our eyes', he wrote in the *Quarterly*, 'to the evidence that an Overruling Intelligence has watched over the action of those [organic] laws, so directing their variations and so determining their accumulation, as finally to produce an organisation sufficiently perfect to admit of, and even to aid in, the indefinite advancement of our mental and moral nature'.[49] It was a shocking disavowal of his own idea and, when Darwin read this, he simply wrote 'No' on his copy of the journal.[50]

Despite dissenting from his own creed, and despite infatuations with quackery and the bogus, Wallace could not resist taking up Hampden's 'flat earth' wager. After seeking the counsel of Charles Lyell, who thought 'it may stop these foolish people to have [the curvature of the Earth] plainly shown them', and trusting that Hampden could 'afford to pay £500', Wallace staked the same sum 'on the understanding to show visibly, and to measure in feet and inches, the convexity of a canal or lake'.[51] There were two preliminary issues to settle. Where would they conduct this experiment? Wallace favoured Llyn Tegid, a glacial lake on the River Dee in Snowdonia, but Hampden prevailed in choosing a stretch of the Bedford River in Norfolk, which was where Rowbotham had conducted his own experiments in the 1830s. And who would referee the contest? Each combatant would have his own champion. Hampden chose a fellow devotee of the flat-earth theory, and Wallace an astronomer from Downham Market.

In the early spring, with each party's stake placed in an escrow, all four men decamped to the Norfolk Fens, their field of inquiry six miles of a dead-straight drainage canal that lay between two bridges.

Wallace's methodology was simple but ingenious. First, he measured the distance between the iron parapet of the first bridge and the surface of the water in the canal: it was thirteen feet and three inches. After this, six miles downstream at the second bridge, he draped over its brickwork a large, white cotton sheet which had 'a thick black band' running horizontally across it, and he positioned the lower edge of that band at precisely the same distance from the water – thirteen feet, three inches – as the parapet on the first bridge. Finally, Wallace went to the midpoint between the bridges, three miles from each, and erected a long pole upright in the canal; adorning that pole, thirteen feet and three inches above the waterline, was a bright red disc. It was a brilliant device: if the Earth were flat, then Wallace, when peering through his telescope upon the parapet of the first bridge, would see

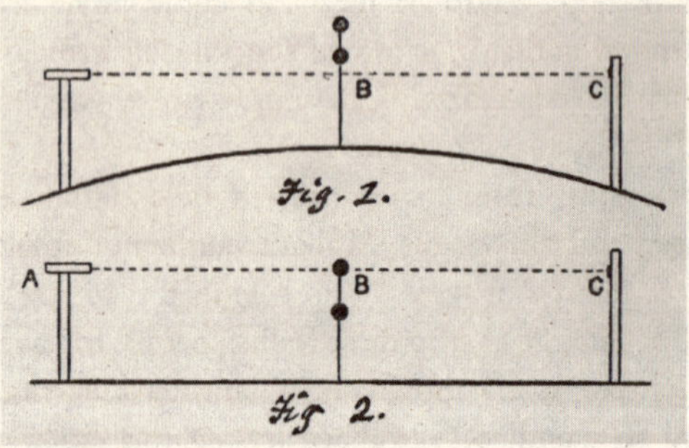

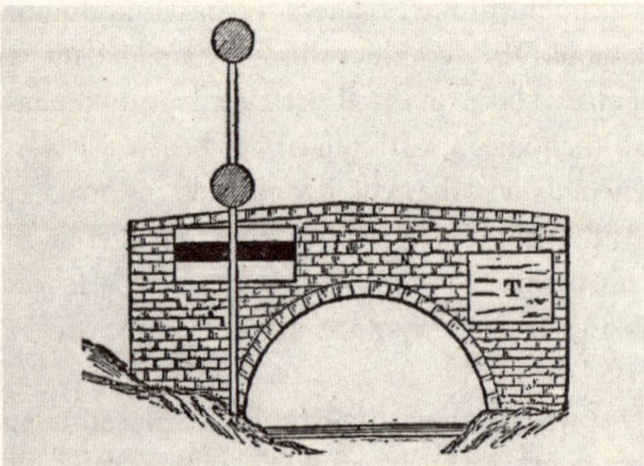

Diagrams of Wallace's Bedford Level Experiment

the red disc aligned with the black band on the sheet downstream, all of them being thirteen feet and three inches above the water. Yet if there was a curvature to the Earth, and if consequently Wallace had placed the mid-way pole on higher ground, the red disc would appear *above* the black band on the distant cotton sheet, despite being the same distance from the water. If the second scenario occurred, Wallace would win the wager. Of course, he was correct. Hampden's referee, on looking through the telescope, described 'a perfect straight line' through the mid-way pole that went beneath the red disc and then to the black cotton on the second bridge: 'Beautiful! Beautiful!' he cried and even 'jumped for joy'.[52]

Yet when the referees made their sketches and reported to the surveying magazine *The Field*, Hampden refused to keep his part of the bargain. Now claiming that Wallace had manipulated his telescope and that its lack of a crosshair rendered it unreliable, he refused to concede. Even when the dispute went to an arbitration at which the editor of *The Field* decided in Wallace's favour, Hampden would not pay; instead, he demanded the return of his stake and went on the attack. In one pamphlet, he claimed to expose 'the Bedford Canal Swindle' and 'the notorious rascality of the scientific world'; referring to 'the triumphing of the wicked' and 'the joy of the hypocrite' by way of the Book of Job, he denounced the arbitration as an example of 'pitiable ignorance' and 'despicable cowardice'.[53] The editor of *The Field* sued for libel, and the courts bound Hampden to keep the peace, but keep the peace he did not.

Instead, Hampden commenced a campaign of sustained abuse and harassment against Wallace. He wrote to the presidents and secretaries of every learned society to which his victim belonged, asking them, 'Do you know that Mr. A. R. Wallace is allowing himself to be posted all over England as a cheat and a swindler?' More disturbingly, he wrote lengthy threats to Wallace's wife: 'If your infernal thief of a husband is brought home some day on a hurdle', he warned, 'with every bone in his head smashed to a pulp, you will know the reason. Do tell him from me he is a lying infernal thief, and as sure as his name is Wallace he never dies in his bed.' Despite several further convictions for libel, Hampden's hostility endured. In 1885, fully fifteen years after the experiment, he would damn Wallace as a 'degraded blackleg' before turning up, unannounced, to the Wallaces' cottage in the town of Godalming. In his final broadside, a pamphlet which he handed out at an exhibition of the Royal Geographical Society, Hampden defended himself as 'the well-known champion of the Mosaic cosmogony' against the impious overtures of infidel geographers. Wallace never did receive his winnings and he later reflected on the futility of engaging 'paradoxers' such as Hampden in rational debate, concluding that the whole affair was 'the most regrettable incident in my life'.[54]

21
The Angel and the Ape

> There was a time, when the earth was to all appearance utterly destitute of both animal and vegetable life, and when according to the opinion of our best philosophers it was simply a hot round ball with a crust gradually cooling. Now if a human being had existed while the earth was in this state . . . and if at the same time he were entirely ignorant of all physical science, would he not have pronounced it impossible that creatures possessed of anything like consciousness should be evolved from the seeming cinder which he was beholding?[1]
>
> Samuel Butler, *Erewhon* (1872)

Charles Darwin was almost sixty now. It was old enough for the time, and older still for a man who had spent so long in pain, with chills and indigestion and whatever else his chronic condition presented. His Harley Street consultant had diagnosed 'suppressed gout', which Hooker lampooned as 'every ill [the doctors] cannot name', and there were weeks on end in bed, with Emma reading novels to her husband. Darwin thought himself 'too dried up' to enjoy even Handel's *Messiah*, and his in-laws lamented: 'What a life of suffering his is!' But he kept busy when he could. After the *Origin* there had come the orchids and the climbing plants, then the Royal Medal that the X Club had won for him. And though the 'big book' on natural selection had not yet appeared, the refinement and exposition of the theory consumed his attention and whatever energy he could muster. 'I am a withered leaf', he accepted, 'for every subject except Science'.[2] He welcomed pilgrims from Germany, Haeckel among them, who knelt before the sage of Down House; he spluttered at the Duke of Argyll's claim that, in nature, 'mere ornament or beauty is in itself a purpose'; and he turned his focus back to domestic breeds, to the vexed question of just *how* selected traits could pass from parent to child.[3]

The manuscript that Darwin delivered to his publisher in 1867, *The Variation of Plants and Animals under Domestication*, had grown wildly in the writing: 'I cannot tell you how sorry I am', he wrote to John Murray, 'to hear of the enormous size of my Book'.[4] Here, over two volumes, each of more than 400 pages, and with tens of thousands of words packed into dense footnotes, Darwin looked for examples of speciation in the farmyard and the field, in ponds, aviaries, and greenhouses. As he remarked in the introduction, he sought to explain 'the amount and nature of the changes which animals and plants have undergone whilst under man's dominion'. How, he asked, had the feral dog become man's best friend? Why could the Persian cat, 'the most distinct in structure and habits', breed with a moggie whereas the mule was infertile?[5] How could botanists induce hybridity – that is, variation – into their orchids?

Towards the end of the second volume, Darwin set out his mechanism for heredity, for the transmission of selected traits. He called it 'pangenesis', an idea that had been with him for years. In the spring of 1865, he had sent a draft to Huxley and asked his opinion, calling it 'rash & crude' but 'a considerable relief to my mind' because he could 'hang on it a good many groups of facts'.[6] And while Huxley's first impression appears not to have been favourable – his reply, which has been lost, moved Darwin to consider not publishing anything at all – he saw some value in it: 'Somebody rummaging among your papers half a century hence', he advised, 'will find Pangenesis & say "See this wonderful anticipation of our modern Theories – and that stupid ass, Huxley, prevented his publishing them"'.[7] So, pangenesis came forth. But how did it work?

In Darwin's hypothesis, which relied heavily on the cell theory that German botanists and physiologists were developing at the time, every single element of a parent organism cast off tiny particles that he called 'gemmules'. These gemmules carried the selected, 'fitter' traits of each part of the parent's body and, because they had a 'mutual affinity for each other', congregated in the parent's reproductive organs. Thereafter, when the parent mated, the reproductive process passed on these gemmules to the child, in whom the gemmules determined the characteristics of the corresponding parts of the child's body.[8] Pangenesis was, and remains, nonsense. It also annoyed some readers because it appeared to resurrect the old Lamarckian theory of transmutation, whereby a man who performed hard labour would bequeath broad shoulders to his offspring. Others thought it too perverse, too complicated. 'Would the *skin & mucous* linings be a single part, or two?' asked one of Darwin's old schoolmates: 'Should the secretions be considered merely as *products* of the organs? Or must there be a *bile* gemmule as well as a *gall* gemmule?'[9] Neither Darwin nor his colleagues had read the work of Gregor Mendel. A Czech biologist and botanist, Mendel had spent years studying the characteristics of pea plants and had proposed, in a minor German-language journal, that where the parents of children possessed contrasting traits, it was the dominant rather than the recessive trait that the child would inherit.[10] When genetics eventually flourished at the turn of the twentieth century, the English biologist William Bateson coined the term 'allele' to describe these Mendelian traits.

Notwithstanding his ignorance of Mendel's breakthrough – and it remains a myth that a copy of Mendel's paper was in the library at Down House – Darwin's work went on and the accolades came with it. He joined Alexander von Humboldt, John Herschel, Lord Macaulay, and Richard Owen in the Prussian ranks of Pour le Mérite; he ascended to the companionship of Peter the Great's Imperial Academy at St Petersburg; and doctorates arrived from Cambridge and Oxford, the latter because Edward Pusey, the intractable Tractarian theologian, refused to tolerate the nomination of seven of the 'blackest heretics' whom he thought even worse than Darwin, among them Thomas Huxley.[11] Even so, there was one last river to cross, one subject that had been missing from Darwin's books so far, even if it had cast a shadow over everything. It was the development of humankind. Over the winter of 1870–71, therefore, as the industrial violence of the Franco-Prussian War drew a 'Thundercloud' over the uncertain future, Darwin looked back into the uncertain past; he called this latest book *The Descent of Man*.[12]

In its introduction, Darwin made clear his goals. 'The sole object of this work', he wrote, 'is to consider, firstly, whether man, like every other species, is descended from some pre-existing form'. Beyond that, *Descent* would look to identify 'the manner of [man's] development' and 'the value of the differences between the so-called races of man'. He admitted freely that there would be little original research in its pages; rather, *Descent* would apply his theory – which for the first time in a publication he described as 'evolution' – to the facts adduced in the 'admirable treatises' of others including Huxley and Haeckel. He would not accept that such endeavours were useless or merely interesting; they were essential. 'It has often and confidently been asserted', he declared, 'that man's origin can never be known: but ignorance more frequently begets confidence than does knowledge: it is those who know little, and not those who know much, who so positively assert that this or that problem will never be solved by science.' It was the boldest declaration of expertise and the strongest repudiation of incurious orthodoxy that Darwin ever made.[13]

There were probably five main points to the *Descent*. First, Darwin showed that humankind had evolved from a 'lower form' in the same way as other animals: it shared a rudimentary structure with primates, who could suffer from the same diseases, and it 'developed from an ovule ... which differs in no respect from the ovules of other animals'; Darwin thought it 'wonderful', meaning nonsensical, that anyone should think of man and monkey as the children of separate creations. Second, he argued that humans shared mental and emotional faculties with 'lower animals': they all possessed curiosity, memory, imagination, reason, communication, even kindness in common, with the effect that 'we may trace a perfect gradation from the mind of an utter idiot, lower than that of the lowest animal, to the mind of a Newton'. Third, that humankind had probably risen from 'the extinct apes closely allied to the gorilla and chimpanzee' in Africa, and then through stages of development into the sentient, civilised creature of the Victorian age. On this note, tellingly, Darwin reminded his readers that 'the old Dinosaurians are intermediate in many important respects between certain reptiles and certain birds', and that 'the Ichthyosaurians ... present many affinities with fishes'. Fourth came the issue of race and, here, Darwin denounced polygenism, which was the idea – increasingly popular in the 1860s – that the human 'races' were separate species which had separate creations. 'The most weighty of all the arguments against treating the races of man as distinct species', he wrote, 'is that they graduate into each other, independently . . . of their having intercrossed'. Finally, but at the greatest length, Darwin explained how the most important element of natural selection was *sexual* selection. By this, he meant that females of a species would look for and choose their preferred characteristics in males, who consequently competed against one another – as when peacocks displayed their plumage, or when stags rutted – for female attention.[14]

As Desmond and Moore relate, the press was reading *Descent* during the worst violence of the Paris Commune, *la semaine sanglante*. In keeping with the bloodshed of the season, when the French national army killed as many as 7,000 working-class rebels, Darwin expected his own 'execution'.[15] Some reviewers obliged. The *Edinburgh Review* warned that if Darwin was right, that if human morality was merely an instinct 'identical in kind with those of ants or bees', and if 'the revelation of God' had simply been 'invented for the good of society', then such a truth would 'shake society to its very foundations by destroying the sanctity of the conscience and the religious sense'.[16] The Duke of Argyll likewise chided Darwin for publishing such an argument at this delicate moment in European history: a 'man incurs a grave responsibility', he wrote, 'when, with the authority of a well-earned reputation, he advances at such a time the disintegrating speculations of this book'. Besides, asked the duke, was not human history proof that a progressive, improving evolution was claptrap? 'No poetry surpasses Homer', he protested; 'no art is more perfect than that of Greece; no specimens of the human form are more beautiful than the models which Greek sculptors have preserved for us'.[17] So there, Mr Darwin.

St George Mivart then perfected his betrayal. He had already complained that Darwinian theory was 'beset with certain scientific difficulties', and in *On the Genesis of Species* (1871) he sniped that Darwinism was popular only because of its utility 'as a weapon of offence [for] irreligious writers'.[18] But in reviewing the *Descent* for the *Quarterly*, Mivart made fresh enemies. In a vicious polemic, he judged Darwin to have 'utterly failed', to have adopted 'a radically false metaphysical system', and to have 'set at naught the first principles of both philosophy and religion'.[19] Although he wrote anonymously, it was not difficult to divine Mivart's authorship. The review 'mortified' Darwin, and Huxley turned bulldog on the spot: Mivart had been 'insolent', he told Hooker, '& I mean to pin him out'.[20] Some American readers were even more violent in their censure. The *New York Times* reprinted Argyll's review under the headline 'The Man-from-Monkey Theory: Absurdity of Mr Darwin's Hypothesis', and one of its correspondents described the book as 'a hard dose to swallow'.[21] A religious newspaper in North Carolina lashed this 'latest form of infidelity', a book which had dared to contradict 'the legitimate conclusions of the Scriptures'.[22] A reader of the *Chicago Tribune* meanwhile protested that 'Adam was a Man, not a monkey' and urged his brethren not to 'suffer reproach to be cast on our Ancestry'.[23]

Others took greater pleasure in the *Descent*. Alfred Russel Wallace left behind the Bedford Level controversy to praise 'one of the most remarkable works in the English language', a book by which 'Mr Darwin's genius' had illuminated 'a new and inner world of animal life'.[24] There were letters of congratulation from Jena, where Darwinism was 'openly confessed and publicly taught', and there was a request for a translation into Hungarian.[25] Even the *Spectator*, which had gaily mocked Huxley and the Darwinists, conceded that 'the theory of evolution will, as far as we can see, be proved to be really true, in the sense that man is the lineal descendant of animals far his inferior'.[26] And this was right. Even if certain passages of the *Descent* on race and gender might strike the modern reader as outmoded and offensive, and even if subsequent developments showed other passages simply to be wrong, it remains a landmark in the history of science, a book which synthesised anatomy, zoology, botany, physiology, palaeontology, and evolutionary biology into a compelling, readable account of human development.[27]

Darwin had considered appending an essay to *Descent*. The anatomist Charles Bell, who had described the eponymous palsy and written a Bridgewater Treatise on the divine design of the human hand, had long ago declared that humankind alone was divinely 'endowed with certain muscles solely for the sake of expressing his emotions'. This, as Darwin related in the preface to the *Descent*, was an idea 'obviously opposed to the belief that man is descended from some other and lower form'.[28] But with the *Descent* running to more than 900 pages, he and his editors at John Murray decided to reserve this discussion for a separate book, which went on sale the next year. But *The Expression of the Emotions in Man and Animals* (1872) was no mere appendix. Rather, it was one of the first books to feature photographic illustrations, which Darwin used alongside standard engravings to demonstrate the range and similarity of mammalian feelings: there are children crying in sorrow, an old man shrieking at electrocution, a chimpanzee 'disappointed and sulky', and a 'cat in an affectionate frame of mind'; it was also a pioneering effort to infuse biology into the study of the human mind.[29] For a man like Darwin, who often derived misery from the disorders of his stomach, it was a deeply personal project.

There were, Darwin argued, several great principles which underlaid the expression of emotions. First, that under certain states of mind 'there is a tendency through the force of habit and association for the same movements to be performed', such as 'the wagging of a dog's tail' or 'the shrugging of a man's shoulders', which give accurate expression to the emotions being felt. Second was 'the principle of antithesis', which provided that a 'state of mind' that was 'directly opposite' to the habitual may induce 'the performance of movements of a directly opposite nature', even though such movements were 'of no use'. Finally, there was the involuntary, reflexive action of the nervous system, which could produce in both humans and animals a general expression of an emotional state. Working from responses to a questionnaire that he had dispatched around the world, a paper which had asked for detailed information on the emotional habits of various nations, Darwin traced the operation of these principles across human society and the animal kingdom. He recounted how cats and dogs both flattened their ears before fighting; how man and monkey 'likewise grin and make a chuckling sound' in the expression of pleasure and joy; how eyebrows rise in surprise, the pupils dilate in horror, and the cheeks blush in shame. 'The young and the old of widely different races', he concluded, 'both with man and animals, express the same state of mind by the same movements'.[30] If the *Descent* had situated the physical aspects of humankind squarely within the animal kingdom, the *Expression* did the same with the mental.

As Darwin wrote of men and monkeys, Huxley was expanding his empire. Now in his mid-forties, with more bush in the eyebrows and some grey in his hair, he was sitting on several royal commissions, directing the affairs of as many scientific societies, and building up the college at South Kensington in his own image. His house in St John's Wood, adorned with oil paintings and marble busts of its master, had become a salon that Matthew Arnold and the Russian writer Ivan Turgenev often attended.[31] Huxley was also serving on the new London School Board, pressing for a rounded, wide-ranging syllabus for the children of the capital. Effete classicism would make way for practical, useful science, but Huxley was agnostic, not an atheist, so he preserved the moral teachings of the Bible: 'I do not advocate burning your ship', he explained, 'to get rid of the cockroaches'.[32] It was all too much, and too much at a frantic pace, so that by the winter of 1871–2 he was as 'tired as two dogs' and 'grumbling like an ungreased block'. When the fall came, it 'came all at once'. He was sick, incapable of lecturing, even of writing. And when he would not help himself, his wife took charge: Nettie got him two months of leave and booked him a berth on a steamer bound for Egypt, while Darwin and friends raised a subscription worth £2,100 to cover the family's expenses in his absence. Hailed as a hero by the British consulate at Cairo, as one of the great champions of British science, Huxley received a guided tour of the Bulaq Museum, where a sculpture from old Memphis showing 'man leading the apes' caught the eye. He returned to Nettie 'brutal in health' and 'burnt to copper colour' by the desert sun.[33]

Before all this, however, Huxley had gone back to dinosaurs. For more than twenty years, fossilised remains from the Isle of Wight, which naturalists including Richard Owen had ascribed to a young *Iguanodon*, had lain almost forgotten in the British Museum. But when the clergyman William Fox announced that he had found a 'new' species of *Iguanodon* that he could distinguish from previous

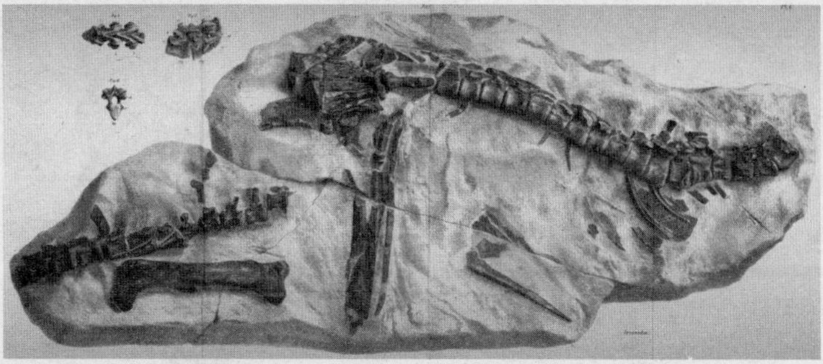

The remains of the *Hypsilophodon*, which Huxley distinguished from the *Iguanodon* in 1867

specimens, Huxley had reason to look again at the Isle of Wight fossils.[34] Now, comparing them to Fox's bones, he discerned distinctive features that others had not. There were none of the serrations that crowned an *Iguanodon*'s teeth but 'strong ridges of enamel' instead; the vertebrae were 'quite different'; and this animal had 'at least four well-developed toes', one more than the *Iguanodon*. These dinosaurs might look 'very similar', but he concluded they were 'perfectly distinct'. At a meeting of the Geological Society, therefore, Huxley announced that the long-neglected bones in the British Museum belonged to a novel species. Borrowing the name that an Austrian zoologist had given to certain species of iguanas, Huxley called this new dinosaur, which was 'mainly herbivorous' and 'about 4½ feet' long, the *Hypsilophodon*.[35]

On the very same evening, Huxley declared a second great finding. Resuming the theme of his seminal paper on the *Archaeopteryx*, he presented 'further evidence of the affinity between the Dinosaurian reptiles and birds'. And whereas the previous examination had focused on the pelvis and clavicle, Huxley now looked at the tibia, at the shin-bones of the *Megalosaurus* and the *Iguanodon*. Referring to the work of French colleagues and of Edward Drinker Cope in America, and building on the 'discovery' of the *Hypsilophodon*, he acclaimed 'the ornithic affinities of the *Dinosauria*' once more. With immaculate

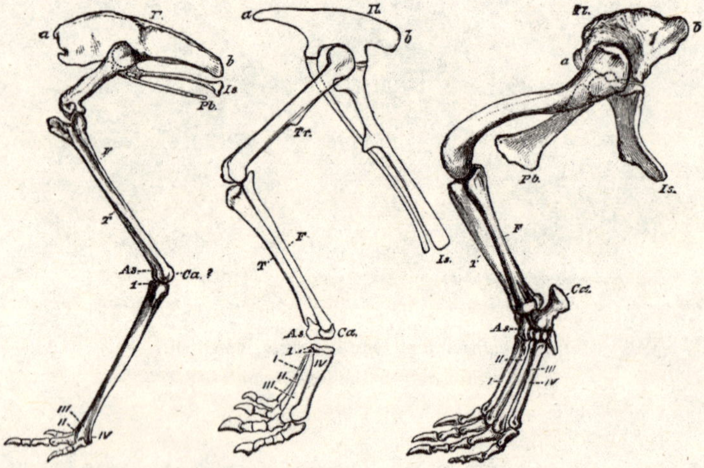

Huxley's comparison of the hip-bones of birds, dinosaurs, and crocodiles

plates illuminating his mastery of these ancient anatomies, Huxley called on the palaeontologists and zoologists of the 1870s to look at the legs of a fowl and ask 'by what test they could be distinguished from the bones of a Dinosaurian'. If the hind-quarters of a chicken could be 'enlarged, ossified, and fossilized', he declared, 'they would furnish us with the last step of the transition between Birds and Reptiles; for there would be nothing in their characters to prevent us from referring them to the *Dinosauria*'.[36]

Darwin was buying into the argument too. Dinosaurs had appeared in the *Origin* by the third edition, where he deployed the extinction of these 'great monsters' as proof that 'bodily strength' alone would not guarantee 'victory in the battle of life', since the process of natural selection was always more complex than mere brutality.[37] But in the sixth and final edition that appeared in 1872, Darwin turned to their evolutionary importance. He noted how Huxley had used 'the ostrich and extinct Archeopteryx' and 'the Compsognathus, one of the Dinosaurians', to bridge 'even the wider interval between birds and reptiles'. And while he recognised that 'some writers have objected to any extinct species, or group of species, being considered as intermediate between any two living species', he thought that the fossil record was clear: 'many fossil species certainly stand between living species, and some extinct genera between living genera'.[38] The difference from 1859's first edition, in which Darwin had expressed serious scepticism of the fossil record, was testament to the discoveries of the 1860s.

Even opponents of Darwinian theory were beginning to accept the evolutionary significance of dinosaurs. John Phillips, for instance, had coined the term 'Mesozoic' to describe that great age of reptiles, and in *Life on the Earth* (1861) he had criticised natural selection in detail, arguing that the Lord had intervened repeatedly in natural history to create each new species. 'Nature', he wrote, was 'the expression of a DIVINE IDEA', and he ridiculed the notion that species were 'the fortunate offspring of a few atoms of matter, warmed by . . . a spark of electricity, or an accidental ray of sunshine'.[39] By 1871, however, Phillips had not only announced the discovery near Oxford of a massive thigh-bone that belonged to the *Cetiosaurus*, but even conceded the anatomical similarities between birds and dinosaurs. In his *Geology of Oxford and the Valley of the Thames*, he observed that the *Megalosaurus* was 'essentially reptilian; yet not a ground-crawler, like the alligator, but moving with free steps chiefly, if not solely, on the hind limbs, and claiming a curious analogy, if not some degree of affinity, with the ostrich'.[40] Inviting Darwin to Oxford 'to come & see my Cetiosaurus' was almost his last act: only a few weeks later, Phillips achieved 'the ne plus ultra of academic distinction' when, after a presumably generous dinner at All Souls College, he fell down the stairs to his death.[41]

With dinosaurs figuring ever more prominently in these fundamental debates over the origins and development of life, scientific journals were devoting greater space to their study. The *Geological Magazine* teemed with new interpretations of the fossil record, while the Palaeontographical Society published dozens of monographs on its members' work. All the while, Richard Owen, now approaching seventy but never mellowing, spewed venom at the radical researchers who would carry off his science into secular terrain; *Vanity Fair* would call him 'Old Bones', a decrepit, possessive, and bitter old man.[42] In the wake of this drama, there was also a growing general readership for dinosaurs. In *Life in the Primeval World* (1872), the popular science writer William Henry Davenport Adams waxed lyrical about eleven great saurians. With striking, dramatic illustrations – many of them hilariously inaccurate, even for the time – he described the 'awful jaws' of the crocodilian *Teleosaurus*, the 'formidable' muzzle of the *Megalosaurus*, the armour of 'great osseous plates' that marked the spine of the *Hylaeosaurus*, and a *Plesiosaurus* that was 'monstrous . . . according to the human ideal of beauty'.[43]

British palaeontologists were also making new discoveries. Harry Seeley – no relation to the author of *Ecce Homo* – was the son of a bankrupt goldsmith who had 'ruined himself with scientific experiments' and so, like Huxley, he had raised himself through education, listening at the School of Mines and studying at the Working Men's College.[44] There followed jobs at the British Museum and with Adam Sedgwick at Cambridge and, by the late 1860s, despite a serious mental breakdown during the middle of the decade, he had become one of Britain's leading experts on dinosaurs: it was Seeley who would soon realise that dinosaurs fell into one of two great groups, the reptile-hipped Saurischia and the bird-hipped Ornithischia.[45] During his time at Cambridge, Seeley not only went fossil-hunting himself, but at the Woodwardian Museum he received specimens that amateurs had uncovered elsewhere in the earth. East Anglia had played host to a great 'rush' for coprolites in the 1850s but now, as Cambridge caught up with its ancient rival to the west, the fens and marshes were yielding more exotic fossils.[46] From the greensand that lay around the university there was some 'horrid' spinal plating that Huxley had elsewhere ascribed to a Cretaceous ankylosaur that he called *Acanthopholis*, meaning 'spiny scales'.[47] There were the jawbones of ichthyosaurs, the teeth of iguanodons, and pterosaurs besides. And from the Kimmeridge clay that surrounded the city of Ely, dug from the earth by the rector of Cottenham, there was a series of bones – vertebrae from the tail, broken shins, fore-limbs, and digits – that in Seeley's view were 'such as might pertain to one animal'.[48] Their provenance is hotly disputed now, but then, in honour of the supposed size of the beast, Seeley christened it the *Gigantosaurus*.

In these ways, by the early 1870s, dinosaurs had embedded

themselves within the practice and theory of British science at universities, in museums, and on the printed page. In turn, Britain's palaeontologists and zoologists had deployed the fossil record and the evident relationship between dinosaurs and birds to construct a system of evolution and descent that biblical apologists – despite the best efforts of Owen, Thomson, and Argyll – were finding more difficult to refute. Now, new questions arose. For most of the nineteenth century, reaction to radical science had been less explosive where that science was esoteric, and where debate was confined within cloistered walls that could limit the spread of the more dangerous ideas. But with Darwin discoursing freely on humankind's fraternity with the animals, and with a booming popular market for scientific literature, how far could Huxley and John Tyndall now carry their culture war? How many more institutions might fall to their secular campaign? There was also a question of international science because, for all its discoveries and discussions of dinosaurs, and for all the inches and ink that publishers now devoted to them, Britain was no longer the centre of the palaeontological world. Huxley, Owen, and Seeley had not ceded their primacy to the new German Empire, not even after the profusion of Bavarian fossils. Instead, British chalk and limestone were surrendering to the valleys and canyons of the American Badlands.

PART FIVE

Entente

DRAMATIS PERSONAE

Charles Bradlaugh, journalist and politician, London
Edward Drinker Cope, palaeontologist, USA
Charles Darwin, naturalist, Kent
Thomas Huxley, naturalist, London
Othniel Charles Marsh, palaeontologist, USA
Richard Owen, naturalist, London
John Tyndall, physicist, London
Alfred Russel Wallace, zoologist, London

22
Learned Baboons

> No man was there. Huge elephantine forms, the mastodon, the hippopotamus, the tapir, antelopes of monstrous size, the megatherium, and the mylodon – all, for the moment, in juxtaposition. Farther back, and overlapped by these, were perched huge-billed birds and swinish creatures as large as horses. Still more shadowy were the sinister crocodilian outlines – alligators and other horrible reptiles, culminating in the colossal lizard, the iguanodon. Folded behind were dragon forms and clouds of flying reptiles; still underneath were fishy beings of lower development; and so on, till the life-time scenes of the fossil confronting him were a present and modern condition of things.[1]
>
> Thomas Hardy, *A Pair of Blue Eyes* (1873)

In 1874, the relationship between science and religion bore the strain of three declarations of war. The first shot rang out from New York, and John William Draper fired it. Fourteen years before, Draper had given the keynote speech at the Oxford meeting of the British Association that provoked the infamous debate between Huxley and Samuel Wilberforce. And although Draper appeared to care little for his own speech, not even mentioning it in his autobiography, the memory of that debate had stayed fresh in his mind. His academic career had flourished. He had been a professor at, and then president of, New York University's School of Medicine; he had given his name to the temperature at which all solids glow red; and he had even found time to write a three-volume history of the American Civil War. But it was the intellectual history of Europe, and the supposed antipathy between science and religion, that enthused Draper most of all. In 1874, therefore, provoked in part by Pius IX's proclamation of papal infallibility, Draper published his *History of the Conflict between Religion and Science*, a sweeping polemic which depicted the two great disciplines as partners in an age-old dance for supremacy.[2]

Draper moved from the library at Alexandria to the 'pillared halls of Persepolis', from the Arabian schools of mathematics to the courts of the Inquisition, and from Renaissance universities to the learned societies of London. There was no question of his favour: 'The history of Science', he declared, 'is not a mere record of isolated discoveries; it is a narrative of the conflict of two contending powers, the expansive force of the human intellect on one side, and the compression arising from traditionary faith and human interests on the other'. Of that 'traditionary faith' he was unsparing, describing the development of Catholicism as an 'intellectual night' which settled on Europe, during which spiritual affairs passed from the control of classical philosophers 'into the hands of ignorant and infuriated ecclesiastics, parasites, eunuchs, and slaves'. At last, however, that night was lifting, and civilised society had recognised the truth: 'that Roman Christianity and Science are recognized by their respective adherents as being absolutely incompatible; they cannot exist together; one must yield to the other; mankind must make its choice – it cannot have both'.[3]

Joining Draper in combat was Andrew Dickson White. A historian, the collector of more than 30,000 books, and the founder and president of Cornell University, White had already lectured on 'the battle-fields of science', arguing that 'interference with Science in the supposed interest of religion . . . has resulted in the direst evils' and that 'all untrammelled scientific investigation . . . has invariably resulted in the highest good'.[4] Now, writing in June 1874 for the *Popular Science Monthly*, a new magazine which had attracted contributions from Huxley and Spencer among others, White opined that the charge of godlessness had been made 'against every great step in the progress of science or education': that Luther had damned Copernicus as a blasphemer, and that the Dean of York had once 'declared the new science of Geology a study invented by the devil, and unlawful for Christians'. These objections, for White, were dogmatic nonsense, since history showed that 'every study which tends to improve the industry of mankind makes a man nobler and better'.[5] He would publish an expanded exposition of this theory as *The Warfare of Science* (1876), which promised 'an outline of the great, sacred struggle for the liberty of science, a struggle which has lasted for so many centuries, and which yet continues'. This was 'a war waged longer, with battles fiercer, with sieges more persistent, with strategy more shrewd than in any of the comparatively transient warfare of Caesar or Napoleon or Moltke'.[6]

When the London publisher Henry King brought out a British edition of White's *Warfare of Science*, he asked the physicist and senior X clubman John Tyndall to write the foreword. Over the previous decade, Tyndall had devoted himself to the Royal Institution on Albemarle Street, working closely, even reverently, with the great Michael Faraday. When Faraday died in 1867, it was Tyndall who took command of the Institution, Britain's great bastion of public education in the sciences. Outside of term time, Tyndall had scaled other heights. Having joined Huxley on his Alpine sojourns in the 1850s, he became one of the first people to climb the deadly Matterhorn, having been the very first to ascend the nearby Weisshorn a few years before; glaciers and mountains around the world bear his name as testimony to these feats. Yet as Tyndall pushed human limits in the mountains and scientific boundaries in the laboratory, breaking new ground in the fields of diamagnetism and thermal radiation, his hatred of dogma – especially Catholic dogma – endured. In his preface to White's pamphlet, he pleaded for the attention of 'intelligent Catholics both in England and Ireland'. He foreswore the idea of conquest, nor did he come to preach 'a creed hostile to their own' or to force a change in 'their spiritual masters'.[7]

Rather, Tyndall begged that they should read, consider, and understand White, who would show them that 'against the benefits which religious associations have conferred upon humanity stands a vast debit of committed wrong'. Tyndall directed this invective at 'the Roman Church', an enemy which aimed 'at the revival and perpetuation of the wrong, striving after a domination which she never fitted herself to exercise'. Should this enemy prevail, he warned, and should societal ignorance allow the Church's malevolence to work unchallenged, it would 'only bring calamity upon the human race'. He closed by appealing to 'the intelligence within her own pale', pleading that White's readers would not 'place the sciences that have given us all our knowledge of this universe at the mercy of their hereditary foe'.[8] It was an eloquent and emotional prayer for science; it was also a declaration of scientific fraternity, a sign that the cause of secular science had become a transatlantic phenomenon. Yet it was a mere postscript to Tyndall's most stirring intervention in the apparent conflict with religion.

In the summer of 1874, the British Association had planned to hold its grand annual meeting in Belfast and, with Tyndall the leading Irish scientist of the day, he was the obvious candidate for the presidency. This meant, of course, that Tyndall would deliver a presidential address. Typically, these speeches would recount the achievements and discoveries of the past year, assess the state and health of scientific inquiry in the British Isles, and identify the points and issues of interest to which the nation's chemists, biologists, physicists, palaeontologists, and general naturalists could devote themselves in the coming twelve months. But even Huxley, the great agitator, worried over what Tyndall could ignite in Belfast, a politically fractious city and a bulwark of 'hot' Protestant literalism. 'I wonder', he wrote to his friend, 'if you are going to be as wise and prudent as I was at Liverpool [in 1870] . . . I declare I have horrid misgivings of your kicking over the traces.'[9] Hooker and his fellow X clubman William Spottiswoode worried too, for they thought it 'impossible either officially or unofficially to advise Tyndall'.[10]

These fears were not groundless. As the members of the Association gathered in the Ulster Hall, a fine new venue that was 'all but unrivalled [in Belfast] as an edifice for the production of musical works', Tyndall unfurled the standard of secular science, declaring war on religious interference and demanding 'the freedom' to discuss even the most radical ideas 'whether right or wrong'.[11] Beginning with Democritus, Epicurus, and the naturalists of ancient Greece, men who battled a 'mob of gods and demons', Tyndall swept through the history of all science. From Copernicus and Galileo to Newton and Descartes, he laid out the basis of scientific materialism: that 'the infinity of forms under which matter appears were not imposed upon it by an external artificer'. Then came Kant, Clerk Maxwell, Huxley, and Darwin: 'All great things', Tyndall explained, 'come slowly to the birth'. From Darwin, he argued, came two choices only: 'Either let us open our doors freely to the conception of creative acts, or, abandoning them let us radically change our notions of Matter'.[12] There was God in nature, or there was not.

As Tyndall approached the close, he ventured to define the border between science and religion and the strict conditions under which they could engage. Religion? Art? Poetry? These belonged to 'the region of *emotion*'. This is not to say that he wished to divorce science from literature; indeed, he presented Huxley as a man of science whose 'clearness and vigour of literary style' was the match of any great man of letters. Nor was it his belief that an educated person could honour Newton but not Shakespeare, Darwin but not Carlyle; it was simply that the latter of these, where they walked with the poetic and the beautiful, had perfected '*creative* faculties of men' that were distinct and separate from the '*knowing*'. Appealing to members of his audience who had 'escaped from these religions into the high-and-dry light' of the intellect, Tyndall proposed the terms on which the sublime and the spiritual could meet the empirical: 'All religious theories, schemes and systems, which embrace notions of cosmogony, or which otherwise reach into the domain of science, must, in so far as they do this, submit to the control of science, and relinquish all thought of controlling it.'[13] It was a declaration of scientific independence and a speech that, ever after, historians and scientists alike would call simply 'the Belfast Address'.[14]

The outrage was immediate. The Presbyterian elders of Belfast, scandalised by this profession of infidelity within their jurisdiction, took to their pulpits to denounce Tyndall and his allies. The minister at the church on Rosemary Street, for instance, thundered against this 'covert assault on Christianity', accusing Tyndall of an 'abuse of office', and over the coming winter that church would hold a series of lectures to save the city's soul and to denounce the 'repulsive moral implications of a materialistic naturalism'.[15] Belfast's Catholic papers were no less savage. The *Ulster Examiner* called the British Association 'a collection of learned baboons' and lamented that Tyndall should have concerned himself with demolishing the 'great plan of creation'.[16] Other critics dragged Huxley into their obloquies on Tyndall, resolving that the two men were 'ignoring the existence of God, and advocating pure and simple materialism'.[17] Neither of them objected to that description, and in the preface to the printed version of his address Tyndall confessed that he did not have 'any real reason to complain of the revilings addressed to me'.[18]

Despite their fears over what he could provoke, Tyndall's friends delighted in the speech. Darwin, who read the proofs while on holiday at Southampton, thought it 'excellent, & as clear as light'. He was 'most deeply gratified' by the passages which touched upon his own work, though confessed that he could not 'even fancy surviving such a week of excitement'.[19] Charles Lyell, now in his late seventies, thought that 'whatever criticisms may be made o[f] Tyndall, [it] cannot be denied that it was a manly & fearless outspeaking of his opinions'. He also took stock of the changing times. 'I feel certain', he reflected, 'that at one of our early meetings of our association, such an address as that of Tyndall & such free discussions as have lately been welcomed on the nature of matter & force would not have been tolerated'.[20] In those early years of the Association, in the halcyon years of clerical and gentlemanly science, would an agnostic Irishman such as Tyndall have been permitted to speak at all? And if not Tyndall, what chance for an atheist such as Charles Bradlaugh?

Since sparring with the government in the 1860s, Bradlaugh's war on religion had burned brighter. There had been new editions of his inquiries into the holy scriptures, where he deplored how the 'nurse and pedagogue' had forced the moralities of Genesis upon 'the child's brain'. What was more 'fearfully horrible', he wondered, than all the men of Sodom demanding to rape the two angels who were visiting the city, and Lot offering up his two daughters as preferable victims with the words, 'Do ye to them as is good in your eyes'? Borrowing from Bishop Colenso of Natal, he savaged the nonsense that '600,000 armed' Israelites, a mighty host that was 'nearly nine times as great as the whole of Wellington's army at Waterloo', would ever have suffered the indignity of enslavement under the pharaoh.[21] And he lampooned the idea that trumpets and the Ark could have collapsed the walls of Jericho, regretting 'that our Christian friends did not try the experiment at Sebastopol' (during the Crimean War).[22] Later, he 'revered [the] memory of those who, with the blood-drops from their wounded feet, had softened the rugged path along which all must travel'. Atheism had its martyrs – 'from Vanini to Carlile, from faggot to dungeon' – and Bradlaugh would honour them.[23]

There was now a sharper, ideological aspect to Bradlaugh's politics and in the early 1870s his sympathies with the Irish Fenians and the Parisian Communards had blossomed into the full embrace of radical republicanism. In *The Land, The People, and the Coming Struggle* (1871), he urged British landlords to generosity and clemency towards poachers lest, in the conflict they would otherwise incite, 'the landless . . . claim political power and use it as a weapon, as did their French brethren' some time before.[24] The next year, in *The Impeachment of the House of Brunswick* (1872), which was 'a vitriolic review of the careers of the Hanoverian monarchs', he argued that 'the right to deal with the throne is inalienably vested in the English people' and called for 'the repeal of the Acts of Settlement and Union'.[25] What would it cost the people, Bradlaugh asked, to rid the land of these German carpetbaggers? He did not advocate violence, instead urging change 'by the school, the brain, the pen, and the tongue, and not heralded by the cannon's roar', but the language was dangerous.[26] Working with G. W. Foote, a fellow secularist, Bradlaugh also established the National Republican League to pursue the creation of 'a

Three of G.W. Foote's scandalous *Comic Sketches* of biblical history

state which guarantees the fullest individual liberty compatible with the general security'.[27] They did not get far.

The National Secular Society therefore remained Bradlaugh's primary vehicle for agitation, and he was recruiting effectively. Annie Besant was twenty-six, an avowed atheist, and already a figure of minor scandal. Having married the evangelical churchman Frank Besant, the brother of the historian Walter, she and her husband had separated when, for fear of disgrace in his parish, Frank forced her to choose: 'I should either resume attendance at the Communion, or should not return home.' In the choice between 'hypocrisy or expulsion', Besant chose the latter, taking their daughter Mabel with her. Bradlaugh first met Besant in 1874 after a lecture on 'the ancestry and birth of Jesus' at the Hall of Science; on that evening, she 'felt the sway of the orator, [and] the power that dwells in spoken words' for the first time. She and Bradlaugh spoke briefly, then again a few days later on the street: 'We found that there was little real difference between our theological views.' Within weeks, Bradlaugh had offered Besant work on the *National Reformer* and, adopting the *nom de guerre* of 'Ajax', she became his most reliable lieutenant. It would be a friendship and collaboration of mutual respect: 'It will be a good thing for the world', she wrote, 'when a friendship between a man and a woman no longer means protective condescension on one side and helpless dependence on the other'.[28]

Besant rivalled Bradlaugh as a presence on the platform. Her first public speech was on 'the political status of women', her second on 'the true basis of morality', and by early 1875 she had made up her 'mind to lecture regularly'.[29] Though nerves would strike Besant before each talk – she compared those moments to crossing the threshold of a dental surgery – the crowds would flock to listen. The academic Beatrice Webb thought she had 'the voice of a beautiful soul', while T. P. O'Connor, the Irish nationalist MP, hinted at other attractions: 'What a beautiful and ... irresistible creature she was,' he reminisced, 'with her slight but full and well-shaped figure, her dark hair, her finely chiselled features ... [and] with that short upper lip that seemed always in a pout'.[30] She toured from Birkenhead to Glasgow and from Aberdeen to Durham, taking postal trains overnight and making friends among the 'shrewd and hard-headed' miners of the north-east. By 1877, she and Bradlaugh had gone as business partners into the Freethought Publishing Company, vowing 'all that we do publish we shall defend'.[31] This was no idle comment, for Besant and Bradlaugh had agreed to republish *The Fruits of Philosophy*, an 1830s manifesto for birth control by the American doctor Charles Knowlton.

Here, Knowlton had argued for Malthusian population management 'without sacrificing the pleasure that attends the gratification of the reproductive instinct'. He had described the reproductive anatomy in explicit detail, prescribed herbal remedies for impotence and infertility, and then advised on a remarkable 'check' to pregnancy: 'syringing the vagina immediately after connection, with a solution of sulphate of zinc, of alum, pearl-ash, or any salt that acts chemically on the semen'.[32] With the previous British publisher of the book having been convicted of obscenity, its reprinting would almost certainly provoke the wrath of the censors, but Bradlaugh declared at the Hall of Science that 'he had fully counted the cost; he knew all he might lose'. An advertisement in the *National Reformer* therefore announced that the book would go on sale from 4 p.m. on the last Saturday in March 1877, and that Besant and Bradlaugh would be there in person, at their premises on Stonecutter Street, to hand over the first hundred copies. It was masterful marketing: customers crowded the shop and within twenty minutes they had sold 'over 500 copies'.[33] The advertisement was also a test of prosecutorial zeal, and the authorities did not disappoint. The inevitable arrest of Besant and Bradlaugh on charges of obscenity soon excited attention across the country: the *Edinburgh Evening News* observed the readiness of the Bow Street magistrates 'to convict . . . all boys or persons selling the book', while the *Leicester Daily Post* noted that 'those who keep watch and ward over the morals' of the capital had provoked a 'jolly row', with 'Mr Bradlaugh and Mrs Besant [doing] their best to shew the Government that their new Gospel of sparse population has as much right to flourish as any other creed'.[34]

With a trial date set for mid-June, Bradlaugh had in mind a star witness for the defence. 'I want to subpoena you', he wrote to Charles Darwin: 'Would you kindly say when it will be convenient for me to send & serve you?'[35] Besant explained that the defendants wished 'to use passages from his works' but it was a hopeless request, for there was no chance of Darwin leaving Kent for the courts.[36] 'I have been for many years much out of health & have been forced to give up all Society', he protested. Besides, the Knowlton pamphlet appalled him. 'I suppose that it refers to means to prevent conception', he replied to Bradlaugh, and 'if so I shd be forced to express in court a very decided opinion in opposition to you'. But why should Darwin, on whom Malthusian ideas had exerted such an influence, repudiate the *Fruits of Philosophy*? It was probably because, even though he had spent twenty years exploring the borderlands of acceptable discourse, Darwin was still socially conservative, even prudish: 'I believe that any such [contraceptive] practices', he explained, 'would in time spread to unmarried women & wd destroy chastity, on which the family bond depends; & the weakening of this bond would be the greatest of all possible evils to mankind'.[37] It was a brutal lesson for Bradlaugh that, despite regarding himself as Darwin's ally in a culture war against the old religious assumptions, there were other conditions – differences within secular moralism, perhaps class as well – that still held sway in Britain.

The trial itself was a formality, for although the jury 'exonerate[d] the defendants from any corrupt motives in publishing' the *Fruits*, it was 'unanimously of opinion that the book in question [was] calculated to deprave public morals'. There followed fines of £200 apiece and a sentence of six months, with only a technicality – the Court of Appeal ruled that the original arrest warrants had been unlawful – sparing Besant and Bradlaugh from prison. There would be no further luck for Besant, however, when her clergyman husband Frank sued for permanent custody of his daughter, Mabel. 'The said Annie Besant', he stated in his petition to the courts, 'is endeavouring to propagate the principles of Atheism'. The judge who heard the case quite agreed, concluding that atheist motherhood was 'not only reprehensible but detestable, and likely to work utter ruin to the child'. He therefore placed Mabel, only eight years old, in the exclusive custody of her father. 'The year 1878', Besant reflected, 'was a dark one for me'.[38] Thus did these events suggest that, regardless of the progress that Huxley and Tyndall made in carving a secular niche for science, and no matter that practising science now appeared – almost incredibly, given the past several decades – to be a tolerable excuse for undermining religious assumptions, there endured a real and present danger for those Britons who would blaspheme for the sake of blasphemy, or merely to test what was socially acceptable to say.

The fate of G. W. Foote, a fervent colleague of Besant and Bradlaugh, was another case in point. In 1881 he would found the *Freethinker*, a newspaper that he proclaimed as 'an anti-Christian organ' which would 'wage relentless war against Superstition in general, and against Christian Superstition in particular'.[39] The *Freethinker*'s unique quality – and Foote's particular crime – was the production of weekly cartoons which mocked miraculous scenes from the Bible without mercy. It was a novel and effective medium and, to the end of persuading his subscribers that mere words had fooled Christians, Foote quoted Herbert Spencer approvingly: 'the image of the Almighty hand launching worlds into space is very fine until you try to form a mental picture of it, when it is found to be utterly unrealisable'. In the same way, Foote thought the narrative of Creation as passable 'until you image the Lord making a clay man and blowing up his nose'. And so there followed weekly 'comic sketches' of the prophet Samuel, as a barber, pouring oil over the head of his customer Saul; of Elisha summoning the two she-bears to tear asunder the children of the town of Bethel; and of Adam taking fright as the fearsome beasts of Genesis sought to challenge his dominion over natural life.[40] Foote would serve twelve months in prison for these impious images of the ancient Earth, of which more and more evidence – as opposed to biblical descriptions – was emerging from the rocks of the American West.[41]

23
Badlands

> The truth seems to be that a long line of disillusive centuries has permanently displaced the Hellenic idea of life, or whatever it may be called. What the Greeks only suspected we know well; what their Aeschylus imagined our nursery children feel. That old-fashioned revelling in the general situation grows less and less possible as we uncover the defects of natural laws, and see the quandary that man is in by their operation.[1]
>
> Thomas Hardy, *The Return of the Native* (1878)

The spring of 1870 had been 'filled with promise and good news', when thirteen men – professors, students, and alumni of Yale – embarked for the American West. This was the university's inaugural Scientific Expedition, the first time that it had funded extended periods of research in the field, and its mission was clear: to scour Wyoming, Utah, and Montana, American territories which had not yet become states of the Union, for the fossils of dinosaurs and whatever other prehistoric creatures had once resided in those vast and rocky Badlands. Their destination was hostile country, dangerous to the unarmed man, and unknown to most Americans. Although the Homestead Act had opened the West to settlers, and though federal forces were pursuing the 'manifest destiny' of the nation by securing these forbidding expanses for the United States, several Indian peoples had mounted stout resistance. 'I don't like the idea of your campaign to the west', urged the friend of one explorer; 'There is no great necessity to exposing one's hair when [the army] will quiet the rascals in a month or two.'[2] The leader of the expedition was Othniel Charles

Marsh, Yale's professor of palaeontology. At the time that he and his colleagues left New England, Marsh was thirty-eight, sturdily built, balding from the front, and possessed of a magnificent beard and moustache that tended 'toward red'. He was the very image of Charles Kingsley's muscular Christian, someone 'strong in body, able to haul a dredge, climb a rock, turn a boulder, walk all day uncertain where he shall eat or rest; ready to face sun and rain, wind and frost'.[3]

Marsh hailed from New York farming stock, growing up only a few miles from Niagara Falls. His immediate family was far from rich, often coming close to failure, and young Othniel spent little time at home. He preferred to hunt and fish and, in the earth thrown up by the Erie Canal, to search for minerals and fossils. At the age of nineteen, money from his late mother's family – his uncle was the financier George Peabody – allowed him to study formally. Spending his winters at Phillips Academy and Yale, and his vacations among the rocks of New England and Nova Scotia, Marsh made his first significant discovery in the summer of 1855 when, in the coalfields of eastern Canada, he found fossilised vertebrae that he ascribed to a new marine lizard, an *Eosaurus*.[4]

Like many young Americans who could, Marsh had spent the early 1860s in Europe, avoiding the draft while honing his craft at Breslau and Berlin. There, by chance, he met another American, Edward Drinker Cope, for the first time. It was an encounter which bred friendship and, for the rest of the 1860s, the men were sincere in signing letters to each other with 'best wishes'; Cope would name a lizard after Marsh, who returned the favour by naming a species of *Mosasaurus* after Cope. While in Europe there were trips to England, too. Marsh paid homage to Darwin in Kent and took keenly to Huxley, whom he lauded as 'a guide, philosopher, and friend, almost from the time I made [the] choice of science as my life['s] work'.[5] He returned to the United States and to Yale in 1866, lured home by the end of the Civil War and his uncle's endowment of the Peabody Museum of Natural History. There was little debate over who would direct the new museum's affairs and spend its new funds, some $3 million in today's money.

In these early years there was work enough for Marsh, both at Yale and around New England. In 1869 he debunked the 'Cardiff Giant', a ten-feet-tall fossil of a 'petrified man' that workmen had found in upstate New York. Far from the alleged proof that biblical giants, the Nephilim, had once walked the Earth, it was a cheap ruse, a carving made from an Iowan block of gypsum. It took Marsh only a few moments to conclude that it was 'a most decided humbug'.[6] He knew that nothing made of gypsum, a soft and soluble form of calcium, could have survived from antiquity; besides, 'tool marks', evidence of recent craftsmanship, were 'still very distinct on different parts of the statue'.[7] When the showman P. T. Barnum failed in his efforts to purchase the 'Giant', he made a fake of the fake and earned a small fortune by putting it on display in New York City: 'There's a sucker born every minute', he is alleged to have said. By the next year, Marsh was throwing Yale's money at the quarries and workmen of New Jersey, offering payment in cash for the first news of fresh finds in the fossil-beds. Cope joined these auctions, and it drove less wealthy fossil-hunters to despair: Joseph Leidy, who had described the *Hadrosaurus*, regretted that where 'formerly every fossil one found in the State came to me ... Professors Marsh and Cope, with long purses, [now] offer money ... and in that respect I cannot compete with them'.[8] A more important incident in Marsh's development occurred in Nebraska. During a brief trip that served as reconnaissance for the later Yale expeditions, he found 'a new and diminutive species of fossil Horse'. Naming it *Equus parvulus*, he described a creature that was 'scarcely a yard in height' and whose 'slender legs' were 'terminated by three toes'.[9] It was, he thought, an ancestor of the modern horse and further fossil evidence of evolution.

Marsh was not an easy person. A landlady from his student years described him as 'very odd' and thought that getting to know him was like 'running against a pitchfork'.[10] Another acquaintance remarked that, having spent so little time at home, 'he developed wholly without the influence of family', something that profoundly informed his adult life. One such consequence was his inability to trust another person, and a close ally observed that 'even where perfect confidence existed, he seldom revealed more about any particular matter than seemed to him necessary or than the circumstances really demanded'.[11] It was not perhaps surprising that Marsh, whom the

Yale president Timothy Dwight described as 'a man quite by himself', remained a bachelor all his life, or that he could bring rancour to academic debate.[12]

Before he led the Yale expedition west on the Union Pacific railroad, the new artery of American expansion that Irish navvies had been laying across the plains, Marsh had procured a letter of introduction from General Philip Sheridan of the United States Army. A veteran of the Appomattox campaign that had closed the Civil War, Sheridan bade 'all officers of the Army' to grant safe passage to the scientists. And so, as Marsh and his colleagues crossed the Mississippi, and as the vastness of the American West stretched out before them 'in every direction as far as the eye could see', they collected armed escorts from US Army forts along the way.[13] The expedition 'greatly puzzled' the soldiers and Marsh endeavoured in vain 'to explain to them the mighty changes of geology', but it was necessary company: 'this was the country of the Sioux, and that warlike tribe', reported one of the travellers, 'was now in a state of unusual excitement'.[14] Among their guardians was William Frederick Cody, the young cavalryman who in later life, as 'Buffalo Bill', became the famous showman of the rodeo.

As the Yale men explored the bluffs and boulders, armed with their revolvers, knives, and geological hammers, they thought of themselves as the heroes of an ancient odyssey. In the baking heat of the American summer 'while the thermometer stood at 110 degrees in the shade of wagons', they marched over burning sand hills from dawn until dusk, mirages of lakes mocking their thirst. 'What *did* God Almighty', asked one of the soldiers, 'make such a country as this for?' When night closed over, when the empty darkness of the plains fell upon them, they 'felt "in for" something more than science'. In a land 'infested by hostile Indians . . . they saw around them the tents, the bivouac fires, the soldiers standing in picturesque groups, the horses cropping in the twilight, the corral of wagons, and pacing sentinels beyond'. These tales of the Badlands, of danger and daring, found homes in the pages of New York's *Weekly Herald* and *Harper's Magazine*.[15] Yet in spite of this heroic imagery and the sense of adventure, the expedition remained a mission of science, the fulfilment of one man's ambition; and Marsh, in the words of one rival, was 'more ambitious than Cope ever was; he [was] raging ambitious'.[16]

That ambition, of course, was to find, name, and describe fossils. And over several more expeditions during the early 1870s, in the reports that flashed back from the West to the newsrooms of New York, Marsh and his colleagues proclaimed a bonanza. The most important of these discoveries were the first American fossils of pterosaurs and of two prehistoric birds. The first of these, the *Hesperornis*, was a large, cormorant-like creature that Marsh found near the Smoky Hill River in Kansas; battling against 'hostile Indians' and 'extreme heat' which caused 'sunstroke and fever', a 'nearly complete skeleton' was 'ample reward' for the danger and hardship. The second creature, the *Ichthyornis*, was a 'delicate, long-winged bird, standing about eight inches high, that probably resembled a modern tern in appearance and habits'.[17] Its discovery was not Marsh's work alone – he was alerted to the first specimen by the Kansas pharmacist Benjamin Mudge – but it was even more sensational than the first: it had reptilian jaws and even teeth. Marsh declared that, between them, the two new animals were 'an important gain to palaeontology, [which] does much to break down the old distinction between Birds and Reptiles, which the Archaeopteryx has so materially diminished'.[18] These fossils would excite the attention of British science too. 'The richness of the western parts of the U. States in fossils', wrote Darwin to Marsh, 'seems quite unparalleled'.[19] Huxley quite agreed and thought there was 'no collection of fossil vertebrates in existence which can be compared to' Marsh's museum.[20]

The Yale man was not, however, the only prospector making tracks in the West. Since encountering Marsh in Berlin, Edward Drinker Cope had established himself as one of America's leading naturalists. He had announced the terrifying carnivore *Laelaps*, which he named after the mythological dog which always caught its prey; he became the first American to illustrate the time of the dinosaurs, depicting the Cretaceous period (which had followed the Jurassic) for the *American Naturalist*; and it was Cope who showed that the famous three-toed footprints in Massachusetts, long ascribed to birds, in fact belonged to dinosaurs. As we have seen, he had also been one of the first scientists anywhere to observe the affinity between birds and dinosaurs such as the *Compsognathus*. There had been missteps – he had, for instance, placed the skull of the *Elasmosaurus*, a cousin of the *Plesiosaurus*, at

Edward Cope's own illustration of the Cretaceous landscape

the end of its tail instead of its neck – but he and Marsh were a pair, the two great titans of American palaeontology. And by 1871, Cope was in Kansas too: 'I have been traveling about in this country', he wrote to his young daughter, 'looking for fossil bones of huge animals such as I bring home . . . to Haddonfield sometimes'.[21]

This meant, however, that Cope was now in direct competition with Marsh, and so their friendship frayed into enmity. They were bidding against each other for the favours of quarrymen, Cope having poached the services of one of Marsh's most valued assistants, and they were both taking 'the liberty of alluding to each [other's] "great mistakes"'. By the winter of 1872–3, this rivalry had degenerated into open warfare. When Cope wrote to Marsh on New Year's Day, informing his rival that he had acquired several specimens from Kansas that had been 'abstracted from one of [Marsh's] boxes', the Yale man exploded with rage and with accusations of theft, even after Cope had returned the fossils. 'Where [are] the rest?' he demanded, 'and how about those from Wyoming?' Had Marsh learned of this alleged larceny at any other time, he warned Cope, he would have 'gone for [him], not with pistols or fists, but in print'. When the two men met again, at a reception for the visiting John Tyndall in New

York, Marsh stuck to Cope 'like a leech' until satisfied that no robbery had occurred. This truce would not last.[22]

The pages of America's leading journals now filled, overflowing into appendices, with allegations of fraud, fabrication, and skulduggery. When Marsh accused Cope of gross malpractice by publishing papers before presenting them to a learned society, Cope denied any wrongdoing and blamed the journal for being 'careless and indifferent'. In return, he sought to belittle Marsh as the pedantic, obsessive 'Professor of Copeology at Yale'; in sorrow, he lamented receiving 'another lesson of the weakness and depravity of human nature'. As the historian of science Url Lanham relates, that education would continue when Marsh persuaded the American Philosophy Society to censure Cope, who recoiled at 'the meanness of the proceedings'. By the next summer, Marsh was mocking Cope's inconsistent descriptions of the prehistoric mammal *Eobasileus* as changing 'more rapidly than Darwin himself ever imagined for the most protean of species'; if Cope were right, Marsh laughed, 'such an animal belongs in the Arabian Nights, and not in the records of modern science'. As the tit-for-tat continued, Cope would protest that Marsh's criticisms were 'misrepresentations', that his rival's innovations were 'untenable', and that his allegations of impropriety were 'either criminally ambiguous or untrue'. He would even beg for peace, hoping that Marsh would be as glad as he to bury their disputes 'in oblivion'.[23] But the most infamous feud in the history of palaeontology, the Bone Wars, had only begun.

Though by the mid-1870s the Peabody Museum at Yale was occupying more of his time, Othniel Charles Marsh had as many as twelve crews in the field, and he was driving them hard: 'Get all the bones of every good specimen,' he ordered his men, 'if it takes a week to dig them out'.[24] And if early reports from his expeditions had hammed up the threat of Native American violence, Marsh was now on friendly terms with the Lakota and their chief, Red Cloud. Marsh, the chief was told, was 'a friend of the Great Father at Washington [the American president]', and 'if he was allowed to hunt for stone bones . . . he would be a good friend of the Sioux people'. It was a risky bargain to strike, potentially placing Marsh and his men at odds with the American government, but he kept his word. On returning to the east coast,

Marsh complained to the administration of Ulysses S. Grant about federal conduct towards the Lakota and even launched a press campaign to condemn the government's handling of 'the Indian question generally'. For Red Cloud, it was an uncommon display of American decency. 'I thought [Marsh] would be like all white men', he recalled, 'and forget me when he went away. But he did not. He told the great Father everything, just as he promised he would, and I think he is the best white man I ever saw.'[25]

As before, Marsh did not have the West to himself. Edward Drinker Cope had inherited $250,000 from his father – almost $7 million in today's money – and he was exhausting this fortune in searching for dinosaurs. Still in his thirties, Cope was vigorous, fit, and brave, and when the Sioux rode off to rout Custer at Little Bighorn, he and his crews stole into their lands to plunder virgin soil. And while he had allegedly earned 'the contempt of all the scientific men of the country', with, moreover, 'no woman ... safe within five miles of him', he inspired devotion in others.[26] When the fossil-hunter Charles Sternberg begged him for work in 1876, Cope had replied with a cheque for $100 and the instruction: 'Go to work'. In turn, Sternberg resolved 'to endure immeasurable hardships and privations in the barren fossil fields ... in the service of the greatest naturalist [that] America has produced', pledging that he would 'make a collection of [the West's] wonderful fossils no matter what it might cost me in discomfort and danger'.[27]

By now there was more than simple rivalry between Marsh and Cope; there was also intellectual difference. For Cope the Quaker, Darwinian theories and dinosaur fossils belonged to different fields of enquiry. As early as 1868, although he conceded that 'descent with modifications' was 'exceedingly probable', he had complained that natural selection was nowhere near powerful enough to explain the diversity of species. Instead, Cope put forward the Lamarckian argument that 'the will of the animal, applied to its body' wrought changes in a context where 'the will of the Creator' remained the most important factor.[28] Thus, despite having played a key role in identifying the kinship of dinosaurs to birds, Cope would not use Darwin to explain it.

Marsh, however, had become an avowed disciple of Darwin and Huxley. At the American Association for the Advancement of Science in August 1877, delivering the keynote speech at its Nashville meeting, he proclaimed: 'I am sure I need offer here no argument for evolution; since to doubt evolution to-day is to doubt science, and science is only another name for truth.' He took enormous pride in having helped to construct that truth as well: 'Birds have come down to us through the Dinosaurs,' he told his colleagues, 'and the close affinity of the latter with recent Struthious Birds will hardly be questioned. The case amounts almost to a demonstration if we compare, with Dinosaurs, their contemporaries, the Mesozoic Birds.' Invoking the authority of his British heroes, Marsh explained that '*Compsognathus* and *Archaeopteryx* of the Old World, and *Ichthyornis* and *Hesperornis* of the New, are the stepping stones by which the evolutionist of to-day leads the doubting brother across the shallow remnant of the gulf'.[29] But Cope doubted still. 'The doctrines of "selection" and "survival"', he insisted, 'plainly do not reach the kernel of evolution'; the development of species, he maintained, required 'the Creator of all things [who] has set agencies at work which will slowly develop a perfect humanity'.[30]

Thomas Huxley was witness to this fight. 'The whole nation', exclaimed a Kentuckian banker, had been 'electrified with the announcement' that Darwin's bulldog was to visit the United States in 1876.[31] When his ship arrived, New York's reporters raced aboard to grab the first interview; the governor of Connecticut begged for an audience; and as Huxley toured the north-east that summer, he became as famous as any attraction at Philadelphia's Centennial Exposition, even Alexander Graham Bell's new telephone. Marsh had the honour to guide him around the Peabody Museum at Yale, and the pair got on famously. Huxley wondered at the range and depth of the American's collection: 'Professor Marsh would simply turn to his assistant', Huxley's son recorded, 'and bid him fetch box number so and so, until [my father] turned upon him and said, "I believe you are a magician; whatever I want, you just conjure up"'.[32] By September, when Huxley delivered three lectures at Chickering Hall on New York's Fifth Avenue, he had won the guarded admiration of even the conservative, Catholic-owned *Herald*. 'There are few English writers in any department of letters', ran one editorial, 'who possess such rich gifts of expression'.[33]

The lectures themselves, which Huxley delivered every other night, turned on the history of nature, evolution, and the 'demonstrable' evidence for it. 'Fossils', he told the sell-out crowds, 'furnish us with a record, the general nature of which cannot be misinterpreted, of the kinds of things that have lived upon the surface of the earth . . . [Even] the circumstantial evidence absolutely negatives the conception of the eternity of the present condition of things.' The matters of creation and evolution, he went on, were nothing to do with metaphysics; rather, they were historical questions just as 'when the Angles and the Jutes invaded England, whether they preceded or followed the Romans'. Adverting to ichthyosaurs and plesiosaurs, to the *Compsognathus* and the *Archaeopteryx*, and the fossil birds that his new friend Marsh had uncovered in the West, Huxley hammered away at the relationships between the oldest reptiles and dinosaurs, and between dinosaurs and birds. 'It is, in fact, quite possible', he suggested, 'that all these . . . avi-form reptiles of the Mesozoic epoch are not terms in the series of progression from birds to reptiles at all but simply the more or less modified descendants of Palaeozoic forms through which that transition was actually effected'.[34] On the great ladder of the chain of being, Huxley was joining together the ancient and the modern. His trip to the United States was a signal moment in the history of palaeontology, perhaps of science. Here was the great man, the terror of the Old World's old guard, embracing the materials and resources of the New.[35]

Had he travelled a little later, Huxley would have had so much more to see, for 1877 was the banner year for hunting fossils. From just outside Denver, Colorado, the teacher Arthur Lakes would report on bones 'so monstrous that I could hardly believe my eyes', and Marsh would ascribe them – probably wrongly – to a sauropod that he called *Atlantosaurus*.[36] On the other side of the city, at a place now known as 'Dinosaur Ridge', diggers found the *Allosaurus*, a terrifying Jurassic carnivore with serrated teeth and hooking claws that was thought to have hunted down prey at thirty miles an hour. And about eighty miles to the south, in a quarry outside Cañon City, the palaeontologist Samuel Wendell Williston found the first bones of the *Diplodocus*. Yet if Denver now became one great site of the Dinosaur Rush, with prospectors and palaeontologists hunting for fossils as if they were gold, even more fertile ground was being broken at Como Bluff in Wyoming. There, along a ridge which protruded from the earth outside the town of Medicine Bow, teams belonging to Marsh and to Cope erected camps for the battles to come.

As news of all these discoveries crossed the Atlantic, Charles Darwin rejoiced. 'What progress Palaeontology has made during the last 20 years', he wrote to Alfred Russel Wallace; 'if it advances at the same rate in the future, our views on the migration & birth place of the various groups [of animals] will I fear be greatly altered'.[37] It was all enough that, by 1878, Huxley was able to credit the fossil record, and the palaeontologists who had compiled it, with proving the reality of evolution. Paying tribute to colleagues from France, Switzerland, and Russia and to Othniel Charles Marsh's adventures in the American West, all while arrogating a little of the credit to himself, he cheered that science no longer obsessed over 'the hodman's work of making "new species" of fossils' but was 'tracing out the succession of the forms presented by any given type in time'. By now, Huxley and his brethren had worked out 'the successive forms of the Equine type', of ungulate mammals and 'the *Carnivora* . . . through the tertiary deposits'. Most crucially, they had established 'the gradations between birds and reptiles' and 'the modifications undergone by the *Crocodilia*, from the Triassic epoch to the present day'. He therefore concluded that, 'on the evidence of palaeontology, the evolution of many existing forms of animal life from the predecessors is no longer an hypothesis, but an historical fact'; it was only the *cause* of evolution that was 'still open to discussion'.[38] Within another two years, Huxley was sufficiently confident to state that Darwin's initial hesitancy over the fossil record, as expressed in the first edition of the *Origin*, was now irrelevant: 'the amount of our knowledge . . . [had] increased fifty-fold, and in some directions even approaches completeness'.[39] If dinosaurs had once been regarded as impossible monsters, they were now – at least in Huxley's considered opinion – the means of understanding life itself.

Of course, the Americans had not monopolised the discovery of new dinosaurs. The geologist George Mercer Dawson, a former student of Huxley's, was reporting on fossils from the Canadian side of the Lake of the Woods and along the Red Deer River; from India, the surveyor Richard Lydekker had announced the arrival of the *Titanosaurus*, a gargantuan sauropod, from 'the Lameta group of rocks' near Jabalpur; and in Britain, though it was now a junior partner to the American West, palaeontologists both professional and amateur had coaxed new creatures from the earth.[40] William Davies, for instance, was a geologist at the British Museum and, in 1874, the directors of the Swindon Brick and Tile Company had summoned him to Wiltshire to inspect their clay pits. In 'a large septarian nodule', essentially a fossilised mud bubble, he found the bones of what he called 'another huge dragon'. Though the exhumation of the beast proved hazardous, with many of the fossils 'wet, rotten, and crumbling', Davies rescued enough material for Richard Owen to pronounce more sensibly on the discovery.[41] In honour of 'the formidable weapon with which [its] fore limb appears to have been armed', he proposed the name *Omosaurus armatus*, a massive cousin to the stegosaurs that Marsh and Cope were finding in the Badlands.[42] (A subsequent dispute would see the *Omosaurus* renamed as the *Dacentrurus*, by which it is known today.)

Keeping Londoners abreast of these new discoveries was Harry Seeley, who had spent the 1860s working with Adam Sedgwick at Cambridge. A fervent advocate of public education, a man who wanted 'to be to the people of England in Science what Cobden was to his Country in procuring the adoption of free trade', Seeley would spend the summer of 1875 delivering lectures at the Royal Institution on 'fossil flying animals'.[43] Taking the *Pterodactyl* and *Dimorphodon* as his subjects, and 'detailing the characteristic points of resemblance between the pterodactyles and existing flying animals', Seeley employed these lectures in service of a broader mission: 'to show how a naturalist does his work', to show that an honest enquiry of the earth could yield 'a solid return for labour . . . without exhausting the treasury of Nature's truths'.[44] These lectures, rousing and popular, designed to inspire as much as to educate, were of the kind that Huxley had made his own. The *Morning Post* published weekly reports on Seeley's progress, and local newspapers across the British Isles reprinted these in the weeks that followed; soon enough, provincial readers were demanding that Seeley should make a tour of theatres beyond

Harry Seeley's illustrations of the *Dimorphodon* and the *Pterodactyl*, as reprinted in the *Illustrated London News* in 1875

London.[45] Two years later, the *Evening Mail* reported on a similarly triumphant course of lectures on evolution where Seeley took his audience to the 'ornithosaurs, which in essential points of organization resemble birds [and] . . . make a nearer approximation to the reptiles than do the surviving orders of birds'.[46]

There was also a new, visual dimension to this public engagement with prehistoric fossils. Of course, there had been previous depictions of prehistoric life: there was Henry De la Beche's *Duria Antiquior* in aid of Mary Anning, John Martin's engravings for Gideon Mantell, the dinosaurs at the Crystal Palace, and majestic illustrations by Édouard Riou for the novels of Jules Verne and Louis Figuier's *La Terre avant le Déluge* (1863).[47] But with Seeley making some 'remarkable' new drawings of these flying creatures, images that the *Illustrated London News* reproduced for tens of thousands of weekly readers, these lectures marked a major milestone in the cultural history of dinosaurs and their ancient neighbours: these drawings did not feature in expensive books or elite monographs, but in a bestselling weekly paper.[48] Even so, if the British public wished to see these fossils for themselves, where could they go? Despite the Crystal Palace, and despite the collections at the British Museum and Oxford and Cambridge, there was no great venue for the display and the appreciation of dinosaurs and prehistoric life. At least not yet, for ever since the late 1860s, Richard Owen had been working assiduously in the service of public science, for the construction of London's own Natural History Museum.

24
The Citadels Fall

> We must bear in mind that every organised being now in existence represents the last link of an inconceivably long series of organisms, which come down in a direct line of descent, and of which each has inherited a part of the acquired characteristics of its predecessor. Everything, furthermore, points in the direction of our believing that at the beginning of this chain there existed an organism of the very simplest kind, something, in fact, like those which we call organised germs.[1]
>
> Samuel Butler, *Unconscious Memory* (1880)

For *The Times*, it was 'a great day with the young people of the metropolis, and . . . an epoch in many lives'. It was a day which compared to the morning 'when the first Noah's Ark came into the nursery; or . . . when the faithful patriarch saw his floating menagerie duly tenanted'. On that day in April 1881, more than a thousand Londoners made pilgrimage to South Kensington, where they worshipped in 'a true Temple of Nature' that celebrated 'the Beauty of Holiness'. Hitherto, museums had been 'very dull and wearisome places' and, for *The Times*, there was force in the saying that 'a living dog is better than a dead lion'. All that had changed now. If the British Museum had improved upon the Ashmolean, and if that museum at Oxford had improved upon the old curiosity shop, this new monument to science was something else entirely.[2] The Natural History Museum was a masterpiece of the Romanesque revival. In the colour of buff terracotta, its façade ran for 200 yards down Exhibition Road. From the Thames gravel that lay beneath 'lawns, flower-beds, and broad walks', two great towers of brick reared up into the London sky.[3] Everything here was of a piece, nothing without meaning. If, as one historian has observed, the frontage evoked medieval guilds and marketplaces, the secular and the exact, the towers and the museum's great vestibule were monuments to 'sacred, metaphysical principles'.[4] On the walls inside and out, where graven images of saints and the Stations of the Cross might once have looked down upon the laity, there were 'mammals, reptiles, birds, fishes, crustaceans, Echinodermata, and the shells of molluscs' carved into the stone.[5] It was the triumph of its architect, the Liverpudlian maestro Alfred Waterhouse, and of endless campaigning by Richard Owen.

In his own telling, Owen had conceived of these new premises almost as soon as he joined the British Museum in 1856. In the old and cramped quarters at Bloomsbury, the specimens of the natural history department had lain unloved in 'damp vaults' where the fossils suffered decay and erosion. This was no place for good science. By 1859, Owen had prepared a report for the Museum's trustees on the unfitness of the present site, pleading for additional space and for 'a hall, or exhibition-space . . . adapted to convey an elementary knowledge of . . . natural history to the large proportion of public visitors'.[6] He recommended a site at South Kensington, on land that the government had purchased with profits from the Great Exhibition. Well versed in the demands of scientific bureaucracy, Owen knew the rivers yet to cross. The trustees passed his report to the Treasury, who passed it in turn to a committee of the House of Commons that would consider 'how far, and in what way, it may be desirable to find increased SPACE for the Extension and Arrangement of the various COLLECTIONS of the BRITISH MUSEUM'.[7] Over ten weeks the next summer, the committee interviewed Owen, Huxley, Roderick Murchison, the zoologist William Carpenter, and the librarian Anthony Panizzi, who told the MPs that 'if there was to be added a mastodon, or a megatherium, or a great saurian' to the Museum's collection, the fossil department 'certainly would not have sufficient space'.[8]

At the height of their dispute over the fraternity of mankind to the gorilla, it was almost inevitable that Owen and Huxley would provide conflicting testimony. Owen believed that a new museum was necessary to exhibit the whole of creation all at once and, naturally, that he should

Comic News's mockery of Richard Owen's proposal for a new natural history museum

direct it; Huxley, however, thought that removing the Museum's collection to west London would cause terrible inconvenience to anyone living elsewhere in the city. (That would not stop him raising Imperial College in the same place.) With the new Chancellor of the Exchequer William Gladstone pushing policies of retrenchment, the committee declined to ask government 'to expend a very large sum of money' on a new museum. It concluded that the construction of a new museum over ten acres of floorspace would lead only to 'the needless bewilderment and fatigue of the public, and the impediment of the studies of the scientific visitor'.[9]

Delighting in his defeat, Owen's critics pursued his 'mortification'. The chairman of the committee regretted that 'a man whose name stood so high should connect himself with so foolish, crazy, and extravagant a scheme'. In 1862, by which time Owen had persuaded Gladstone to reconsider the project, Disraeli took his turn to denounce the new museum; another MP sneered at Owen's plan as 'a bad way of doing business', and another still revolted at the price of £600,000 (almost £60 million today) for nothing more than 'a new collection of birds, beasts, and fishes'.[10] The *Comic News*, a scurrilous weekly paper, meanwhile poked fun with a cartoon that showed monkeys and giraffes riding the Brompton omnibus on the road to 'their new house in South Kensington'.[11] Although Owen's grief for the project was considerable, it did not trouble him for long; after all, this was a man who made enemies before breakfast. Later in the decade, for instance, he earned the special hatred of Joseph Dalton Hooker for manoeuvring against the gardens at Kew, drafting a secret report for the Commission of Works that despised the whole science of botany as 'attaching barbarous binomials to dried foreign weeds'.[12] He therefore did not cower before this hostility and, as his son explained, Owen remained faithful to the idea of a museum for 'the proper exhibition of the natural history treasures of the nation'.[13]

Still the grey eminence of London science, living at the favour of Queen Victoria in a Richmond Park mansion that even the Brazilian emperor would visit, Owen soon had better luck. Gladstone led a motion through the Commons, committing the government to purchase a more modest five acres for a new museum; a competition for designs drew entries from the best of British architects; and by the early 1870s, working from plans by Francis Fowke, Waterhouse was directing an army of masons, carpenters, and surveyors in South Kensington. As the museum took shape, Thomas Carlyle expressed his approval: 'It seems to me that a nation ready to spend any amount of millions on any foolery that turns up', he told Owen, 'really might as well take counsel of its chief naturalist, and build such a museum as will satisfy him'. John Ruskin, typically waspish, was less enthusiastic: 'It would be misery to me', he wrote, 'to see your new abode'. Despite Ruskin's misgivings, the collections of the British Museum now made their way from Bloomsbury to South Kensington, where these fossils fulfilled Owen's great plan to exhibit 'our national treasures ... in a manner worthy of the greatest commercial and colonial empire of the world'.[14]

Yet if Owen had raised his towers in tribute to both natural history and the role of the Lord in directing its course, he would share the west London skyline with his greatest foe. Since opening his school for science on Exhibition Road, only yards from Owen's new fiefdom, Huxley had collected the disciplines of a scientific university under his roof. And though his domain was nowhere near so grand, it had by

1881 relaunched as the Normal School of Science, the name an imitation of the *École Normale* at Paris. Soon changing again to the Royal College of Science – and later still, as we have seen, to Imperial College – this institution would stand as Huxley's watchtower, the keeper of a secular tradition in defiance of Owen's great palace. A young H. G. Wells would describe his time in this modest building, under Huxley's tutelage, as 'the most educational year' of his life; Huxley himself, now 'a yellow-faced, square-faced old man, with bright little brown eyes', was simply 'the master'.[15] Even so, opening the Natural History Museum had not only allowed Richard Owen to satisfy the ambition of a lifetime, but it had helped Britain to keep pace with its rivals. By the early 1880s, there were specialist museums for natural history in Vienna, Paris, Berlin, St Petersburg, and Madrid. In the United States, Ulysses S. Grant had in 1874 laid a cornerstone in New York for the American Museum of Natural History, which Rutherford B. Hayes opened three years later amid scenes of great pageantry. And for this, as for museums around the world, it was the acquisition of American fossils – specimens from the Badlands of the West – that became their driving mission.

As Othniel Charles Marsh was stocking the halls of the Peabody Museum at Yale, Edward Drinker Cope had found his own buyers along the eastern seaboard. The value that now attached to dinosaur bones meant that offers came in floods to America's premier fossil-hunters. In late 1877, two men calling themselves 'Harlow and Edwards' wrote to Marsh. They were apparently 'desirous of disposing of what fossils we have, and also the secret of the others', yet because they were 'working men and not able to present them as a gift', they would only 'sell the secret of the fossil bed, and [look for] work in excavating others'.[16] They sent a shoulder blade and a single vertebra 'as proof of our sincerity and truth', and Marsh, ever desperate to prevent Cope stealing a march, dispatched an agent to Wyoming to find his correspondents. Outside Medicine Bow, the town of saloon bars and shootouts that Owen Wister would memorialise in the classic western novel *The Virginian* (1902), bribes from 'a freshly opened box of cigars' procured knowledge of their whereabouts.[17] Of course, 'Harlow' and 'Edwards' were pseudonyms of sorts – the men, William Harlow Reed and William Edwards Carlin, had used their middle names because they had been hunting for fossils when they should have been working on the railway – but they had not lied. Marsh would offer them each $90 a month to excavate the bones of Como Bluff.

This was the promised land, where three formations of Jurassic rock erupted into the sky. As Marsh's team quarried to depths of forty feet, men worked with picks and ropes to hew the fossils from the earth. 'I doubt not that many thousands of tons of bones will yet be exhumed', reported Samuel Wendell Williston to Yale.[18] Yet by the time that Williston had gone back east that winter, Carlin and Reed had made even greater discoveries: 'I wish you were hear [*sic*] to see the bones roll out,' they wrote to him, 'and they are beauties ... It would astonish you to see the holes we have dug.'[19] One of the first new dinosaurs that Marsh's men removed from the bluff was the *Apatosaurus*, a stupendous four-legged herbivore that, in honour of the warrior of Greek mythology, he gave the epithet 'Ajax'. After the *Apatosaurus*, from fossils found 'in the most inextricable confusion, and in every conceivable position', came several dinosaurs that have become household names: the stunning 'thunder lizard' that Marsh named *Brontosaurus excelsius*, and the armour-plated, spike-tailed *Stegosaurus*.[20] There were new species of *Allosaurus* and the light, fleet-footed *Laosaurus* besides. It was impossible to keep word of the windfall from spreading and, by the end of the first winter at Como, Williston had sore news: 'I am sorry to say', he told Marsh, 'that Cope knows of this locality and the general nature of the fossils'.[21]

Soon enough, Marsh's great rival had four quarries of his own at Como Bluff. Cope had also stolen the services of Mr Carlin, and it was an ingenious act of theft: Carlin was the local stationmaster and so, during the winter, he could package and guard fossils for Cope in the warmth of the waiting room; Reed, still working for Marsh, was made to suffer without shelter on the platform. Spite now met with spite. Whenever Reed finished excavating the prime fossils from a site, his men would smash the remainder into dust to prevent Cope's men from scavenging; Marsh meanwhile spread the slander that Cope was lower than a 'damned thief'. By 1879 the business of the crews was espionage as much as excavation, with scouts in the pay of each great fossil-hunter tracking the movement of their enemies. Even in the worst of winter, with the men shuddering in tents and furs as the temperature fell twenty below zero, the rivalry burned: 'Peace and harmony does not prevail here to any extent', it was reported.[22]

In the grander sweep of the Bone Wars, however, Marsh was winning. He had collected more fossils, named more new dinosaurs, and he had the awesome resources of Yale University and his uncle George Peabody at his disposal; the US Geological Survey soon chipped in as well, providing Marsh and his men with an annual fund of $15,000, almost half a million dollars in today's money. Cope, in contrast, had exhausted much of his inheritance, having invested badly in mining operations, and he had nothing like the institutional support that his rival enjoyed; by the mid-1880s, he recognised that his 'entire future in a financial sense' would depend on selling copies of his fossils. As the two men wrote up their finds for learned journals, it became clear that Marsh was making the more incisive contributions to science as well. In 1881 he gave the name 'theropods' to the two-legged, meat-eating dinosaurs that he had discovered; two years later, he gave a brilliant reconstruction of long-necked, four-legged sauropods such as brontosaurs. This is not to say that Marsh was the hail-fellow of his colleagues. Samuel Williston had spent ten years by his side but griped that Marsh had 'never been known to tell the truth when a falsehood would serve the purpose as well'; he even claimed to have written many of Marsh's papers.[23]

Still, it was Marsh the enthusiastic evolutionist, not Cope the recalcitrant Quaker, who would more profoundly shape our understanding of dinosaurs and the prehistoric world. As late as 1887, Cope was preaching a 'theology of evolution' which emphasised 'the great truth of the control of mind over matter', where there was copious evidence for 'the existence of a Supreme Being', and where 'Science . . . had little to say on the subject of immortality'. For all that Cope had done and seen, he believed until his death in 1897 that there was 'a great Mind now invisible to us', that 'the energy of living things' manifested the 'design' of that great mind, and that mental energy had 'directive power . . . over the process of Evolution'.[24] Cope was wrong, and his ideas on the development of species had little influence, but there was a lasting consolation to his memory. At Princeton in the 1870s, Cope had mentored Henry Fairfield Osborn, a wealthy student from Connecticut. On graduating with honours in geology and archaeology, Osborn studied anatomy at New York and then comparative anatomy under Huxley in England and, by the 1890s, he was chief of the department of vertebrate palaeontology at the American Museum of Natural History. There, in 1905, a host of skeletal fragments came before him. There were teeth from Colorado, pieces of cranial matter from Wyoming, and shards of a backbone that Cope himself had found. Most stunningly of all, there were dozens of bones that his assistant, Barnum Brown, had recovered from the Badlands. The dinosaur that Brown and Osborn reconstructed from these pieces was majestic. It was twenty feet tall, forty feet long, and it had teeth like the tines of a pitchfork. And that year, in his museum's bulletin, it was Cope's student Osborn who named the dinosaur that has haunted nightmares ever since: *Tyrannosaurus rex*.[25]

In the spring of 1880, as Marsh and Cope warred in the West, British voters went to the polls. Benjamin Disraeli's Conservatives were in power, but the winds were against them. The 1870s had seen a global economic downturn but, more damaging, Disraeli's foreign policy was thought to be weak, indecisive, and immoral. It was to the latter of these issues that William Gladstone had addressed his Midlothian campaign. Riding Britain's railways from city to city, speaking for hours at pre-arranged venues to undecided voters and the local press, Gladstone had spent eighteen months assailing Disraeli and the Tories for failing to answer the Irish Question and for abandoning Bulgaria's Orthodox Christians to the Ottomans. One of Gladstone's biographers has described this campaign in almost messianic terms: the Grand Old Man of British politics was reminding the public of 'moral forces they had forgotten' and waking 'the whole spirit of civil duty'. It worked. As Gladstone 'pounded the ministerial citadel into the ground', the Liberals won a majority of all votes cast across Britain.[26] Remarkably, it remains the last time that any party other than the Conservatives has done so. From late April, therefore, more than 350 men arrived at Westminster to take their seat as Liberal MPs, with numerous household names among them. There was the banking magnate Nathan Rothschild; the master of Birmingham, Joseph

Chamberlain; the historian George Otto Trevelyan; and the radical, one-time republican Charles Dilke. Joining them all was the new MP for Northampton, Charles Bradlaugh.

Upon their acquittal in 1878, he and Annie Besant had returned to their work with the National Secular Society, which had threatened to splinter under the pressure of the obscenity trial. Besant had maintained her focus on contraception, arguing that it was 'clearly useless to preach the limitation of the family, and to conceal the means whereby such limitation may be effected'.[27] Bradlaugh had meanwhile resumed his attacks on the perversions of religion. At Nottingham, he debated whether it was 'reasonable to Worship God' and he decided it was not; his *National Reformer* continued to harry the dogmatic and the intolerant; and he published new editions of his execration of the gospels.[28] 'To discuss the credibility of the miraculous conception and birth', he proclaimed anew, would be 'to insult the human understanding ... A miraculous birth would be scouted to-day as monstrous; antedate it 2,000 years and we worship it as a miracle.'[29] And despite Darwin's refusal to testify during the obscenity trial, the influence of the old man on Bradlaugh's thinking was increasingly clear. 'Between the cabbage and the man', he now proclaimed, 'I know no break. Between the highest of which I know and the lowest of which I know ... I trace step by step down the ladder until I come where it is impossible for me ... to distinguish between animal and vegetable life.'[30]

Bradlaugh had also turned his attention to politics, to demonstrations for peace between the Russians and the Ottomans, and to running for Parliament in the Midlands town of Northampton. It was familiar terrain. He had campaigned there in support of the Second Reform Act of 1867, lecturing as 'Iconoclast', and there had been two failed runs for Parliament as a radical Liberal, losing out in both 1868 and 1874 to candidates from the moderate wing of the party; indeed, Liberal grandees such as Gladstone and John Bright had worked actively against him. Yet when two of the Liberals' prospective candidates fell gravely ill in early 1880, Bradlaugh had secured the nomination. He electioneered as he lectured, magnificently: 'He raises his voice', reflected a fellow candidate, 'until the very walls re-echo with it, and winds up with a fierce appeal to the electors to do their duty'. Though Bradlaugh could easily subscribe to the general Liberal platform of 'Peace, Retrenchment, and Reform', he brought his own peculiar concerns to the hustings: if elected, he would work to reform the House of Lords, disestablish the Church of England, and end British wars against the Zulu. His Conservative rivals made much of Bradlaugh's atheism, denouncing him as 'THE GREATEST ENEMY OF GOD AND OF HIS TRUTH NOW LIVING AMONGST MEN', but Northampton 'swallowed' this radical candidate.[31] Bradlaugh secured the vote of 3,827 electors, and even the New York papers noticed his victory.

There was, however, one last obstacle in Bradlaugh's way, because Parliament's arcane rules did not automatically allow the MP who had been elected by voters to take his seat in the House of Commons. This problem had arisen several times over the course of the nineteenth century, most notably when Daniel O'Connell had been forbidden from entering Parliament on account of his Catholicism. A similar issue had confronted Lionel Rothschild and David Salomons in the 1850s: before taking part in the business of the Commons, MPs were obliged to swear an oath of allegiance which included the phrases 'So help me God' and 'On the true faith of a Christian'. This was impossible for Rothschild and Salomons, who were Jewish, but a compromise permitted them to replace the word 'God' with 'Jehovah' and to leave out the second phrase altogether.[32] Bradlaugh's case was more complicated. He was probably the most famous atheist in the world, so swearing any such oath – however mildly phrased – would represent the public and humiliating abjuration of the beliefs by which he lived. What could he do?

As the clerk and Speaker of the House of Commons worked through that year's intake, swearing in dozens of MPs each day, Bradlaugh laid out his plan. Recent legislation had allowed British subjects to 'affirm' their honesty before the law courts instead of swearing a religious oath, and Bradlaugh himself had done this on several occasions. He would therefore seek to affirm before Parliament and placed a notice in *The Times* to make known his intention. At two o'clock on the appointed afternoon, as another batch of MPs 'took and subscribed the Oath according to Law', Bradlaugh demurred. Although several reports suggested that he 'refused to take the oath' and even 'obtruded his opinions' on the House, the truth was more polite: he merely approached the table and begged permission to affirm 'in virtue of the provisions of the Evidence Amendment Acts'. With the Speaker harbouring 'grave doubts' about Bradlaugh's construction of the relevant legislation, the issue devolved to a select committee, a special panel of MPs who would decide whether this right applied in Parliament as well as in the courts.[33]

The events of the following weeks and months were a tawdry spectacle. The committee found against Bradlaugh so the matter went back to the Commons, where a gang of staunch Tories – Randolph Churchill was one of the ringleaders of this 'Fourth Party' – rallied colleagues against the atheist. Had the Liberals won a majority, they asked, so that 'Christian religion might be safely derided, and the existence of God publicly and with scorn denied'? Churchill begged Gladstone not to use his new-found power 'for the purpose of placing on those benches opposite an avowed Atheist and a professedly disloyal person'.[34] He also read to the House salacious extracts from Bradlaugh's pamphlets, painting him as a danger to public safety: 'I loath[e] these small German breast-bestarred wanderers', he had written about the monarchy, 'whose only merit is their loving hatred of one another'.[35] It was enough to animate the queen herself, and Victoria's diaries record her contempt for this 'horrible, immoral Atheist'. She thought Bradlaugh 'dreadful' and her son Leopold reported from Parliament that he was 'horrid & repulsive looking'.[36]

With the Tories taking the field 'in this great battle', mere association with Bradlaugh became a dangerous thing.[37] Edward Aveling had embarked on a promising medical career, first under an agnostic physiologist at Cambridge, then as a comparative anatomist in London. But when he took up the vice-presidency of the National Secular Society, he aroused the suspicion of the medical censors and lost his job at the London Hospital, with Aveling complaining that 'the real reason for his dismissal was his avowal of certain religious and political views'.[38] In the bedlam, even plausible allies held firm against Bradlaugh. The Irish liberal Robert Lyons, for instance, predicted only calamity if Parliament should yield to unbelief: 'Woe betide the day', he told the Commons, 'when the ship of State should be left to go on the waste of waters with her bulwarks of religion and morality stove in'. The Tories, of course, condemned Bradlaugh outright, with Lord Norton suggesting that he should be publicly flogged in Trafalgar Square. Either way, the case was the cause of the day, and a Dublin newspaper reported that 'home and foreign policy . . . all fade into insignificance . . . in front of the problem whether or not [he] is to be admitted'.[39] As for Bradlaugh himself, he was by high summer in prison. When the Tories whipped a motion to force his withdrawal, and when Bradlaugh refused to obey, the serjeant-at-arms had escorted him from the House of Commons and confined him to a cell in the clocktower beneath Big Ben.

The authorities released him soon enough, fearing that Bradlaugh would become a martyr to his cause, but the ordeal was far from over. A lawsuit before the High Court brought judgment that Bradlaugh could not affirm, and there was a by-election called for Northampton, where his Tory opponent sounded 'a clarion blast against . . . the champion advocate of iniquity and lawlessness'.[40] But when Bradlaugh triumphed in this rematch, albeit with a smaller margin, the whole farce began again: he tried to affirm, Parliament denied him, and the electoral authorities called for another poll. The pattern repeated in 1881, 1882, 1884, and 1885, while Gladstone failed three times to pass a bill that would have given him entry. It was not until 1886, two years after the last great Reform Act of the nineteenth century, that a new and exasperated Speaker of the House at last let Bradlaugh take his seat. This was not the signal triumph of atheism or the delivery of a knock-out blow to religion; rather, like so many of the changes that had been wrought over the past seventy years, and perhaps like the very processes of geology and biology that humankind now understood, it had been a work of relentless attrition. Even so, after so many years of trying, the godless had finally breached the walls of the citadel.

It was a Tuesday evening in April 1882 that another bastion fell, when, in 'the dim light [of] two old-fashioned lanterns', they placed the coffin in the Chapel of St Faith.[41] At twenty to twelve the next morning, as family and colleagues assembled in their grief, Thomas Huxley helped to bear the casket from the chapel and through the cloisters of Westminster Abbey. In that day's edition of *Nature*, he had eulogised 'an intellect which had no superior, and . . . a character which was even nobler than the intellect'. He remembered how his friend had found 'a great truth, trodden under foot, reviled by bigots, and ridiculed by all the world', but how, with 'ready humour' and through 'the sovereignty of reason', he had prevailed.[42] At the Abbey, Joseph Hooker and Alfred Russel Wallace went with him as pallbearers, as did John Lubbock and William Spottiswoode of the X Club. The others who shared in this solemn duty were men of stature, men of state: the linguist Canon Frederick Farrar, the Dukes of Devonshire and Argyll, the Earl of Derby, even the American ambassador to the United Kingdom. Black velvet edged with white silk covered the coffin; wreaths of 'beautiful white flowers', some of them from 'scientific societies in Liverpool', lay there too.[43] When the procession met the family, the service began. They sang Handel's *Israel in Egypt* – 'His body is buried in peace, but his name liveth evermore' – and the mourners passed around the grave. Charles Darwin would rest here, at Scientists' Corner, beside the astronomer John Herschel and Isaac Newton, and not far from James Ussher.

Though Darwin's faith had failed upon the death of his daughter Annie, and though he confessed in later years that he did not 'believe in the *Bible* as a divine revelation, & therefore not in Jesus Christ as the Son of God', he had never attacked religion or endorsed the works of atheism. Writing to the clergyman John Fordyce in 1879, he had suggested that 'generally (& more and more so as I grow older) but not always . . . agnostic would be the most correct description of my state of mind'. He had, moreover, refused the dedication of a young atheist's translation from the German of *The Student's Darwin* (1881). 'Though I am a strong advocate for free thought on all subjects', he wrote to Bradlaugh's disciple Edward Aveling, 'it appears to me . . . that direct arguments against christianity & theism produce hardly any effect on the public . . . It has, therefore, been always my object to avoid writing on religion.'[44]

In the last decade of his life, in the years since the *Descent*, Darwin had turned back to botany. Milling around the house and garden in Kent, the laps on the Sandwalk ever slower, he was a man who knew his age. 'I have little strength', he wrote to the German publisher Victor Carus, '& feel very old'.[45] Despite these infirmities, and still working with John Murray, there had come studies of insectivorous plants, fertilisation 'in the vegetable kingdom', variations of flowers

FIG. 62.—*Funeral of Charles Darwin, Esq., in Westminster Abbey.*

A wood engraving of Charles Darwin's funeral at Westminster Abbey in 1882

within the same species, and 'the great classes of movements' among plants.[46] The last of all his books returned to the subject of his very earliest papers, the formation of mould and the behaviour of earthworms. As his greatest biographers relate, Darwin's admirers could not understand this late obsession with such a subject, but these worms were a leitmotif for his life's work: small creatures working great changes in the earth by 'minute, incremental events . . . evolving life, and transforming the soil'.[47] Like the worms, Darwin had worked his own changes. When he disembarked from the *Beagle* in 1836, his ideas about 'descent' and changes in animal species were so dangerous, so inimical to the orthodoxy, that fear kept him quiet for more than twenty years. Yet by the time of his death in 1882, ideas about natural selection and evolution were commonplace, an atheist was on the verge of admission to the House of Commons, and the author of the Devil's gospel was regarded as a natural tenant of Westminster Abbey.

Epilogue: Rex

All the world's a stage
And all the men and beasties merely players
. . . Then Dinosaurs,
In fair round belly, with food well lined,
Their eyes severe, erect on great hind legs,
The Lords of Mesozoic times.
And so they played their part. The sixth age shifts
Into the Bird, a diver, six feet high,
Hesperornis it is called, with teeth in jaws,
Large skull and reptile-like affinities,
And yet a Bird![1]

Henry Neville Hutchinson (1889)

In the seventy-odd years that followed Mary Anning's discovery of the *Ichthyosaurus*, awesome changes had taken place in the politics and even the appearance of the British Isles. Thanks to the Reform Acts of 1832, 1867, and 1884, the relative size of the electorate had trebled; new laws had removed the civil disabilities which applied to Roman Catholics and Dissenting Protestants; colonial slavery had been abolished; and the pink of the British Empire now coloured the maps of much of Africa, the Asian subcontinent, and Australasia. The people themselves had gone forth and multiplied: despite the devastation of the Great Famine and the mass emigration from Ireland that it compelled, the population of the British Isles had doubled. More strikingly, three times as many Britons were living in towns and cities, drawn in from the shires by the wages of factory-work. London was now the world's metropolis, the fulcrum of global power and finance that almost four million people called home; and with great industrial cities such as Manchester, Birmingham, Leeds, and Sheffield having swollen in size over the course of the nineteenth century, Britain was the most urbanised country in the world.

The rising skylines of these cities said something about the modern, industrial character of Victorian Britain: there was Waterhouse's new Town Hall at Manchester, the rebuilt Palace of Westminster, George Gilbert Scott's railway station at St Pancras, and the Natural History Museum at South Kensington. Still, these great monuments to progress – these churches of the modern – would share the sky with older concerns. By the turn of the century there were new cathedrals at Bristol and Southwark, another under construction at Truro, and a renovated York Minster; until the erection of the Blackpool Tower, the tallest building in Britain remained Salisbury cathedral. British churches had extended their bases too. The 1851 census of religious worship had appeared to reveal a pattern of disengagement, but muscular Christianity and the games-playing cult that swept across the Empire soon allowed British churches to welcome thousands of prodigals back into the fold: boxing gyms became centres of missionary outreach; cricket developed the self-consciously Christian qualities of discipline, honesty, and fair play; and several of this season's Premier League football clubs – Manchester City, Everton, and Fulham among them – spawned from the extra-curricular activities of Victorian churches. It was no coincidence that the Earl of Shaftesbury, who had denounced J. R. Seeley's *Ecce Homo* as 'the most pestilential book ever vomited from the jaws of Hell', was the inaugural president of the Young Men's Christian Association, the worldwide YMCA movement of today. Of course, simply associating with a church did not mean that Britons of the late Victorian period were sincerely religious; nor did it make them more likely to believe the Bible than a secular source of information. As these pages have shown, Christianity could no longer command automatic or exclusive authority in public life, and the verses of the scriptures no longer formed the essential core of knowledge with which many Britons approached questions of morality, politics, and especially science. Few if any transformations in intellectual history have been more profound.

When Anning, Mantell, and Buckland made their breakthrough

discoveries in the 1810s and 1820s, the bones of the *Megalosaurus*, *Iguanodon*, and *Dimorphodon* astonished – as Charles Kingsley later put it, they were 'impossible monsters' – but they did not disturb the ages-old assumption that the Lord had made each species perfectly and separately, with humankind his special creation. Yet over the coming decades, evidence of another world accumulated. There came the bones of bipedal *Hadrosaurus* from the eastern United States, the fossilised feathers of the *Archaeopteryx* from the limestone of Bavaria, and the tiny, chicken-like creature called the *Compsognathus*. To all of these Othniel Charles Marsh would add the prehistoric remains of the birds *Ichthyornis* and *Hesperornis*. By the late 1860s, and certainly by the time of his tour of the United States in the late 1870s, it was clear to Thomas Huxley that there was a close, consecutive affinity between prehistoric dinosaurs and modern birds; these fossils were proof, he declared, that Charles Darwin's mechanism of evolution by natural selection was no mere theory. And if such a process required millions of years to have the effects which Huxley and his allies contended, Charles Lyell had long made the case that, because geological change occurred only incrementally, the Earth was much, much older than anyone once had thought.

As for dinosaurs and birds, so for humankind. Genesis 1:26 told that on the sixth day the Lord said: 'Let us make man in our image, after our likeness'. To be clear on the matter, the next verse was a refrain: 'So God created man in his *own* image, in the image of God created he him'. It followed that a Lamarckian suggestion of historical human development – that is, any suggestion that humankind had evolved from a lesser form – was a dangerous, sacrilegious theory. Yet with Darwin's examination of orangutans in London Zoo, and the discovery of the Neanderthal in the Rhineland and of further remains in Gibraltar, the evidential base was shifting. By the time that Darwin published *The Descent of Man* (1871), it had become possible and indeed plausible to argue not only that humankind had evolved from a more primitive form of life, but also that there was an intrinsic biological fraternity between humans and primates. These remained controversial ideas, and Huxley's 'gorilla war' against Richard Owen epitomised the rhetorical violence of the mid-Victorian period, but it was to evolution and to natural selection, not to the guidance of the Old Testament, that Britons now looked for explanation of the natural world.[2] As for the dinosaurs, the impossible monsters whose fossils had played such a vital role in the development of these new disciplines, they had become a phenomenon of popular culture.

Dinosaurs had fascinated Arthur Conan Doyle ever since childhood, 'ever since coming across a fossil dig while out on a walk' at Stonyhurst College in the early 1870s.[3] The 'romance of palaeontology' would never leave him. On his second honeymoon, in Greece in 1907, he claimed to see 'a creature which has never ... been described by Science. It was exactly like a young ichthyosaurus, about 4 feet long, with thin neck and tail, and four marked side-flippers'; in his autobiography he reflected that, if these fruits of the Aegean Sea were anything to notice, the 'old world has got some surprises for us yet'. Shortly afterwards, on moving to Crowborough in Sussex, he and his wife would explore the same Wealden rocks that had surrendered treasures to Gideon Mantell in the 1820s. 'We have been getting some curious objects', he reported to the Natural History Museum, 'in the Green Sand formation near my house ... At first I thought they were marks of the hind feet of Iguanodon or some other big lizard, but as they never cover each other I begin to wonder ... [whether] Iguanodon walked on his hind legs only.'[4] He was not wrong.

As he crafted the Sherlock Holmes stories for the *Strand Magazine*, Conan Doyle drew further inspiration from the adventures of his friend Percy Harrison Fawcett, who had spent years looking for the lost Amazonian city of 'Z'. During periodic trips back to Britain, Fawcett had described how 'time and the foot of man had not touched' the Richard Franco Hills of the Bolivian highlands. 'They stood like a lost world,' he mused, 'forested to their tops, and the imagination could picture the last vestiges there of an age long vanished'. Fawcett thought that 'monsters from the dawn of man's existence' could still have roamed in those mountains, 'imprisoned and protected by unscalable cliffs'.[5] Conan Doyle quite agreed. From April to November 1912, in the pages of the *Strand*, he told of the pugnacious Professor Challenger and his quest to prove that dinosaurs and other prehistoric animals had survived in the remote reaches of South America.

In perhaps the purest and most accomplished expression of Edwardian adventuring, *The Lost World* sees Challenger and his companions – a sceptical scientific rival, an intrepid reporter, an aristocratic explorer –

voyage into the unknown from the Brazilian city of Manaus. Cannibals beat war drums that stalk their progress upriver. A *Pterodactyl* with 'monster wings' and 'fierce, red, greedy' eyes descends upon their camp like the harpies in Virgil. And in the soft mud of a plateau, Challenger spies the fresh footprints of a dinosaur. 'Nothing else could have left such a track', he explains: 'They puzzled a worthy Sussex doctor some ninety years ago; but who in the world could have hoped – hoped – to have seen a sight like that?' From here, through brushwood and thick trees, the party emerges into an open glade where a family of iguanodons, their slate-coloured skin shimmering in the sunlight, stuns them into 'motionless amazement'. Photographs taken and journals kept, the travellers make their way back to humankind. They survive an attack from another dinosaur, either a *Megalosaurus* or an *Allosaurus*, and they skirt a war between indigenous humans and a race of ape-men. Back in London, at a meeting of the Zoological Institute, the sceptics laugh and scorn and accuse Challenger of lying, all before he unveils a specimen, proof beyond proof of their quest: a living *Pterodactyl*. At the sight of the creature, which had the face of 'the wildest gargoyle', there is chaos. Ladies faint; the Duke of Durham collapses from his seat; and the *Pterodactyl* escapes, last seen flying out to sea.[6]

The Lost World is probably the most enduring depiction of dinosaurs from this period, a staple of science fiction that has inspired countless adventures on page and screen; it is also, of course, the name of the sequel to *Jurassic Park*. But at the turn of the twentieth century, it was far from alone in deploying dinosaurs for dramatic purpose. In *Beyond the Great South Wall* (1901), Frank Mackenzie Savile sends a band of explorers to Antarctica in pursuit of a lost civilisation. They find these polar tribes, Mayans who had fled south from the conquistadors, but also 'a Beast . . . like unto nothing known outside the frenzy of delirium'. Worshipped by the lost Mayans as 'the great god Cay', it had 'the lithe neck of a boa-constrictor', a 'coarse, heavy, serrated tail', and 'horny eyelids [that] winked languidly over . . . deep-set wicked eyes'.[7] Apparently, it was a *Brontosaurus*. Even more speculatively, the short story 'The Monster of Lake LaMetrie' (1899) by Wardon Allan Curtis sees an *Elasmosaurus*, 'a colossal, indescribable thing', receive a brain transplant and devour its human master before the US Army shoots it to death.[8]

There was a burgeoning non-fiction market, too. Throughout the 1890s, the clergyman Henry Neville Hutchinson had published prolifically on dinosaurs and the 'antique world', introducing the British public to the most recent discoveries of the Badlands such as the *Stegosaurus* and the *Triceratops*. Beguiling his readers with tales from a dinosaurian world that was 'quite as strange as the fairy-land of Grimm or Lewis Carroll', Hutchinson recalled the early work of William Buckland in reconciling the evidence of the earth with biblical verse.[9] He thought not only that humanity was 'the final product of countless ages of Evolution', but even that 'the account of the creation in the opening chapters of Genesis *implies* evolution'.[10] Finding Hutchinson's books 'literature more than science', the palaeontologist Harry Seeley offered a more prosaic but equally popular account of pterosaurs in *Dragons of the Air* (1901).[11]

All the while, British newspapers continued to pay minute attention to the discoveries of the American West. In May 1899, for instance, under the headline 'Concerning Dinosaurs', the *St James's Gazette* informed its readers of the precise process by which Henry Fairfield Osborn had excavated four 'interesting fossils of gigantic animals' from a quarry to the west of Cheyenne, Wyoming.[12] The *Daily Telegraph*'s New York correspondent reported that, in the wake of yet another mammoth discovery in the Badlands, 'old and young' were flocking to the American Museum 'to see the dinosaur', and that 'dinosaur tea-parties now promise to become fashionable'.[13] Even local newspapers such as the *Abergavenny Chronicle* kept up with American palaeontology, with one article relating how 'the Peabody Museum at Newhaven (Connecticut) now possesses one of the only two Dinosaurs known to exist in a complete form'.[14]

The early screen played host to dinosaurs too. In *Prehistoric Peeps* (1905), the four-minute adaptation of a cartoon strip that ran in *Punch*, a scientist falls asleep 'surrounded by fossils' and, in his dream, actors dressed as dinosaurs and other 'prehistoric monsters' chase him around a cave, 'undeterred by his revolver shots'.[15] By 1914, the great filmmaker D. W. Griffith had released *Brute Force*, where the same device of a dream sees two tribes of cavemen go to war with each other while a mechanical dinosaur lurks ominously in the background. That same year, in a seminal moment in the history of animation, the cartoonist

Winsor McCay's *Gertie the Dinosaur* (1914)

and vaudeville star Winsor McCay produced *Gertie the Dinosaur*. While the cartoon played on the screen, McCay would perform the role of a ringmaster. Holding a whip, and presenting 'the only dinosaur in captivity', he commanded the brontosaurus to leave her cave and bow to the audience, to raise her right then her left foot, and to catch and eat an apple. Sometimes Gertie obeyed, sometimes she did not, but with each instruction synchronised to the animation, she wowed the crowds. In November that year, McCay sold the rights to the film to William Fox: for just one week at the Palace Theater, he earned $350, 'the biggest figure ever paid for a single reel attraction'.[16]

These cinematic adventures reflected a booming market for visual representations of dinosaurs and the prehistoric world, and one of the artists who met this demand was Charles R. Knight, the Brooklyn-born heir to Henry De la Beche, John Martin, and Édouard Riou. Knight was blind, at least legally: born with astigmatism, he suffered a childhood accident when 'a boy about [his] own age carelessly tossed a small pebble which struck [him] directly in [his] right eye'. Losing all sight in it, Knight would need special lenses to see, and he would keep his one good eye only inches from the canvas when he painted. Still, he trained at various New York institutions before a friendship at the American Museum of Natural History brought a breakthrough: a commission to paint the prehistoric pig known as the *Elotherium*. There followed the rare sight of nimble, vicious dryptosaurs fighting one another in *Leaping Laelaps* (1897) and museums across the United States were soon paying Knight to 'restore' their specimens to life.

Working closely with eminent palaeontologists 'for the edification of the general public', Knight's most famous works were the giant murals that adorned museums in New York, Chicago, and Los Angeles.[17] There, even now, in bright and vivid colours, allosaurs tear meat from the bones of fallen prey, tyrannosaurs spar with a *Triceratops*, apatosaurs roam through green pastures, and pterosaurs soar above the sea. Millions of Americans have since stopped to gaze upon Knight's murals, which came to inspire the animator Ray Harryhausen and the author Ray Bradbury among others; in time, they wrote forewords for the posthumous publication of his autobiography.

By the early twentieth century, therefore, dinosaurs had become a cultural phenomenon. A hundred years earlier, before Joseph and Mary Anning had liberated the *Ichthyosaurus* from the limestone of the Dorset coast, the idea that dinosaurs had once ruled the Earth was laughable, unthinkable. But now it was common knowledge; dinosaurs were an industry. They featured in serialised novels, short stories of science fiction, news reports, scientific monographs, silent films, animated films, and vast murals of the prehistoric world. None of this, however, compared to the excitement – even the mania – that attended the Carnegie *Diplodocus*.

Andrew Carnegie and his parents had emigrated from Scotland to the United States in the 1840s. Taking work for a cotton magnate, then as a messenger boy for a telegraph company, the young Carnegie paid forensic attention to the operations of industry. By the 1850s, at the age of twenty-four, he was running a division of a railway company; after the Civil War, having invested in Pennsylvanian oilfields, he went into steel, and here he made his fortune. Crucially, Carnegie found a way to apply the Bessemer process cheaply and effectively; he brought every element of production and distribution under the same roof; and he ploughed profits back into the business. As workmen laid his steel under the railroad, and as the same metal formed the spine of America's early skyscrapers, Carnegie acquired extraordinary wealth: in 1901, when J. P. Morgan bought him out, his steel business alone was worth more than $300 million, almost $11 billion in today's money. Carnegie gave much of his money back to cities and communities, funding libraries, opera houses, schools, and universities; indeed, he preached 'a gospel of wealth', urging his fellow tycoons to indulge in philanthropy, not avarice. But for the museum that he founded at Pittsburgh, a place synonymous with steel, he wanted the most majestic dinosaur that money could buy. An article in the *New York Journal* had heralded the 'most colossal animal ever on earth' and Carnegie asked his museum director openly: 'Can't you buy this for Pittsburgh?' He thought they could. 'We will get His Imperial Majesty,' he promised, 'the Prince of Lizards!'[18]

The first bones of the *Diplodocus* had been found by Samuel Williston in Colorado in 1877. The next year, in his account of 'the principal characters of American Jurassic dinosaurs', Othniel Charles Marsh had named and described a dinosaur 'of very large size, and herbivorous in habit ... about fifty feet in length'.[19] But as further *Diplodocus* specimens came to light in the 1880s, and by the time that Carnegie had funded a permanent expedition to the fossil-beds, it was clear that the *Diplodocus* was much, much bigger than Marsh had once thought. One Wyoming official proclaimed it was 'the biggest thing on earth', while the *New York Post* carried exaggerated reports of its dimensions: 'a monster 130 feet long, 35 feet high at the hips, 25 feet at the shoulders, [and] an animal that in life weighed 120,000 pounds'. With a full skeleton being excavated from the rocks of the Morrison Formation, abducting such a prize became a political matter: if the beast escaped to Pittsburgh, noted one fossil-hunter, 'a howl would go up over the whole State of Wyoming'. Over these objections, pragmatism prevailed. 'It will do the University in Wyoming a great deal more good', argued Carnegie's agent in the West, 'to have this skeleton ... mounted and exhibited ... at Pittsburgh, where three or four hundred thousand people will see it every year ... than it will if they should have some [Wyoming] bone butcher go in and grub the thing up'.[20] Within months, the dinosaur was making the journey east to Pennsylvania.

It was not just the discovery of Carnegie's *Diplodocus* which cemented American primacy in 'dinosaur science', it was the presentation. In 1902, when visiting Carnegie at Skibo Castle in the Scottish Highlands, Edward VII spied a sketch of the dinosaur on the wall. In Carnegie's own telling, 'His majesty ... expressed the hope that the Diplodocus before him might some day be seen here', and though the original fossil skeleton would not leave Pittsburgh, Carnegie would provide the next best thing, a life-sized plaster of Paris copy.[21] Paying four men to work on the project for eighteen months, Carnegie spent £2,000 on replicas of all 292 of the dinosaur's bones. These were shipped to London in thirty-six crates, and his best men then supervised the reconstruction of the *Diplodocus*, which by May 1905 was ready for exhibition at the Natural History Museum. At one o'clock in the afternoon, more than 200 dignitaries gathered in the Gallery of Reptiles. There was the Liberal grandee George Trevelyan, the geologist Archibald Geikie, and Darwin's old friend John Lubbock, now Lord Avebury and one of only two surviving members of the X Club.

'You will have seen from the published accounts', said Carnegie, who had travelled to London for the occasion, 'how it comes about that this gigantic monster makes his appearance and takes up his abode among you'. And despite the king's absence, he paid tribute to Edward VII, who 'even in his recreations ... seems to keep his eyes and mind ever open for opportunities to advance the interests of his country'. In accepting the gift of the *Diplodocus* on behalf of the Museum and the British nation, Lubbock thanked Carnegie and suggested that the dinosaur would electrify public interest in science. 'Size', he declared, 'appealed to the imagination, and he did not doubt that this specimen would excite the wonder and admiration of all who saw it'. The most telling remarks, however, concerned the shifting balance of transatlantic power. Though the Museum's director joked that the *Diplodocus* was 'an improved and enlarged form of an English creature', Carnegie knew that American palaeontology had outstripped anything that Britain could offer. 'It is doubly pleasing', he proclaimed, 'that this [dinosaur] should come from the youngest of *our* museums on the other side to yours, the parent institution of all, for certainly all those in America may be justly considered in one sense your offspring; we have followed you, inspired by your example.'[22] But the son had grown stronger than the father. Here was Carnegie, low-born in Scotland but a titan of America, coming home to claim the kingdom.

London was not his only conquest, for among the great powers of the Old World the *Diplodocus* became a totem of essential cultural prestige. The French president procured a cast for Georges Cuvier's old haunt at the National Museum for Natural History, and Wilhelm II's ministers negotiated over a copy for Berlin; at Vienna, one of the world's great cultural cities at the *fin de siècle*, Franz Josef received another *Diplodocus* in person, while Archduke Vladimir, the uncle of Nicholas II, begged Carnegie 'not to forget Russia' and the Imperial Academy of Science at St Petersburg.[23] In a final act of defiance of Old World customs, Carnegie's team would not consider requests from Italy unless they came from Victor Emmanuel III himself. By 1910, Russian students had composed a ditty in honour of 'Dippy' the dinosaur: 'Crowned heads of Europe', they sang, 'All make a royal fuss / Over Uncle Andy / And his old diplodocus'.[24]

If this book began with a seventeenth-century archbishop poring over scrolls and parchment, piecing together the biblical history of the Earth, and pinning the date of Creation to 23 October 4004 BC, it ends with the process of zircon dating.[25] In 1896, as the Bone Wars of the American West were winding down, two French scientists discovered the phenomenon of radioactivity. Henri Becquerel and Marie Curie were working in their Paris laboratory when Becquerel, whose early work had focused on phosphorescence, found that uranium salts emitted invisible rays that turned photographic plates to black. Within a few years, researchers such as the great New Zealand physicist Ernest Rutherford and Marie Curie's husband, Pierre, had found that other elements had similar emissions, and that some substances were able to 'trap' these rays. As early as 1907, following discussions with Rutherford, the Yale chemist Bertram Boltwood suggested that these processes could reveal the age of rocks that had trapped radioactive material. He tested samples from around the world, and the results were astonishing. Certain minerals from Connecticut were 410 million years old, and some from the Carolinas even older; rocks from a farm at Telemark in Norway were more ancient again, at least 1.7 billion years old. The material from Ceylon, between Colombo and Kandy in present-day Sri Lanka, was older still: Boltwood calculated it had been there for at least 2.2 billion years, ten times what even the most radical Victorian geologist had suggested for the age of the Earth.

Boltwood is rightly regarded as the father of radiochemistry, but he had been working with a mineral known as thorianite, a rare oxide of the metal thorium. A much better 'test' is zircon, a silicate mineral which occurs in crystals of various colours, often reddish-brown, throughout the crust of the Earth. As the scientist Adam Rutherford relates, zircon has two precious qualities that set it apart from and above its rivals. First, it is incredibly tough: it can withstand all the awesome changes in the earth – tectonic shifts, earthquakes, and volcanic eruptions – that have reworked our planet's surface since its creation. The second is the structure of its atoms, which arrange themselves in cubic patterns, and this neat geometry allows the atoms of certain elements such as uranium to infiltrate zircon crystals, where the mineral's 'molecular box' will trap them. It is this trapped uranium which provides the key to the age of the Earth: when radioactive uranium decays over time, it decays into lead, which by happy coincidence cannot penetrate zircon by itself. And since physicists know the precise, uniform rate at which the decay from uranium into lead occurs, and because any lead within the zircon must have decayed from the uranium, this means that the relative proportions of uranium and lead within zircon crystals provide an incredibly accurate guide to the age of the crystals themselves. The more lead, the more decay has occurred; the more decay, the older the crystal.

The Jack Hills lie about 400 miles north-east of Perth in the forbidding desert of Western Australia. Few people live anywhere near them, although the Mitsubishi corporation has spent years mining iron from the ground, then shipping the ore to Japan from the port at Geraldton. But since the 1980s, physicists and geochronologists have found wondrous things in the fastness of these hills. Among them are plaques of rock and crystal that sparkle in blue, yellow, green, and red. And one of these zircon crystals, which trapped its uranium that decayed into lead, is the oldest substance on Earth. When James Ussher constructed his annals of the Old Testament, and when he suggested that Creation had occurred in 4004 BC, it was the most sophisticated estimate that he could make, given the methods and resources that were available to him. Having refined James Hutton's geology into the system of 'uniformitarianism', Charles Lyell suggested an age of the Earth of many millions of years, and in the later nineteenth century William Thomson, now Lord Kelvin, applied thermodynamics to the question and allowed for an age of up to ninety-six million years. But using radiochemistry and geochronology, scientists from Australia, Scotland, and the United States have calculated the age of the Jack Hills rocks, with an error margin of less than one per cent, to something quite staggering: 4.404 billion years. And that is where we are now.

Acknowledgements

Writing this book has proved challenging for several reasons.

First, there is the aged problem of whether mere historians can do justice to the science, but in respect of every scientific paper, speech, or monograph that is available in English, I have gone to the original material and sought to present those sources in their intellectual contexts. Of course, many others have ploughed that furrow before me. Quite aptly for a book concerning science, therefore, I have been honoured to stand on the shoulders of giants, among them Adrian Desmond, James Moore, Nicolaas Rupke, James Secord, Martin Rudwick, Frank Turner, and Peter Bowler. Whether I have kept my balance on those heights is for the reader to decide.

Second, when I began working on this book in 2020, many if not all libraries and archives had closed in response to the coronavirus pandemic, with visiting restrictions enduring late into 2021. This compelled me to abandon plans for extensive research among the manuscripts of the National Archives, the British Library, the Bodleian Library, and any number of other learned institutions; accordingly, this book's bibliography differs considerably from the one I had anticipated. Even so, and even if the Main Library at the University of Manchester once again became my office, I have been able to cite so many primary sources only because of the growing brilliance of digital archives: the Smithsonian's Biodiversity Heritage Library contains almost every nineteenth-century scientific paper; the Darwin Correspondence Project at Cambridge has published more than 15,000 letters and articles; the British Library's Newspaper Archive has uploaded more than 72 million pages of newsprint; and many of the relevant books – whether monographs, autobiographies, novels, or 'life and letters' volumes – have fallen out of copyright, meaning they are freely available through Google Books, Internet Archive, the HathiTrust Digital Library, and John van Wyhe's wonderful Darwin Online website.

Finally, and more personally, I did not cope especially well during the lockdowns of 2020–21. Where once I had relished working from home, or the splendid isolation of days and nights in the library, enforced solitude played havoc with my health. I plunged into lengthy episodes of depression and anxiety, eating too much, drinking too much, and losing any capacity to concentrate. The acquisition of a beautiful, long-haired cat helped in part – it was at least another heartbeat in the house – but I wonder even now whether I have recovered, whether I have truly re-entered society as a functioning person, and how much better this book could otherwise have been. In any event, I must thank my family, friends, colleagues, employers, teammates, and publishers for the kindness and patience they have shown to me over the past few years.

I owe particular thanks to any number of people. For reading the manuscript and offering advice that has done much to improve it, I must thank Simon Skinner, Peter Mandler, David Norman, Adam Rutherford, and Tom Holland; for their guidance with sources, Jon Parry and Martin Hewitt; and for the generosity of their brief reviews that appear on this book's cover and elsewhere, Richard Holmes, Catherine Fletcher, Adrian Desmond, Steve Brusatte, Sathnam Sanghera, and Tom Holland. Further, I thank my agent Donald Winchester for his enduring support; my editor Will Hammond for once again excavating a book from my manuscript; Rhiannon Roy and Laura Reeves at Bodley Head for guiding me through the post-submission process; Joe Pickering for his tireless work in promoting the book; Duncan Heath for his forensic copy-editing; Mandy Greenfield for her equally thorough proof-reading; Michael Hill for the charming 'Dinosaur Map of Britain'; and Amelia Tolley for the compelling cover art, which – of course – owes much to Henry de la Beche's *Duria Antiquior*. Whatever mistakes remain in this book are my own.

Bibliography

PRINTED PRIMARY SOURCES

'Abstract of the Evidence of Members of the Royal College of Surgeons in London, taken before the Parliamentary Medical Committee in 1834', *Western Journal of the Medical and Physical Sciences*, 9 (1836), 353–61.

Adams, William Henry Davenport, *Life in the Primeval World: Founded on Meunier's 'Les Animaux D'Autrefois'* (London, 1872).

Aikin, John, *The Lives of John Selden, Esq., and Archbishop Ussher; with the Notices of the Principal English Men of Letters with whom They Were Connected* (London, 1812).

Arnold, Matthew, *New Poems* (2nd edn, London, 1868).

Auriol, Anne, *Statement and Correspondence consequent on the Ill-Treatment of Lady de la Beche by Major-General Henry Wyndham* (London, 1843).

Aurousseau, M. (ed.), *The Letters of F. W. Ludwig Leichhardt* (3 vols, Cambridge, 1968).

Austen, Jane, *Northanger Abbey: and Persuasion* (4 vols, London, 1817).

Beche, Henry De la, *Notes on the Present Condition of the Negroes in Jamaica* (London, 1825).

———, *Remarks on the Geology of Jamaica* (London, 1827).

———, 'Anniversary Address of the President', *Quarterly Journal of the Geological Society of London*, 4 (1848), xxi–xxxi.

——— and W. D. Conybeare, 'Notice of the Discovery of a New Fossil Animal, forming a Link between the Ichthyosaurus and Crocodile, together with General Remarks on the Osteology of the Ichthyosaurus', *Transactions of the Geology Society*, 5 (1821), 559–94.

Beecher, Charles, 'Memoir of Othniel Charles Marsh', *American Journal of Science*, 7 (1899), 403–28.

Bell, Charles, *The Hand, Its Mechanism, and Vital Endowments as Evincing Design* (London, 1833).

Bell, Thomas, 'Presidential Address', *Journal of the Proceedings of the Linnean Society*, 4 (1860), viii–xx.

Bernard, Nicholas, *The Life and Death of the Most Reverend and Learned Father of Our Church, Dr James Ussher, late Arch-Bishop of Armagh, and Primate of All-Ireland* (London, 1656).

Besant, Annie, *The Law of Population: Its Consequences and Its Bearing upon Human Conduct and Morals* (New York, 1878).

———, *Autobiographical Sketches* (London, 1885).

Bierce, Ambrose, *The Devil's Dictionary* [1906] (Cleveland, 1911).

[Blake, Charles Carter], 'Professor Huxley on Man's Place in Nature', *Edinburgh Review*, 117 (1863), 278–93.

Boswell, James, *Life of Samuel Johnson, LL.D.* [1791] (Chicago, 1972).

Bradlaugh, Charles, *The Bible: What It Is: Genesis, Its Authorship and Authenticity* (London, 1870).

———, *The Land, The People, and the Coming Struggle* [1871] (3rd edn, London, 1880).

———, *The Impeachment of the House of Brunswick* [1872] (2nd edn, London, 1873).

———, *When Were Our Gospels Written?* (4th edn, London, 1881).

[Brewster, David], 'Reflexions on the Decline of Science in England, and on Some of its Causes' [review], *Quarterly Review*, 43 (1830), 305–42.

'A Brief Relation of the Life and Memoirs of John, Lord Belasyse, Written and Collected by His Secretary, Joshua Moore', *Calendar of the Manuscripts of the Marquess of Ormonde, K.P., preserved at Kilkenny Castle*, 2 (1903), 376–99.

'British Association: Section D – Zoology and Botany', *Athenaeum*, 1707 (1860), 64–5.

'The British Association at Cambridge', *Quarterly Review*, 113 (1863), 362–92.

Brookes, Richard, *The Natural History of Waters, Earths, Stones, Fossils, and Minerals, with their Virtues, Properties, and Medicinal Uses* (London, 1763).

Brydone, Patrick, *A Tour through Sicily and Malta in a Series of Letters to William Beckford, Esq., of Somerly in Suffolk* (London, 1806).

Buckland, Francis T., *Curiosities of natural history* (4th edn, New York, 1859).

Buckland, William, *Vindiciae Geologicae; or, the Connexion of Geology with Religion Explained, in an Inaugural Lecture Delivered before the University of Oxford, May 15, 1819, on the Endowment of a Readership in Geology by His Royal Highness, the Prince Regent* (Oxford, 1820).

———, *Reliquiae Diluvianae: or, Observations on the Organic Remains contained in Caves, Fissures, and Diluvial Gravel, and on Other Geological Phenomena, Attesting the Action of an Universal Deluge* (London, 1823).

———, 'Notice on the Megalosaurus or Great Fossil Lizard of Stonesfield', *Transactions of the Geological Society of London*, 1 (2nd ser., 1824), 390–96.

———, 'On the Discovery of a New Species of Pterodactyle in the Lias at Lyme Regis', *Transactions of the Geological Society of London*, 3 (2nd ser., 1829), 217–22.

———, 'On the Discovery of Coprolites, or Fossil Faeces, in the Lias at Lyme Regis, and in Other Formations', *Transactions of the Geological Society of London*, 3 (2nd ser., 1829), 223–36.

———, *Geology and Mineralogy considered with Reference to Natural Theology* (2 vols, London, 1836).

Bugg, George, *Scriptural Geology: or, Geological Phenomena, Consistent Only with the Literal Interpretation of the Sacred Scriptures* (2 vols, London, 1826–7).

Burke, Edmund, *Reflections on the Revolution in France, and on the Proceedings in Certain Societies in London* (3rd edn, London, 1790).

Burnet, Gilbert, *The Life of William Bedell D.D., Lord Bishop of Killmore in Ireland* (London, 1692).

Busk, George, 'On the Crania of the Most Ancient Races of Man. By Professor D. Schaaffhausen, of Bonn', *Natural History Review*, 1 (1861), 155–76.

———, 'On a Very Ancient Human Cranium from Gibraltar', *Report of the Thirty-Fourth Meeting of the British Association for the Advancement of Science; Held at Bath in September 1864* (London, 1865), 91–2.

Butler, Samuel, *Erewhon: or, Over the Range* (London, 1872).

———, *Unconscious Memory: A Comparison between the Theory of Dr. Ewald Hering . . . and the 'Philosophy of the Unconscious' of Dr. Edward von Hartmann* (London, 1880).

Byron, George Gordon, *Don Juan: Cantos IX, X, and XI* (London, 1823).

Campbell, George, 'On Tertiary Leaf-Beds in the Isle of Mull', *Quarterly Journal of the Geological Society of London*, 7 (1851), 89–103.

———, *The Reign of Law* [1867] (5th edn, London, 1871).

———, *Primeval Man: An Examination of Some Recent Speculations* [1869] (3rd edn, London, 1870).

Campbell, John, *The Lives of the Lord Chancellors and Keepers of the Great Seal of England from the Earliest Times till the Reign of King George IV* (3 vols, London, 1845).

'The Cardiff Giant A Humbug', *Proceedings of the Massachusetts Historical Society*, 11 (1871), 161–2.

Carron, William, *Narrative of an Expedition Undertaken Under the Direction of the Late Mr. Assistant Surveyor E. B. Kennedy for the Exploration of the Country Lying Between Rockingham Bay and Cape York* (Sydney, 1849).

Carus, C. G., *The King of Saxony's Journey through England and Scotland in the Year 1844* (London, 1846).

Census of Great Britain, 1851: Religious Worship: England and Wales: Report and Tables (London, 1853).

[Chambers, Robert], *Vestiges of the Natural History of Creation* (London, 1844).

———, *Vestiges of the Natural History of Creation* (10th edn, London, 1853).

Charles Bradlaugh: A Record of His Life and Work, ed. Hypatia Bradlaugh Bonner (2 vols, London, 1895).

Charles Darwin: His Life Told in an Autobiographical Chapter, and in a Selected Series of His Published Letters, ed. Francis Darwin (London, 1892).

Charles Darwin's Beagle Diary, ed. Richard Darwin Keynes (Cambridge, 1988).

Cobbett, William, *Rural Rides* (London, 1830).

Colenso, John William, *Ten Weeks in Natal: A Journal of a First Tour of Visitation among the Colonists and Zulu Kafirs of Natal* (London, 1855).

———, *First Steps of the Zulu Mission (Oct. 1859)* (London, 1860).

———, *Zulu–English Dictionary* (Pietermaritzburg, 1861).

———, *St Paul's Epistle to the Romans: Newly Translated, and Explained from a Missionary Point of View* (Cambridge, 1861).

———, 'On Missions to the Zulus in Natal and Zululand', *Social Science Review and Journal of the Sciences*, 18 (1865), 482–510.

———, *The Pentateuch and Book of Joshua Critically Examined: Part I* (London, 1865).

'Convocation', *Contemporary Review*, 1 (1866), 250–69.

Conybeare, W. D., *Outlines of the Geology of England and Wales, with an Introductory Compendium of the General Principles of that Science* (London, 1822).

———, 'On the Discovery of an Almost Perfect Skeleton of the Plesiosaurus', *Transactions of the Geological Society of London*, 1 (2nd ser., 1824), 381–9.

———, 'Report on the Progress, Actual State, and Ulterior Prospects of Geological Science', *Report of the First and Second Meetings of the British Association for the Advancement of Science* (London, 1833), 365–414.

Conybeare, W. J., *Church Parties: An Essay, reprinted from 'The Edinburgh Review'* (London, 1854).

Cooper, Robert (ed.), *The London Investigator: A Monthly Journal of Secularism*, 1 (London, 1854–5).

Cope, Edward Drinker, '[On the *Archaeopteryx* and *Compsognathus*], *Proceedings of the Academy of Natural Sciences of Philadelphia*, 19 (1867), 234–5.

———, *On the Origin of Genera* (Philadelphia, 1869).

———, *On the Origin of the Fittest: Essays on Evolution* (New York, 1887).

[Copleston, Edward], 'Buckland-*Reliquiae Diluvianae*', *Quarterly Review*, 29 (1823), 138–65.

The Correspondence of Charles Darwin, ed. Frederick Burkhardt (30 vols, Cambridge, 1985–).

The Correspondence of James Ussher, 1600–1656, ed. Elizabethanne Boran (3 vols, Dublin, 2015).

Cox, G. V., *Recollections of Oxford* (2nd edn, London, 1870).

Cumberland, George, 'On [the] Proper Objects of Geology', *Monthly Magazine*, 40 (1815), 130–33.

Curtis, Wardon Allan, 'The Monster of Lake LaMetrie', *Pearson's Magazine*, 7 (1899), 341–7.

Cuvier, Georges, 'Extract from a Memoir on an Animal of Which the Bones are Found in the Plaster Stone around Paris, and which appears no longer to exist alive today' [1798], in Martin J. S. Rudwick, *Georges Cuvier, Fossil Bones, and Geological Catastrophes: New Translations & Interpretations of the Primary Texts* (Chicago, 1997), 35–41.

———, *Recherches sur les Ossemens Fossiles des Quadrupèdes* (5 vols, Paris, 1812–23).

———, *Le Règne Animal, Distribué d'après son Organisation* (4 vols, Paris, 1816).

———, 'Biographical Memoirs of M. de Lamarck', *Edinburgh New Philosophical Journal*, 20 (1836), 1–21.

Darwin, Charles, *Extracts from Letters Addressed to Professor Henslow* (Cambridge, 1835).

———, *Narrative of the Surveying Voyages of His Majesty's Ships Adventure and Beagle between the Years 1826 and 1836, Describing their Examination of the Southern Shores of South America, and the Beagle's Circumnavigation of the Globe, 1832–1836* (3 vols, London, 1839).

———, *The Structure and Distribution of Coral Reefs* (London, 1842).

———, *Geological Observations on the Volcanic Islands* (London, 1844).

———, *A Monograph of the Sub-Class Cirripedia, with Figures of All the Species: The Lepadidae; or Pedunculated Cirripedes* (London, 1851).

———, *A Monograph on the Fossil Lepadidae, or Pedunculated Cirripedes of Great Britain* (London, 1851).

———, *A Monograph of the Sub-Class Cirripedia, with Figures of all the Species. The Balanidae (or Sessile Cirripedes); the Verrucidae, etc.* (London, 1854).

———, *A Monograph on the Fossil Balanidae and Verrucidae of Great Britain* (London, 1854).

——— and Alfred Russel Wallace, 'On the Tendency of Species to Form Varieties; and on the Perpetuation of Varieties and Species by Natural Means of Selection', *Journal of the Proceedings of the Linnean Society*, 3 (1859), 45–62.

———, *On the Origin of Species by Means of Natural Selection, or the Preservation of Favoured Races in the Struggle for Life* (2nd edn, London, 1860).

——, *On the Origin of Species by Means of Natural Selection, or the Preservation of Favoured Races in the Struggle for Life* (3rd edn, London, 1861).

——, *On the Various Contrivances by which British and Foreign Orchids are Fertilised by Insects and on the Good Effects of Intercrossing* (London, 1862).

——, *On the Movements and Habits of Climbing Plants* (London, 1865).

——, *On the Origin of Species by Means of Natural Selection, or the Preservation of Favoured Races in the Struggle for Life* (4th edn, London, 1866).

——, *The Variation of Animals and Plants under Domestication* (2 vols, London, 1868).

——, *The Expression of the Emotions in Man and Animals* (London, 1872).

——, *The Origin of Species by Means of Natural Selection, or the Preservation of Favoured Races in the Struggle for Life* (6th edn, London, 1872).

——, *The Effects of Cross and Self Fertilisation in the Vegetable Kingdom* (London, 1876).

——, *The Power of Movement in Plants* (London, 1880).

Darwin Correspondence Project (www.darwinproject.ac.uk).

Darwin, Francis (ed.), *The Foundations of the Origin of Species: Two Sketches Written in 1842 and 1844* (Cambridge, 1909).

'Darwin on *The Descent of Man*', *Edinburgh Review*, 134 (1871), 195–235.

Davies, William, 'On the Exhumation and Development of a Large Reptile (*Omosaurus Armatus*, Owen), from the Kimmeridge Clay, Swindon, Wilts', *Geological Magazine*, 3 (1876), 193–7.

Deluc, Jean-André, *Geological Travels* (3 vols, 1810–11).

'Denarius' [Henry Cole], *Shall We Keep the Crystal Palace, and Have Riding and Walking in all Weathers among Flowers, Fountains and Sculptures?* (London, 1851).

Dickens, Charles, *The Mudfog Papers and Other Sketches* [1837–8] (New York, 1880).

——, *Dealings with the Firm of Dombey and Son: Wholesale, Retail, and for Exportation* (2 vols, New York, 1848).

——, *Bleak House* (London, 1853).

——, *Hard Times: For These Times* (Leipzig, 1854).

'Discussion on Captain S. Osborn's Paper', *Proceedings of the Royal Geographical Society*, 15 (1871), 37–9.

'Donations to the Cabinet of Minerals', *Transactions of the Geological Society*, 5 (1819), 640–50.

Doyle, Arthur Conan, *The Lost World* (New York, 1912).

'Dr Buckland's *Bridgewater Treatise*', *Quarterly Review*, 46 (1836), 31–64.

Draper, John William, 'On the Intellectual Development of Europe, Considered with Reference to the Views of Mr. Darwin and Others', *Report of the*

Thirtieth Meeting of the British Association for the Advancement of Science; Held at Oxford in June and July 1860 (London, 1861), 115–16.

———, *History of the Conflict Between Science and Religion* (New York, 1874).

Du Chaillu, Paul B., 'The Geographical Features and Natural History of a Hitherto Unexplored Region of Western Africa', *Proceedings of the Royal Geographical Society of London*, 5 (1860–61), 108–12.

Duncan, P. B., 'On the Woman in Paviland Cavern', in C. G. B. Daubeny (ed.), *Fugitive Poems connected with Natural History and Physical Science* (Oxford, 1869).

Dwight, Timothy, *Memories of Yale Life and Men, 1854–1899* (New York, 1903).

Eliot, George, 'The Lifted Veil', *Blackwood's Edinburgh Magazine*, 86 (1859), 24–47.

———, *The Mill on the Floss* (2 vols, Leipzig, 1860).

———, *Silas Marner: The Weaver of Raveloe* (Leipzig, 1861).

Elrington, Thomas, *A Letter to the Clergy of the United Diocese, on the Church Temporalities Act* (Dublin, 1833).

Engels, Friedrich, *The Condition of the Working Class in England in 1844*, trans. Florence Kelley Wischnewetzky [1845] (London, 1892).

Essays and Reviews [1860] (4th edn, London, 1861).

'Essays and Reviews' [review], *Edinburgh Review*, 113 (1861), 461–500.

'The Feathered Reptile of Solenhofen', *Intellectual Observer*, 1 (1862), 367–8.

Feuerbach, Ludwig, *The Essence of Christianity*, trans. Mary Ann Evans [1841] (2nd edn, New York, 1855).

Figuier, Louis, *La Terre avant le Déluge* (Paris, 1863).

FitzRoy, Robert, 'Sketch of the Surveying Voyages of His Majesty's Ships Adventure and Beagle, 1825–1836', *Journal of the Royal Geographical Society of London*, 6 (1836), 311–43.

———, *Narrative of the Surveying Voyages of His Majesty's Ships Adventure and Beagle between the Years 1826 and 1836, Describing Their Examination of the Southern Shores of South America, and the Beagle's Circumnavigation of the Globe. Proceedings of the Second Expedition, 1831–36, under the Command of Captain Robert Fitz-Roy, R.N.* (London, 1839).

Flower, William Henry, 'On the Posterior Lobes of the Cerebrum of the Quadrumana', *Philosophical Transactions of the Royal Society of London*, 152 (1862), 185–203.

Foote, G. W., *Secularism: The True Philosophy of Life: An Exposition and a Defence* (London, 1879).

———, *Comic Bible Sketches reprinted from 'The Freethinker'* (London, 1885).

———, *Prisoner for Blasphemy* (London, 1886).

———, *Reminiscences of Charles Bradlaugh* (London, 1891).

['Foulke and Leidy on the *Hadrosaurus*'], *Proceedings of the Academy of Natural Sciences of Philadelphia*, 10 (1859), 213–18.

Fox, 'On the Skull and Bones of an Iguanodon', *Report of the Thirty-Eighth Meeting of the British Association for the Advancement of Science; Held at Norwich in August 1868* (London, 1869), 64–5.

Froude, J. A., *Shadow of the Clouds* (London, 1847).

———, *The Nemesis of Faith* (London, 1849).

Gall, Franz Josef, *On the Functions of the Brain and of Each of Its Parts: On the Organ of the Moral Qualities and Intellectual Faculties, and the Plurality of the Cerebral Organs*, trans. Winslow Lewis (6 vols, Boston, MA, 1835).

Gaskell, Elizabeth, *North and South* (Leipzig, 1855).

Geoffroy Saint-Hilaire, Étienne, *Philosophie Anatomique: Des Organes Respiratoires sous le Rapport de la Détermination et de l'Identité de Leurs Pièces Osseuses* (Paris, 1818).

'Geology and Mineralogy', *Edinburgh Review*, 45 (1837), 1–39.

George Douglas Campbell, Eighth Duke of Argyll, K.G., K.T. (1823–1900): Autobiography and Memoirs, ed. Dowager Duchess of Argyll (2 vols, London, 1906).

Gibbon, Charles, *The Life of George Combe: Author of 'The Constitution of Man'* (2 vols, London, 1878).

Gleig, George (ed.), *Supplement to the Third Edition of the Encyclopaedia Britannica, or, a Dictionary of Arts, Sciences, and Miscellaneous Literature* (2 vols, Edinburgh, 1801).

Gosse, Philip Henry, *Omphalos: An Attempt to Untie the Geological Knot* (London, 1857).

Grant, Robert, 'Observations on the Nature and Importance of Geology', *Edinburgh New Philosophical Journal*, 1 (1826), 293–302.

———, *An Essay on the Study of the Animal Kingdom: Being an Introductory Lecture delivered in the University of London* (London, 1829).

Greenhill, William, 'To the Judicious Reader, and All Desirous to Understand the Chronology of Sacred Scriptures', in Thomas Allen, *A Chain of Scripture Chronology; from the Creation of the World to the Death of Jesus Christ* [1659] (London, 1668).

Haeckel, Ernst, *Generelle Morphologie der Organismen: Allgemeine Grundzüge der Organischen Formen-Wissenschaft, Mechanisch Begrundet Durch Die von Charles Darwin* (Berlin, 1866).

Hake, Gordon, *Memoirs of Eighty Years* (London, 1892).

Hampden, John, *Is Water Level or Convex After All? The Bedford Canal Swindle Detected & Exposed* (Swindon, 1870).

Hardy, Thomas, *A Pair of Blue Eyes* (3 vols, London, 1873).

———, *The Return of the Native* (3 vols, London, 1878).

Hawkins, Thomas, *The Book of the Great Sea-Dragons, Ichthyosauri and Plesiosauri, Gedolim Taninim of Moses, Extinct Monsters of the Ancient Earth* (London, 1840).

Holyoake, George Jacob, *The History of the Last Trial by Jury for Atheism in England: A Fragment of Autobiography* (London, 1851).

———, *Self-Help by the People: Thirty-Three Years of Co-Operation in Rochdale* [1858] (9th edn, 2 vols, London, 1882).

———, *The Principles of Secularism Illustrated* (3rd edn, London, 1871).

———, *Sixty Years of an Agitator's Life* (2 vols, London, 1891).

———, *English Secularism: A Confession of Belief* (London, 1896).

Home, Everard, 'Some Farther Account of the Fossils Remains of An Animal, of which a Description was Given to the Society in 1814', *Philosophical Transactions of the Royal Society*, 106 (1816), 318–21.

———, 'Additional Facts respecting the Fossil Remains of an Animal, on the Subject of which Two Papers have been printed in the Philosophical Transactions', *Philosophical Transactions of the Royal Society of London*, 108 (1818), 24–32.

———, 'An Account of the Fossil Skeleton of the Proteo-Saurus', *Philosophical Transactions of the Royal Society*, 109 (1819), 209–11.

———, 'Reasons for Giving the Name Proteo-Saurus to the Fossil Skeleton which has been Described', *Philosophical Transactions of the Royal Society of London*, 109 (1819), 212–16.

Hone, William, *The Late John Wilkes's Catechism of a Ministerial Member* (London, 1817).

———, *The Sinecurist's Creed, or Belief* (London, 1817).

———, *The Political Litany, Diligently Revised* (London, 1817).

Hood, Thomas, 'A Geological Excursion to Tilgate Forest, A.D. 2000', in *Relics from the Wreck of a Former World; or, Splinters Gathered on the Shores of a Turbulent Planet* (New York, 1847), 53–6.

Hooke, Robert, 'Lectures and Discourses of Earthquakes, and Subterraneous Eruptions', in *The Posthumous Works of Robert Hooke … containing his Cutlerian Lectures and Other Discourses*, ed. Richard Waller (London, 1705), 277–450.

Hooker, Joseph Dalton, 'Introductory Essay', in *Flora Tasmaniae* (2 vols, London, 1860), 1:i–cxxviii.

Hughes, Thomas, *Tom Brown's School Days* (London, 1857).

———, *Memoir of a Brother* [1859] (2nd edn, London, 1874).

———, *Tom Brown at Oxford* [1859] (3 vols, Boston, MA, 1868).

Humboldt, Alexander von, *Personal Narrative of Travels to the Equinoctial Regions of the New Continent During the Years, 1799–1804*, trans. Helen Maria Williams (7 vols, London, 1829).

Hurrell Froude: Memoranda and Comments, ed. Louise Guiney (London, 1904).

Hutton, James, *Abstract of a Dissertation read in the Royal Society of Edinburgh, upon the Seventh of March, and Fourth of April, MDCCLXXXV, concerning the System of the Earth, its Duration, and Stability* (Edinburgh, 1785).

———, 'Theory of the Earth; or an Investigation of the Laws Observable in the Composition, Dissolution, and Restoration of Land upon the Globe', *Transactions of the Royal Society of Edinburgh*, 1 (1788), 209–304.

Huxley, Thomas Henry, 'On a Hitherto Undescribed Structure in the Human Hair Sheath', *Medical Gazette*, 1 (1845), 1340–41.

———, 'On the Anatomy and the Affinities of the Family of the *Medusae*', *Philosophical Transactions of the Royal Society*, 139 (1849), 413–37.

———, 'The Vestiges of Creation', *British and Foreign Medico-Chirurgical Review*, 13 (1854), 332–43.

———, 'Lectures on General Natural History: Lecture XII', *Medical Times & Gazette*, 15 (1857), 238–41.

———, 'On the Persistent Types of Animal Life' [1859], *Notices of the Proceedings at the Meetings of the Members of the Royal Institution*, 3 (1858–60), 151–3.

———, 'The Darwinian Hypothesis' [1859], in *Collected Essays* (2 vols, New York, 1894), 2:1–21.

———, 'The Croonian Lecture: On the Theory of the Vertebrate Skull', *Proceedings of the Royal Society of London*, 9 (1859), 381–457.

———, 'On the Zoological Relations of Man with the Lower Remains', *Natural History Review*, 1 (1861), 67–84.

———, 'Man and the Apes', *Athenaeum*, 1744 (1861), 433.

———, 'Man and the Apes' (II), *Athenaeum*, 1746 (1861), 498.

———, 'On Species and Races, and Their Origin', *Notices of the Proceedings at the Meetings of the Members of the Royal Institution*, 3 (1862), 195–200.

———, *Evidence as to Man's Place in Nature* (London, 1863).

———, 'On *Acanthopolis Horridus*, a New Reptile from the Chalk-Marl', *Geological Magazine*, 4 (1867), 65–7.

———, 'On the Animals which are Most Nearly Intermediate between Birds and Reptiles', *Annals and Magazine of Natural History*, 2 (1868), 66–75.

———, 'On Some Organisms Living at Great Depths in the North Atlantic Ocean', *Quarterly Journal of Microscopical Science*, 8 (1868), 203–12.

———, 'A Liberal Education; and Where to Find It', *Macmillan's Magazine*, 17 (1868), 367–78.

———, *On the Physical Basis of Life* [1868] (2nd edn, New Haven, CT, 1870).

———, 'Triassic Dinosauria', *Nature*, 1 (1869), 23–4.

———, 'Geological Reform' [1869], in Huxley, *Discourses, Biological and Geological: Essays* (New York, 1896), 305–39.

———, 'The Anniversary Address of the President', *Quarterly Journal of the Geological Society of London*, 25 (1869), xxviii–liii.

———, 'On the Ethnology of Britain', *Journal of the Ethnological Society of London*, 2 (1870), 382–4.

———, *Lay Sermons, Essays, and Reviews* (London, 1870).

———, 'Darwin', *Nature*, 25 (1882), 597.

———, *Science and Culture, and Other Essays* (New York, 1882).

———, *American Addresses: With a Lecture on the Study of Biology* (London, 1886).

———, 'Agnosticism', *Eclectic Magazine*, 49 (1889), 433–50.

'Iconoclast', *The Bible: What It Is* (London, 1861).

'Interesting Fossil Bones in Philadelphia', *Friends' Intelligencer*, 15 (1859), 715–17.

Is It Reasonable to Worship God? Verbatim Report of Debate between R. A. Armstrong and C. Bradlaugh (London, 1878).

Jane Austen's Letters to Her Sister Cassandra and Others, ed. R. W. Chapman (2nd edn, London, 1952).

Jelf, R. W., *Grounds for Laying before the Council of King's College London, Certain Statements contained in a Recent Publication, entitled, 'Theological Essays'* (2nd edn, London, 1853).

Jones, Philip, *Popular Phrenology: Tried by the Word of God, and Proved to be Antichrist, and Injurious to Individuals and Families* (London, 1845).

The Journal of Gideon Mantell, Surgeon and Geologist, Covering the Years, 1818–1852, ed. E. Cecil Curwen (London, 1940).

Keble, *National Apostasy: Considered in a Sermon Preached in St Mary's Oxford, before His Majesty's Judges of Assize, on Sunday, July 14, 1833* (Oxford, 1833).

Kidd, John, *A Geological Essay on the Imperfect Evidence in Support of a Theory of the Earth, Deducible either from its General Structure or from the Changes Produced on its Surface by the Operation of Existing Causes* (Oxford, 1815).

———, *An Answer to a Charge against the English Universities contained in the Supplement to the Edinburgh Encyclopaedia* (Oxford, 1818).

Kingsley, Charles, *Westward Ho! Or the Voyages and Adventures of Sir Amyas Leigh* (London, 1855).

———, 'The Wonders of the Shore', *North British Review*, 22 (1855), 1–56.

———, *Two Years Ago* (3 vols, London, 1857).

———, *The Water-Babies: A Fairy Tale for a Land Baby* [1862–3] (London, 1880).

——, 'Speech of Lord Dundreary in Section D, on Friday Last, on the Great Hippocampus Question', in *Charles Kingsley: His Letters and Memories of His Life*, ed. Frances Kingsley [1877] (New York, 1889), 322–5.

Kirby, William, *On the Power and Goodness of God as Manifested in the Creation of Animals and in Their History, Habits, and Instincts* (2 vols, London, 1835).

Kirwan, Richard, 'Hutton's Theory of the Earth', *British Critic*, 8 (1796), 468–80.

Knight, Charles R., *Autobiography of an Artist*, ed. Jim Ottaviani (New York, 2005).

Knowlton, Charles, *Fruits of Philosophy: An Essay on the Population Question* [1832] (Rotterdam, 1877).

The Lambeth Conferences of 1867, 1878, and 1888: With the Official Reports and Resolutions, together with the Sermons Preached at the Conferences (London, 1889).

Lee, Mrs R., *Memoirs of Baron Cuvier* (London, 1833).

Lehmann, Rudolf Chambers, *Memories of Half a Century* (London, 1908).

Leidy, Joseph, *Cretaceous Reptiles of the United States* (Philadelphia, 1864).

Lesley, J. P., 'Obituary Notice of Wm. Parker Foulke', *Proceedings of the American Philosophical Society*, 10 (1868), 481–510.

Letters and Correspondence of John Henry Newman during His Life in the English Church with a Brief Autobiography, ed. Anne Mozley (2 vols, London, 1891).

Letters and Exercises of the Elizabethan Schoolmasters John Conybeare, with Notes and a Fragment of Autobiography by the Very Rev. William Daniel Conybeare, ed. Frederick Cornwallis Conybeare (London, 1905).

The Life and Correspondence of William Buckland, D.D., F.R.S.., Sometime Dean of Westminster, ed. Elizabeth Oke Gordon (London, 1894).

The Life and Letters of Benjamin Jowett, M.A., Master of Balliol College, Oxford, ed. Evelyn Abbott and Lewis Campbell (2nd edn, 2 vols, London, 1897).

Life and Letters of George Jacob Holyoake, ed. Joseph McCabe (2 vols, London, 1908).

The Life and Letters of the Reverend Adam Sedgwick, ed. John Willis Clark and Thomas McKenny Hughes (London, 1890).

Life and Letters of Sir Joseph Dalton Hooker, O.M., G.C.S.I. (2 vols, London, 1918).

The Life and Letters of Thomas Henry Huxley, ed. Leonard Huxley (2 vols, London, 1903).

Life, Letters, and Journals of Sir Charles Lyell, ed. Mrs Lyell (2 vols, London, 1881).

The Life of Frederick Denison Maurice, Chiefly Told in His Own Letters, ed. Frederick Maurice (2 vols, London, 1884).

The Life of John William Colenso, D.D., Bishop of Natal (2 vols, London, 1888).

The Life of Robert Gray, Bishop of Cape Town, ed. Charles Gray (2 vols, London, 1876).

'Linnean Society', *Annals of Philosophy*, 5 (1815), 70.

Lubbock, John, *Pre-Historic Times as Illustrated by Ancient Remains* (London, 1865).

Lydekker, Richard, 'Notices of New and Other Vertebrata from Indian Territory and Secondary Rocks', *Records of the Geological Society of India*, 9 (1876), 30–49.

Lyell, Charles, *Principles of Geology: or, the Modern Changes of the Earth and Its Inhabitants, Considered as Illustrative of Geology* (3 vols, London, 1830–33).

———, 'Presidential Address to the Geological Society', *Proceedings of the Geological Society of London*, 2 (1837), 479–523.

———, *Elements of Geology* (2nd edn, 2 vols, London, 1841).

———, *Travels in North America; with Geological Observations of the United States, Canada, and Nova Scotia* (2 vols, London, 1845).

———, *A Second Visit to the United States of North America* (2 vols, London, 1849).

———, *The Geological Evidences of the Antiquity of Man with Remarks on Theories on the Origin of Species by Variation* (London, 1863).

Mann, Horace, *Letters on the Existing Law of Registration of Births, Deaths, and Marriages, with Suggestions for its Improvement* (London, 1849).

Mantell, Gideon, *The Fossils of the South Downs; or Illustrations of the Geology of Sussex* (London, 1822).

——— and Rev. T. W. Horsfield, *The History and Antiquities of Lewes and its Vicinity* (2 vols, Lewes, 1824).

———, 'Notice on the Iguanodon, a Newly Discovered Fossil Reptile from the Sandstone of Tilgate Forest, in Sussex', *Philosophical Transactions of the Royal Society of London*, 115 (1825), 179–86.

———, 'The Geological Age of Reptiles', *Edinburgh New Philosophical Journal*, 11 (1831), 181–5.

———, *The Geology of the South-East of England* (London, 1833).

———, 'Observations on the Remains of the Iguanodon, and Other Fossil Reptiles of the Strata of Tilgate Forest in Sussex', *Proceedings of the Geological Society of London*, 1 (1834), 410–11.

———, 'Notice of the Discovery of the Remains of the Iguanodon in the Lower Green Sand Formation of the South-East of England', *American Journal of Science and Arts*, 27 (1835), 355–60.

———, *The Wonders of Geology; Or, A Familiar Exposition of Geological Phenomena* (2 vols, London, 1838).

———, 'Memoir on a Portion of the Lower Jaw of the Iguanodon, and on the Remains of the Hylaeosaurus and Other Saurians, Discovered in the Strata of Tilgate Forest, in Sussex', *Philosophical Transactions of the Royal Society of London*, 131 (1841), 131–51.

———, 'Case of Abscess of the Prostate Gland', *Lancet*, 2 (1841), 301.

———, 'Cultivation of the Canine Brain', *Lancet*, 2 (1841), 907–08.

———, *Thoughts on a Pebble; or, A First Lesson in Geology* (6th edn, London, 1842).

———, *The Medals of Creation; Or, First Lessons in Geology and in the Study of Organic Remains* (2 vols, London, 1844).

———, *Thoughts on a Pebble: or, a First Lesson in Geology* (8th edn, London, 1849).

———, *A Pictorial Atlas of Fossil Remains, Consisting of Coloured Illustrations* (London, 1850).

———, *Petrifactions and Their Teachings; or, a Hand-Book to the Gallery of Organic Remains of the British Museum* (London, 1851).

'Mary Anning, the Fossil Finder', *All The Year Round*, 13 (1865), 60–63.

Marsh, Catherine, *Memorials of Captain Hedley Vicars, Ninety-Seventh Regiment* (London, 1856).

Marsh, Othniel Charles, 'Description of the Remains of a new Enaliosaurian', *American Journal of Science and Arts*, 34 (1862), 1–16.

———, 'Notice of a New and Diminutive Species of Fossil Horse (*Equus Parvulus*), from the Tertiary of Nebraska', *Annals and Magazine of Natural History*, 3 (1869), 95–6.

———, 'On a New Subclass of Fossil Birds (Odontornithes)', *American Journal of Science*, 5 (1873), 233–4.

———, *Introduction and Succession of Vertebrate Life in America* (New Haven, CT, 1877).

———, 'The Principal Characters of American Jurassic Dinosaurs', *American Journal of Science and Arts*, 16 (1878), 411–16.

———, 'Thomas Henry Huxley', *American Journal of Science*, 50 (1895), 177–83.

Martin, John, *A Letter to the Hon. Thomas Erskine, with a Postscript, to the Right Hon. Lord Kenyon, upon their Conduct at the Trial of Thomas Williams for Publishing Paine's Age of Reason* (3rd edn, London, 1797).

Martin, Theodore, *The Life of His Royal Highness the Prince Consort* (2 vols, New York, 1879).

Martineau, Harriet, *Illustrations of Political Economy* (25 vols, London, 1832–4).

———, *The History of England during the Thirty Years' Peace, 1816–1846* (2 vols, London, 1849–50).

Maurice, F. D., *The Epistle to the Hebrews; Being the Substance of Three Lectures Delivered in the Chapel of the Honourable Society of Lincoln Inn* (London, 1846).

———, *The Religions of the World and their Relations to Christianity* [1847] (3rd edn, London, 1854).

———, *The Lord's Prayer: Nine Sermons preached in the Chapel of Lincoln's Inn* (2nd edn, London, 1849).

———, *Theological Essays* (Cambridge, 1853).

———, *The Word 'Eternal' and the Punishment of the Wicked: A Letter to the Rev. Dr. Jelf, Canon of Christ Church, and Principal of King's College* (5th edn, Cambridge, 1854).

Mayhall, John, *The Annals and History of Leeds and Other Places in the County of York* (London, 1860).

Mayhew, Henry, and George Cruikshank, *1851: or, The Adventures of Mr and Mrs Sandboys and Family, Who Came Up to London to 'Enjoy Themselves' and to See the Great Exhibition* (London, 1851).

'A Member of the Council of the Clapham Athenaeum', *A Reminiscence of Gideon Algernon Mantell* (London, 1853).

A Memoir of Miss Frances Augusta Bell, Who Died in Kentish Town, ed. Johnson Grant (London, 1827).

Memoirs of Sir Andrew Crombie Ramsay, ed. Archibald Geikie (London, 1895).

Memoirs of the Life and Correspondence of Mrs Hannah More, ed. William Roberts (2 vols, New York, 1845).

The Memoirs of the Life of Edward Gibbon [1796], ed. George Birkbeck Hill (London, 1900).

Memoirs of William Smith, LL.D., Author of the 'Map of the Strata of England and Wales', ed. John Phillips (London, 1844).

Memories of Old Friends: Being Extracts from the Journals and Letters of Caroline Fox, of Penjerrick, Cornwall, from 1835 to 1871, ed. Horace N. Pym (London, 1882).

Mildert, William van, *An Historical View of the Rise and Progress of Infidelity, with a Refutation of its Principles and Reasonings: in a Series of Sermons Preached for the Lecture Founded by the Hon. Robert Boyle, in the Parish Church of St Mary le Bow, London, from the Year 1802 to 1805* (2 vols, London, 1820).

Mill, John Stuart, *A System of Logic, Ratiocinative and Inductive* (2 vols, London, 1843).

Milman, Henry Hart, *The History of the Jews* (3 vols, London, 1829).

Milner, John, *A Defence of Arch-bishop Usher against Dr Cary and Dr Isaac Vossius* (Cambridge, 1694).

Mivart, St George, 'Difficulties of the Theory of Natural Selection', *The Month*, 11 (1869), 35–53.

———, *On the Genesis of Species* (2nd edn, London, 1871).

———,'Descent of Man', *Quarterly Review*, 131 (1871), 47–90.

Molyneux, Thomas, 'A discourse concerning the large horns frequently found under the ground in Ireland', *Philosophical Transactions of the Royal Society of London*, 19 (1697), 489–512.

More, Hannah, 'The Shepherd of Salisbury Plain', in *The Works of Hannah More* (7 vols, New York, 1835), 1:176–201.

Murchison, Roderick, Édouard de Verneuil, and Alexander von Keyserling, *The Geology of Russia in Europe and the Ural Mountains* (2 vols, London, 1845).

Newman, Francis William, *A History of the Hebrew Monarch from the Administration of Samuel to the Babylonish Captivity* (London, 1847).

———, *The Soul, Her Sorrows and Her Aspirations: An Essay towards the Natural History of the Soul, as the True Basis of Theology* (London, 1849).

———, *Phases of Faith; or, Passages from the History of My Creed* (London, 1850).

Nolan, Frederick, *The Analogy of Revelation and Science Established in a Series of Lectures delivered before the University of Oxford* (Oxford, 1833).

North, Roger, *The Life of the Right Honourable Francis North, Baron of Guildford, Lord Keeper of the Great Seal under King Charles II and King James II* (2nd edn, 3 vols, 1808).

Official Descriptive and Illustrated Catalogue of the Great Exhibition of the Works of Industry of All Nations, 1851 (London, 1851).

'One of the Royal Fusiliers' [Timothy Gowing], *A Soldier's Experience, or A Voice from the Ranks: Showing the Cost of War in Blood and Treasure* (Nottingham, 1892).

Osborn, Henry Fairfield, '*Tyrannosaurus* and other Cretaceous Carnivorous Dinosaurs', *Bulletin of the American Museum of Natural History*, 21 (1905), 259–65.

Owen, Richard, 'A Description of a Portion of the Skeleton of the Cetiosaurus, a Gigantic Extinct Saurian Reptile, Occurring in the Oolitic Formations of Different Portions of England', *Proceedings of the Geological Society of London*, 3 (1842), 457–62.

———, 'Report on British Fossil Reptiles', in *Report of the Eleventh Meeting of the British Association for the Advancement of Science; Held at Plymouth in July 1841* (London, 1842), 60–204.

———, *A History of British Fossil Reptiles* (4 vols, London, 1849–84).

———, 'Dr Mantell', *Literary Gazette and Journal of Belles Lettres, Science, and Art for the Year 1852* (London, 1852), 842.

——, *Geology and Inhabitants of the Ancient World* (London, 1854).

——, 'Address of the President', in *Report of the Twenty-Eighth Meeting of the British Association for the Advancement of Science; Held at Leeds in September 1858* (London, 1859).

——, 'Darwin on the Origin of Species' [review], *Edinburgh Review*, 111 (1860), 487–532.

——, 'The Gorilla and the Negro', *Athenaeum*, 1743 (1861), 395–6.

——, 'The Gorilla and the Negro' (II), *Athenaeum*, 1745 (1861), 467.

——, 'On the Characters of the Aye-Aye, as a Test of the Lamarckian and Darwinian Hypothesis of the Transmutation and Origin of Species', *Report of the Thirty-Second Meeting of the British Association for the Advancement of Science; Held at Cambridge in October 1862* (London, 1863), 114–16.

——, 'On the Zoological Significance of the Cerebral and Pedial Characters of Man', *Report of the Thirty-Second Meeting of the British Association for the Advancement of Science; Held at Cambridge in October 1862* (London, 1863), 116–18.

——, 'On the Archaeopteryx of von Meyer, with a Description of the Fossil Remains of a Long-Tailed Species, from the Lithographic Stone of Solenhofen', *Philosophical Transactions of the Royal Society of London*, 153 (1863), 33–47.

Owen, Richard (Jr), *The Life of Richard Owen by His Grandson, rev. C. Davies Sherborn* (2 vols, London, 1894).

Oxford University Statutes, trans. G. R. M. Ward and James Heywood (2 vols, London, 1851).

Paine, Thomas, *The Age of Reason: Being an Investigation of True and Fabulous Theology* (3 vols, London, 1794–1807).

The Palace and Park: Its Natural History, and its Portrait Gallery, together with a Description of the Pompeian Court (London, 1855).

Parkinson, James, *Organic Remains of a Former World: An Examination of the Mineralized Remains of the Vegetables and Animals of the Antediluvian World* (3 vols, London, 1804–11).

——, *Outlines of Oryctology: An Introduction to the Study of Fossil Organic Remains, Especially of Those Found in the British Strata* (London, 1822).

Parr, Richard, *The Life of the Most Reverend Father in God, James Usher, late Lord Arch-Bishop of Armagh, Primate and Metropolitan of All Ireland* (London, 1686).

Paxton, Joseph, *What is to Become of the Crystal Palace?* (London, 1851).

Penn, Granville, *A Comparative Estimate of the Mineral and Mosaical Geologies: Revised, and Enlarged with Relation to the Latest Publications on Geology* (2nd edn, 2 vols, London, 1825).

Phillips, John, *Life on the Earth: Its Origin and Succession* (London, 1860).

———, *Geology of Oxford and the Valley of the Thames* (Oxford, 1871).

Phillips, Samuel, *Crystal Palace: A Guide to the Palace & Park* (London, 1858).

[Philpotts, Henry], *Socialism. Second Speech of the Bishop of Exeter, in the House of Lords, February 4, 1840* (London, 1840).

Playfair, John, 'Biographical Account of the late James Hutton, M.D.', in *The Works of John Playfair, Esq.* (4 vols, Edinburgh, 1822), 4:33–120.

Plot, Robert, *The Natural History of Oxford-Shire: Being an Essay toward the Natural History of England* (Oxford, 1677).

Porter, George Richardson, *The Progress of the Nation, in its Various Social and Economical Relations, from the Beginning of the Nineteenth Century* (3rd edn, London, 1851).

Poulett Scrope, George, *Considerations on Volcanos, the Probable Causes of their Phenomena, the Laws which Determine their March, the Disposition of their Products, and their Connexion with the Present State and Past History of the Globe* (London, 1825).

———, 'Principles of Geology', *Quarterly Review*, 43 (1830), 411–68.

Queen of Science: Personal Recollections of Mary Somerville, ed. Dorothy McMillan (Edinburgh, 2001).

Queen Victoria's Journal (http://www.queenvictoriasjournals.org).

Renan, Ernest, *The Life of Jesus*, trans. Charles Edwin Wilbour [1863] (New York, 1864).

Religion and Science; the Letters of 'Alpha' on the Influence of Spirit upon Imponderable Actienic Molecular Substances, and the Life-Forces of Mind and Matter (Boston, 1875).

Report from the Select Committee on the British Museum, together with the Proceedings of the Committee, Minutes of Evidence, and Appendix (London, 1860).

Report of the First and Second Meetings of the British Association for the Advancement of Science; at York in 1831, and at Oxford in 1832 (2nd edn, London, 1835).

Ritchie, J. Ewing, *Days and Nights in London: or, Studies in Black and Gray* (London, 1880).

Roberts, George, *The History and Antiquities of the Borough of Lyme Regis and Charmouth* (London, 1834).

Robertson, William, *Sermon preached on the Day of Humiliation on Account of the War, ordered by Her Majesty, on Wednesday, 26 April, 1854, in New Greyfriars' Church, Edinburgh* (Edinburgh, 1854).

Routledge's Guide to the Crystal Palace and Park at Sydenham: with Descriptions of the Principal Works of Science and Art, and of the Terraces, Fountains,

Geological Formations, and Restoration of Extinct Animals, therein Contained (London, 1854).

Rowbotham, Samuel, *Zetetic Astronomy: Earth not a Globe! An Experimental Inquiry into the True Figure of the Earth: Proving it a Plane, without Axial or Orbital Motion* (London, 1865).

Russell, Richard, *A Dissertation on the Use of Sea Water in the Diseases of the Glands, particularly the Scurvy, King's-Evil, Leprosy, and the Glandular Consumption* (4th edn, London, 1760).

Savile, Frank Mackenzie, *Beyond the Great Wall: The Secret of the Antarctic* (New York, 1903).

Secularism: Unphilosophical, Immoral, and Anti-Social: Verbatim Report of a Three Nights' Debate between the Rev. Dr. McCann and Charles Bradlaugh, in the Hall of Science, London (London, 1881).

Sedgwick, Adam, 'Presidential Address', *Proceedings of the Geological Society of London*, 1 (1834), 270–316.

———, 'Natural History of Creation', *Edinburgh Review*, 82 (1845), 1–85.

Seeley, H. G., *Index to the Fossil Remains of Aves, Ornithosauria, and Reptilia, from the Secondary System of Strata, Arranged in the Woodwardian Museum of the University of Cambridge* (Cambridge, 1869).

———, *Dragons of the Air: An Account of Extinct Flying Reptiles* (New York, 1901).

Seeley, J. R., *Ecce Homo: A Survey of the Life and Work of Jesus Christ* [1865] (London, 1866).

———, *The Expansion of England: Two Courses of Lectures* (London, 1891).

Silliman, Benjamin, 'Obituary: Gideon Algernon Mantell', *American Journal of Science and Arts*, 15 (1853), 147–50.

Sotheby, Samuel Leigh, *A Letter Addressed to the Directors of the Crystal Palace Company* (London, 1855).

———, *A Few Words by Way of a Letter Addressed to the Shareholders of the Crystal Palace Company: With Remarks on the Various Subjects Necessarily Brought under the Notice of the Committee, Appointed August 9, 1855, to Investigate the Affairs of the Company* (3rd edn, London, 1856).

Spencer, Herbert, *Social Statics: or, The Conditions Essential to Human Happiness Specified, and the First of Them Developed* (London, 1851).

———, 'The Development Hypothesis' [1852], in Herbert Spencer, *Essays Scientific, Political & Speculative* (London, 1858), 389–95.

———, 'Progress: Its Law and Causes', *Westminster Review*, 67 (1857), 445–65.

———, *The Principles of Biology* (2 vols, London, 1864).

———, *An Autobiography* (2 vols, London, 1904).

Stanley, A. P., 'Two Years Ago' [review], *Saturday Review of Politics, Literature, Science, and Art*, 3 (1857), 176–7.

Sternberg, Charles H., *The Life of A Fossil Hunter* (New York, 1909).

Strauss, David Friedrich, *The Life of Jesus, Critically Examined* [trans. Mary Ann Evans] (4th edn, 3 vols, London, 1846).

Stretton, Hesba, *Jessica's First Prayer* (London, 1866).

Taylor, Robert, *The Diegesis: Being a Discovery of the Origin, Evidences, and Early History of Christianity* [1829] (Boston, MA, 1860).

Temple, Frederick, *The Present Relations of Science to Religion: A Sermon, Preached on Act Sunday, July 1, 1860, before the University of Oxford* (Oxford, 1860).

Tennyson, Alfred, *The Princess: A Medley* [1847] (London, 1860).

———, *In Memoriam, A.H.* (London, 1850).

Thackeray, William Makepeace, *The Newcomes: Memoirs of a Most Respectable Family* [1854–5] (Boston, 1869).

Thomson, William, 'On Geological Time', *Transactions of the Geological Society of Glasgow*, 3 (1868), 1–29.

———, 'Geological Dynamics', *Geological Magazine*, 6 (1869), 472–6.

'Transactions of the Geological Society of London' [review], *Quarterly Review*, 34 (1826), 507–39.

The Trial of George Jacob Holyoake, on an Indictment for Blasphemy, before Mr Justice Erskine, and a Common Jury, at Gloucester, August the 15th, 1842 (London, 1842).

Trial of the Bishop of Natal for Erroneous Teaching (Cape Town, 1863).

The Truth about Church Extension: An Exposure of Certain Fallacies and Misstatements contained in the Census Reports on Religious Worship and Education (London, 1857).

Tuckwell, Rev. W., *Reminiscences of Oxford* (London, 1900).

Turner, Sharon, *The Sacred History of the World: As Displayed in the Creation and Subsequent Events to the Deluge; Attempted to be Philosophically Considered in a Series of Letters to a Son* (3 vols, New York, 1832).

Tyndall, John, 'Introduction', in Andrew Dickson White, *The Warfare of Science* (London, 1876), iii–iv.

———, *New Fragments* (New York, 1892).

The Unpublished Journal of Gideon Mantell, 1819–1852, ed. John A. Cooper (Brighton, 2010).

Ure, Andrew, *A New System of Geology, in which the Great Revolutions of the Earth and Animated Nature, are Reconciled At Once to Modern Science and Sacred History* (London, 1829).

Ussher, James, *The Annals of the World: Deduced from the Origin of Time, and Continued to the Beginning of the Emperour Vespasians Reign, and the Totall Destruction and Abolition of the Temple and Common-Wealth of the Jews* (London, 1658).

Van Wyhe, John (ed.), *The Complete Works of Charles Darwin Online* [http://darwin-online.org.uk].

Verne, Jules, *Journey to the Centre of the Earth* [1864] (London, 1876).

Wallace, Alfred Russel, 'On the Law which has Regulated the Introduction of New Species', *Annals and Magazine of Natural History*, 16 (1855), 184–96.

———, *The Malay Archipelago: The Land of the Orangutan, and the Bird of Paradise: A Narrative of Travel, with Studies of Man and Nature* (New York, 1869).

———, 'The Descent of Man', *Academy*, 2 (1871), 177–83.

———, *My Life: A Record of Events and Opinions* (2 vols, London, 1905).

———, *Letters from the Malay Archipelago*, ed. John van Wyhe and Kees Rookmaaker (Oxford, 2013).

Waterhouse Hawkins, Benjamin, *On Visual Education as Applied to Geology, Illustrated by Diagrams and Models of the Geological Restorations at the Crystal Palace* (London, 1854).

Waterton, Charles, *Essays on Natural History: Third Series* (London, 1857).

Wells, David A. (ed.), *Annual of Scientific Discovery: Or, Year-Book of Facts in Science and Art for 1865* (London, 1865).

Wells, H. G., *Kipps: The Story of a Simple Soul* (New York, 1906).

What is Bishop Colenso's New Book? And Who is Bishop Colenso? (Manchester, 1865).

White, Andrew Dickson, 'Scientific and Industrial Education in the United States', *Popular Science Monthly*, 5 (1874), 170–91.

———, *The Warfare of Science* (New York, 1876).

The Whole Works of the Most Rev. James Ussher, D.D., Lord Archbishop of Armagh, and Primate of All Ireland, ed. Charles Richard Elrington (16 vols, Dublin, 1847).

Wilberforce, Samuel, 'Darwin's Origin of Species', *Quarterly Review*, 108 (1860), 118–38.

———, 'Essays and Reviews' [review], *Quarterly Review*, 248–305.

Wordsworth, William, *The Excursion, Being A Portion of The Recluse, A Poem* (London, 1814).

The Works of Charles Darwin, ed. Paul H. Barrett and R. B. Freeman (29 vols, New York, 1986–92).

The Works of John Ruskin, ed. E. T. Cook and Alexander Wedderburn (39 vols, London, 1903–12).

The Year-Book of Facts in Science and Art: Exhibiting the Most Important Discoveries & Improvements of the Past Year, ed. John Timbs (London, 1852).

NEWSPAPERS, JOURNALS, AND MAGAZINES

Abergavenny Chronicle
Annals of Philosophy
Bath Chronicle and Weekly Gazette
Belfast Mercury
Belfast Newsletter
Bell's Weekly Messenger (London)
Berkshire Chronicle
British Critic
British Medical Journal
Bury and Norwich Post
Buxton Herald
Cardiff Times
Cheltenham Chronicle
Chicago Tribune
Comic News
Daily Telegraph & Courier
Dundee Courier
Dunfermline Saturday Press
The Economist
Edinburgh Evening News
Evening Mail
Exeter Flying Post
The Gardeners' Chronicle and Agricultural Gazette
General Evening Post (London)
Hansard
Hereford Times
Illustrated London News
Illustrated Magazine
Lady's Newspaper and Pictorial Times
Leeds Intelligencer
Leeds Times
Leicester Daily Post
The Leisure Hour: A Family Journal of Instruction and Recreation
The Lion
London Evening Standard

London Investigator
Loughborough Monitor
Madras Courier
Monthly Review or Literary Journal Enlarged
Morning Advertiser
Morning Herald
The Movement; and Anti-Persecution Gazette
National Reformer
New Monthly Magazine
New York Herald
New York Times
Observer
The Oracle of Reason; Or, Philosophy Vindicated
Perthshire Courier
Punch; or the London Charivari
Preston Chronicle
Raleigh Christian Advocate
Reasoner
St James's Gazette
Salisbury and Winchester Journal
Scientific Opinion
Scotsman
Spectator
Statesman (London)
Suffolk Chronicle
The Sunday at Home: A Family Magazine for Sabbath Reading
Sussex Advertiser
Taunton Courier and Western Advertiser
The Times
Tracts for the Times
Vanity Fair
Wiltshire and Gloucestershire Standard
York Herald

SECONDARY SOURCES

Adams, Max, *The Prometheans: John Martin and the Generation that Stole the Future* (London, 2010).

Akin Burd, Van, 'Ruskin and his "Good Master", William Buckland', *Victorian Literature and Culture*, 36 (2008), 299–315.

Anderson, Olive, 'Gladstone's Abolition of Compulsory Church Rates: A Minor Political Myth and its Historiographical Career', *Journal of Ecclesiastical History*, 25 (1974), 185–98.

Appel, Toby A., *The Cuvier–Geoffroy Debate: French Biology in the Decades before Darwin* (Oxford, 1997).

Arnstein, Walter L., *The Bradlaugh Case: A Study in Late Victorian Opinion and Politics* (Oxford, 1965).

Auerbach, Jeffrey A., *The Great Exhibition of 1851: a Nation on Display* (New Haven, CT, 1999).

Balston, Thomas, *John Martin, 1799–1854, His Life and Works* (London, 1947).

Barr, James, 'Why the World was Created in 4004 BC: Archbishop Ussher and Biblical Chronology', *John Rylands Bulletin*, 67 (1985), 575–608.

Barrett, Paul (ed.), 'A Transcription of Darwin's First Notebook on "Transmutation of Species"', *Bulletin of the Museum of Comparative Zoology*, 122 (1960), 245–96.

———, 'Pre-Scientific Chronology: The Bible and the Origin of the World', *Proceedings of the American Philosophical Society*, 143 (1999), 379–87.

Bartley, Paula, *Queen Victoria* (London, 2016).

Barton, Ruth, 'John Tyndall, Pantheist: A Rereading of the Belfast Address', *Osiris*, 3 (1987), 111–34.

———, *The X Club: Power and Authority in Victorian Science* (Chicago, 2018).

Bashford, Alison, *An Intimate History of Evolution: The Story of the Huxley Family* (London, 2022).

Baxter, Stephen, *Revolutions of the Earth: James Hutton and the True Age of the World* (London, 2003).

Bentley, Michael, 'Victorian Historians and the Larger Hope', in Michael Bentley (ed.), *Public and Private Doctrine: Essays in British History presented to Maurice Cowling* (Cambridge, 1993), 127–49.

Benton, Michael J., *The Dinosaurs Rediscovered: How a Scientific Revolution is Rewriting History* (London, 2019).

Brady, Ciaran, *James Anthony Froude: An Intellectual Biography of a Victorian Prophet* (Oxford, 2013).

Bowler, Peter J., 'Darwinism and the Argument from Design: Suggestions for a Reevaluation', *Journal of the History of Biology*, 10 (1977), 29–43.

———, *Evolution, The History of an Idea* (London, 1983).

———, *The Eclipse of Darwinism: Anti-Darwinian Evolution Theories in the Decades around 1900* (Baltimore, MD, 1983).

———, *Reconciling Science and Religion: The Debate in Early Twentieth-Century Britain* (Chicago, 2001).

Bratton, J. S., 'Hesba Stretton's Journalism', *Victorian Periodicals Review*, 12 (1979), 60–70.

Brock, W. H., 'The Selection of the Authors of the Bridgewater Treatises', *Notes and Records of the Royal Society of London*, 21 (1966), 162–79.

Brooks, David, 'Gladstone and Midlothian: The Background to the First Campaign', *Scottish Historical Review*, 64 (1985), 42–67.

Brown, Callum G., *The Death of Christian Britain: Understanding Secularisation, 1800–2000* (Abingdon, 2009).

Brown, Matthew, 'Darwin at Church: John Tyndall's Belfast Address', in James H. Murphy (ed.), *Evangelicals and Catholics in Nineteenth-Century Ireland* (Dublin, 2005), 235–46.

Browne, Janet, *Charles Darwin* (2 vols, London, 1995–2002).

Brusatte, Steve, *The Rise and Fall of the Dinosaurs: A New History of a Lost World* (London, 2018).

Burchfield, Joe D., *Lord Kelvin and the Age of the Earth* (Chicago, 1990).

Burrow, J. W., *Evolution and Society: A Study in Victorian Social Theory* (Cambridge, 1966).

Cadbury, Deborah, *The Dinosaur Hunters: A True Story of Scientific Rivalry & the Discovery of the Prehistoric World* (London, 2000).

Callaway, Jack M., and Elizabeth L. Nicholls (eds.), *Ancient Marine Reptiles* (San Diego, 1997).

Cantor, Geoffrey, *Religion and the Great Exhibition of 1851* (Oxford, 2011).

Carr, J. A., *The Life and Times of Archbishop James Ussher, Archbishop of Armagh* (London, 1895).

Chadwick, Owen, *The Secularization of the European Mind in the Nineteenth Century* (Cambridge, 2011).

Chaffin, Tom, *Odyssey: Young Charles Darwin, The Beagle, and the Voyage that Changed the World* (London, 2022).

Chambers, Paul, *Bones of Contention: The Fossil that Shook Science* (London, 2002).

Chapman, Allan, *Caves, Coprolites, and Catastrophes: The Story of Pioneering Geologist and Fossil-Hunter William Buckland* (London, 2020).

Clark, J. C. D., *English Society, 1660–1832: Religion, Ideology, and Politics during the Ancien Regime* (2nd edn, Cambridge, 2000).

Clark, Michael, 'Jewish Identity in British Politics: the Case of the First Jewish MPs, 1858–87', *Jewish Social Studies*, 13 (2007), 93–126.

Colbert, Edwin Harris, *The Great Dinosaur Hunters and Their Discoveries* (New York, 1984).

Conlin, Jonathan, *Evolution and the Victorians: Science, Culture, and Politics* (London, 2013).

Cook, E. T., *The Life of John Ruskin* (2 vols, London, 1911).

Cooter, Roger, *The Cultural Meaning of Popular Science: Phrenology and the Organization of Consent in Nineteenth-Century Britain* (Cambridge, 1984).

Crafton, Donald, *Before Mickey: The Animated Film, 1898–1928* (Chicago, 2015).

Crais, Clifton C., *Sara Baartman and the Hottentot Venus: A Ghost Story and a Biography* (Princeton, NJ, 2009).

Critchley, E. M. R., *Dinosaur Doctor: The Life of Gideon Mantell* (Stroud, 2010).

Darwin on Evolution: The Development of the Theory of Natural Selection, ed. Thomas F. Glick and David Kohn (Indianapolis, 1996).

Davidson, Jane P., *The Bone Sharp: The Life of Edward Drinker Cope* (Philadelphia, 1997).

Davidson, Nick, *The Greywacke: How a Priest, a Soldier, and a School Teacher Uncovered 300 Million Years of History* (London, 2021).

Davis, David Brion, *The Problem of Slavery in the Age of Revolution, 1770–1823* (Ithaca, NY, 1975).

Dawson, Gowan, 'Literary Megatheriums and Loose Baggy Monsters: Palaeontology and the Victorian Novel', *Victorian Studies*, 53 (2011), 203–30.

———, *Show Me the Bone: Reconstructing Prehistoric Monsters in Nineteenth-Century Britain and America* (Chicago, 2016).

———, 'Dickens, Dinosaurs, and Design', *Victorian Literature and Culture*, 44 (2016), 761–78.

De Beer, Gavin, *Archaeopteryx Lithographica* (London, 2009).

Deacon, Richard, *The Cambridge Apostles: A History of Cambridge University's Elite Intellectual Secret Society* (New York, 1986).

Dean, Dennis R., *James Hutton and the History of Geology* (Ithaca, NY, 1992).

———, *Gideon Mantell and the Discovery of Dinosaurs* (Cambridge, 1999).

Desilva, Jeremy M. (ed.), *A Most Interesting Problem: What Darwin's* Descent of Man *Got Right and Wrong about Human Evolution* (Princeton, 2021).

Desmond, Adrian, *The Hot Blooded Dinosaurs: A Revolution in Palaeontology* (London, 1975).

———, 'Designing the Dinosaur: Richard Owen's Response to Robert Edmond Grant', *Isis*, 70 (1979), 224–34.

———, 'Richard Owen's Reaction to Transmutation in the 1830s', *British Journal for the History of Science*, 18 (1985), 25–50.

——— and James Moore, *Darwin: The Life of a Tormented Evolutionist* [1991] (London, 2009).

———, *The Politics of Evolution: Morphology, Medicine, and Reform in Radical London* (Chicago, 1992).

———, *Huxley* (London, 1997).

———, 'Redefining the X Axis: "Professionals", "Amateurs", and the Making of Mid-Victorian Biology: A Progress Report', *Journal of the History of Biology*, 34 (2001), 3–50.

——— and James Moore, *Darwin's Sacred Cause: How a Hatred of Slavery Shaped Darwin's Views on Human Evolution* (New York, 2009).

———, 'T. H. Huxley's Turbulent Apprenticeship Years: John Charles Cooke and the John Salt Scandal', *Archives of Natural History*, 48 (2021), 215–26.

Di Gregorio, Mario A., *From Here to Eternity: Ernst Haeckel and Scientific Faith* (Gottingen, 2005).

Draper, Jonathan A. (ed.), *The Eye of the Storm: Bishop John Wiliam Colenso and the Crisis of Biblical Inspiration* (London, 2003).

Dunn, Waldo Hilary, *James Anthony Froude: A Biography* (Oxford, 1961).

Edmonds, J. M., 'The Founding of the Oxford Readership in Geology, 1818', *Notes and Records of the Royal Society of London*, 34 (1979), 33–51.

Ellis, Ieuan, *Seven Against Christ: 'A Study of Essays and Reviews'* (Leiden, 1980).

Emling, Shelley, *The Fossil Hunter: Dinosaurs, Evolution, and the Woman Whose Discoveries Changed the World* (New York, 2009).

England, Richard, 'Censoring Huxley and Wilberforce: A New Source for the Meeting that the *Athenaeum* "Wisely Softened Down"', *Notes & Records of the Royal Society of London*, 71 (2017), 371–84.

Fallon, Richard, *Reimagining Dinosaurs in Late Victorian and Edwardian Literature: How the 'Terrible Lizard' Became a Transatlantic Cultural Icon* (Cambridge, 2021).

Farrar, W. V., 'Andrew Ure, F.R.S., and the Philosophy of Manufactures', *Notes and Records of the Royal Society of London*, 27 (1973), 299–324.

Fastovsky, David E., and David B. Weishampel, *Dinosaurs: A Concise Natural History* (4th edn, Cambridge, 2021).

Fawcett, Percy Harrison, *Lost Trails, Lost Cities: From His Manuscripts, Letters, and Other Records*, ed. Brian Fawcett (New York, 1953).

Febvre, Lucien, and Henri-Jean Martin, *The Coming of the Book: The Impacting of Printing, 1450–1800*, trans. David Gerard [1958] (London, 1976).

Field, Clive D., 'Counting Religion in England and Wales: The Long Eighteenth Century, *c.*1680–*c.*1840', *Journal of Ecclesiastical History*, 63 (2012), 693–720.

Figes, Orlando, *Crimea: The Last Crusade* (London, 2010).

Finnegan, Diarmid A., and Jonathan Jeffrey Wright, 'Catholics, Science, and Civic Culture in Victorian Belfast', *British Journal for the History of Science*, 48 (2015), 261–87.

———, *The Voice of Science: British Scientists on the Lecture Circuit in Gilded Age America* (Pittsburgh, 2021).

Fitzwilliams, Duncan C. L., 'The destruction of John Hunter's papers', *Proceedings of the Royal Society of Medicine*, 42 (1949), 871–6.

Forgan, Sophie, and Graeme Gooday, 'Constructing South Kensington: The Buildings and Politics of T. H. Huxley's Working Environments', *British Journal for the History of Science*, 29 (1996), 435–68.

Freeman, Michael, *Tracks to a Lost World: Victorians and the Prehistoric* (New Haven, CT, 2004).

Fry, Hannah, and Adam Rutherford, *Rutherford & Fry's Complete Guide to Absolutely Everything* (New York, 2021).

Fuller, J. G. C. M., 'A Date to Remember: 4004 BC', *Earth Sciences History*, 24 (2005), 5–14.

Galton, David, 'Did Darwin Read Mendel?', *QJM*, 102 (2009), 587–9.

Gardiner, Samuel R., *History of the Great Civil War, 1642–1649* [1886–91, 4 vols] (London, 1898).

Garwood, Christine, *Flat Earth: The History of an Infamous Idea* (London, 2007).

Gillespie, Charles Coulston, *Genesis and Geology: A Study in the Relations of Scientific Thought, Natural Theology, and Social Opinion in Great Britain, 1790–1850* (Cambridge, MA, 1951).

Gilley, Sheridan, and Ann Loades, 'Thomas Henry Huxley: The War between Science and Religion', *The Journal of Religion*, 61 (1981), 285–308.

Good, James I., *History of the Reformed Church of Germany* (Reading, PA, 1894).

Goodhue, Thomas W., 'The Faith of a Fossilist: Mary Anning', *Anglican and Episcopal History*, 70 (2001), 80–100.

———, 'Mary Anning's Commonplace Book' (Lyme Regis Museum, undated) (www.lymeregismuseum.co.uk).

Goodman, Jordan, *The Rattlesnake: A Voyage of Discovery to the Coral Sea* (London, 2005).

Gould, Stephen Jay, 'Foreword', in Jean Chandler Smith (ed.), *Georges Cuvier: An Annotated Bibliography of His Published Works* (Washington, DC, 1993), vii–xii.

———, 'Fall in the House of Ussher' in *Eight Little Piggies: Reflections in Natural History* (New York, 1994), 181–93.

Green, S. J. D., *The Passing of Protestant England: Secularisation and Social change, c.1920–1960* (Cambridge, 2011).

Gregorio, Mario A. di, 'The Dinosaur Connection: A Reinterpretation of T. H. Huxley's Evolutionary View', *Journal of the History of Biology*, 15 (1982), 397–418.

Gruber, Jacob W., *A Conscience in Conflict: The Life of St George Jackson Mivart* (New York, 1960).

——, 'Brixham Cave and the Antiquity of Man', in Tim Murray (ed.), *Histories of Archaeology: A Reader in the History of Archaeology* (Oxford, 2008), 13–45.

Guy, Jeff, *The Heretic: A Study of the Life of John William Colenso, 1814–1883* (Pietermaritzburg, 1983).

Hall, Donald E., 'Muscular Christianity: Reading and Writing the Male Social Body' in Donald E. Hall (ed.), *Muscular Christianity: Embodying the Victorian Age* (Cambridge, 1994), 3–14.

Hanson, Thor, *Feathers: The Evolution of a Natural Miracle* (New York, 2011).

Hardin, Jeff, Ronald A. Numbers, and Ronald A. Binzley (eds.), *The Warfare between Science & Religion: The Idea that Wouldn't Die* (Baltimore, MD, 2018).

Harris, José, *Private Lives, Public Spirits: A Social History of Britain, 1870–1914* (Oxford, 1993).

Headingley, Adolphe S., *The Biography of Charles Bradlaugh* (London, 1883).

Heffer, Simon, *High Minds: The Victorians and the Birth of Modern Britain* (London, 2013).

Herrick, Jim, *Humanism: An Introduction* (Amherst, NY, 2005).

Hesketh, Ian, *Of Apes and Ancestors: Evolution, Christianity, and the Oxford Debate* (Toronto, 2009).

——, 'Behold the (Anonymous) Man: J. R. Seeley and the Publishing of "Ecce Homo"', *Victorian Review*, 38 (2012), 93–112.

——, *Victorian Jesus: J. R. Seeley, Religion, and the Cultural Significance of Anonymity* (Toronto, 2017).

Hewitt, Martin, *The Dawn of the Cheap Press in Victorian Britain: The End of the 'Taxes on Knowledge', 1849–1869* (London, 2014).

——, 'Darwin's Generations: The Reception of Darwinian Evolution in Britain, 1859–1909' (unpublished manuscript, 2023).

Higgitt, Rebekah, and Charles W. J. Withers, 'Science and Sociability: Women as Audience at the British Association for the Advancement of Science, 1831–1901', *Isis*, 99 (2008), 1–27.

Hilton, Boyd, *The Age of Atonement: The Influence of Evangelicalism on Social and Economic Thought, 1795–1865* (Oxford, 1988).

——, *A Mad, Bad, and Dangerous People? England, 1783–1846* (Oxford, 2006).

Holmes, Andrew R., 'Presbyterians and Science in the North of Ireland before 1874', *British Journal of the History of Science*, 41 (2008), 541–65.

Hopes, Jeffrey, 'Dating the World: The Science of Biblical Chronology', *XVII–XVIII*, 71 (2014), 65–83.

Howlett, E. A., W. J. Kennedy, H. P. Powell, and H. S. Torrens, 'New Light on the History of *Megalosaurus*, the Great Lizard of Stonesfield', *Archives of Natural History*, 44 (2017), 82–102.

Hunting, Jill, *For Want of Wings: A Bird with Teeth and a Dinosaur in the Family* (Norman, OK, 2002).

Inglis, K. S., 'Patterns of Religious Worship in 1851', *Journal of Ecclesiastical History*, 11 (1960), 74–86.

Jackson, Patrick Wyse, *The Chronologers' Quest: Episodes in the Search for the Age of the Earth* (Cambridge, 2006).

Jaffe, Mark, *The Gilded Dinosaur: The Fossil War between E. D. Cope and O. C. Marsh and the Rise of American Science* (New York, 2000).

James, Frank A. J. L., 'An "Open Clash Between Science and the Church"? Wilberforce, Huxley, and Hooker on Darwin at the British Association, Oxford 1860', in David M. Knight and Matthew D. Eddy (eds.), *Science and Beliefs: From Natural Philosophy to Natural Science, 1700–1900* (Aldershot, 2005), 171–93.

Jensen, J. Vernon, 'The X Club: Fraternity of Victorian Scientists', *British Journal for the History of Science*, 5 (1970), 63–72.

———, 'Return to the Wilberforce–Huxley Debate', *British Journal for the History of Science*, 21 (1988), 161–79.

———, *Thomas Henry Huxley: Communicating for Science* (Newark, NJ, 1991).

Jones, John, *Balliol College: A History* (2nd edn, 2005).

Jones, Steve, *Darwin's Island: The Galapagos in the Garden of England* (London, 2009).

Kaalund, Nanna Katrine Lurers, 'Oxford Serialized: Revisiting the Huxley–Wilberforce Debate through the Periodical Press', *History of Science*, 52 (2014), 429–53.

Kennedy, Emmet, *A Cultural History of the French Revolution* (New Haven, CT, 1989).

Kennedy, Padraic C., '"Underhand Dealings with the Papal Authorities": Disraeli and the Liberal Conspiracy to Disestablish the Irish Church', *Parliamentary History*, 27 (2008), 19–29.

Keynes, Richard Darwin, *Fossils, Finches, and Fuegians: Darwin's Adventures on the Beagle* (Oxford, 2003).

Klaver, J. M. I., *The Apostle of the Flesh: A Critical Life of Charles Kingsley* (Leiden, 2006).

Knell, Simon J., 'The Road to Smith: How the Geological Society Came to Possess English Geology', in C. L. E. Lewis and S. J. Knell (eds.), *The Making of the Geological Society* (London, 2009), 1–48.

Knox, R. Buick, *James Ussher: Archbishop of Armagh* (Cardiff, 1967).

Kölbl-Ebert, M., 'The Geological Travels of Charles Lyell, Charlotte Murchison, and Roderick Impey Murchison in France and Northern Italy (1828)', *Geological Society of London: Special Publications*, 287 (2007), 109–17.

———, 'Sketching Rocks and Landscapes: Drawing as a Female Accomplishment in the Service of Geology', *Earth Sciences History*, 31 (2012), 270–86.

Kottler, Malcolm Jay, 'Alfred Russel Wallace, the Origin of Man, and Spiritualism', *Isis*, 65 (1974), 144–92.

Lang, W. D., 'Mary Anning (1799–1847) and the Pioneer Geologists of Lyme', *Proceedings of the Dorset Archaeological and Natural History Society*, 60 (1938), 142–64.

———, 'Early Days of Natural History at Charmouth', *Proceedings of the Dorset Archaeological and Natural History Society*, 62 (1940), 97–113.

———, 'Three Letters by Mary Anning, Fossilist of Lyme', *Proceedings of the Dorset Archaeological and Natural History Society*, 66 (1945), 169–73.

———, 'Mary Anning and Anna Maria Pinney', *Proceedings of the Dorset Archaeological and Natural History Society*, 76 (1955), 146–52.

———, 'Mary Anning's Escape from Lightning', *Proceedings of the Dorset Archaeological and Natural History Society*, 80 (1959), 91–3.

———, 'Mary Anning and a Very Small Boy', *Proceedings of the Dorset Archaeological and Natural History Society*, 84 (1962), 181.

Lanham, Url, *The Bone Hunters* (New York, 1975).

Larsen, Timothy, *Contested Christianity: The Political and Social Context of Victorian Theology* (Waco, TX, 2004).

Liddon, Henry Parry, *Life of Edward Bouverie Pusey*, ed. J. O. Johnston and Robert J. Wilson (4 vols, London, 1894).

Lightman, Bernard V., 'Interpreting Agnosticism as a Nonconformist Sect: T. H. Huxley's "New Reformation"', in Paul Wood (ed.), *Science, Technology and Culture, 1700–1945* (Aldershot, 2004), 197–214.

———, *Victorian Popularizers of Science: Designing Nature for New Audiences* (Chicago, 2009).

———, 'Huxley and the Devonshire Commission', in Gowan Dawson (ed.), *Victorian Scientific Naturalism: Community, Identity, Continuity* (Chicago, 2014), 101–30.

Livesley, Brian, and Gillian M. Pentelow, 'The burning of John Hunter's papers: a new explanation', *Annals of the Royal College of Surgeons of England*, 60 (1978), 79–84.

Lycett, Andrew, *The Man Who Created Sherlock Holmes: The Life and Times of Sir Arthur Conan Doyle* (London, 2007).

Macaulay, Catharine, *The History of England from the Accession of James I to that of the Brunswick Line* [1763–83, 8 vols] (Cambridge, 2013).

McCalman, Iain, *Darwin's Armada: Four Voyagers to the Southern Oceans and Their Battle for the Theory of Evolution* (London, 2009).

McCarthy, Steve, and Mick Gilbert, *The Crystal Palace Dinosaurs: The Story of the World's First Prehistoric Sculptures* (London, 1994).

McGregor, Arthur, 'The Tradescants as Collectors of Rarities', in Arthur McGregor (ed.), *Tradescant's Rarities: Essays on the Foundation of the Ashmolean Museum, 1683, with a Catalogue of the Surviving Early Collections* (Oxford, 1983), 17–23.

McGowan, Christopher, *The Dragon Seekers: How an Extraordinary Circle of Fossilists Discovered the Dinosaurs and Paved the Way for Darwin* (London, 2002).

McGowan-Hartmann, John, 'Shadow of the Dragon: The Convergence of Myth and Science in Nineteenth Century Palaeontological Imagery', *Journal of Social History*, 47 (2013), 47–70.

Maclear, J. F. (ed.), *Church and State in the Modern Age: A Documentary History* (Oxford, 1995).

MacLeod, Roy M., 'The X-Club: A Social Network of Science in Late-Victorian England', *Notes and Records of the Royal Society of London*, 24 (1970), 305–22.

McPhee, John, *Basin and Range* (New York, 1981).

Maddox, Brenda, *Reading the Rocks: How Victorian Geologists Discovered the Secret of Life* (London, 2018).

Mandler, Peter, 'Looking around the World', in Adelene Buckland and Sadiah Qureshi (eds.), *Time Travellers: Victorian Perspectives on the Past* (Chicago, 2020), 24–41.

Markowitz, Stephanie, *The Crimean War in the British Imagination* (Cambridge, 2009).

Markus, Julia, *J. Anthony Froude: The Last Undiscovered Great Victorian* (New York, 2007).

Marsh, Joss, *Word Crimes: Blasphemy, Culture, and Literature in Nineteenth-Century England* (Chicago, 1998).

Marshall, Nancy Rose, '"A Dim World, Where Monsters Dwell": The Spatial Time of the Sydenham Crystal Palace Dinosaur Park', *Victorian Studies*, 49 (2007), 286–301.

Mayor, Adrienne, *The First Fossil Hunters: Dinosaurs, Mammoths, and Myth in Greek and Roman Times* (Princeton, NJ, 2011).

Menez, Alex, 'Custodian of the Gibraltar Skull: The History of the Gibraltar Scientific Society', *Earth Sciences History*, 37 (2018), 34–62.

Merkl, Ulrich, *Dinomania: The Lost Art of Winsor McCay, the Secret Origins of King Kong, and the Urge to Destroy New York* (New York, 2015).

Millhauser, Milton, 'The Scriptural Geologists: An Episode in the History of Opinion', *Osiris*, 11 (1954), 65–86.

Mitchell, Austin, 'The Association Movement of 1792–3', *Historical Journal*, 4 (1961), 56–77.

Moore, James R., *The Post-Darwinian Controversies: A Study of the Protestant Struggle to Come to Terms with Darwin in Great Britain and America, 1870–1900* (Cambridge, 1979).

Moore, Randy, *Dinosaurs by the Decades: A Chronology of the Dinosaur in Science and Popular Culture* (Oxford, 2014).

Morley, John, *The Life of William Gladstone* (3 vols, London, 1903).

Morrell, Jack, and Arnold Thackray (eds.), *Gentlemen of Science: Early Correspondence of the British Association for the Advancement of Science* (London, 1984).

Mortenson, Terry, *The Great Turning Point: The Church's Catastrophic Mistake on Geology – Before Darwin* (Green Forest, AR, 2004).

Moss, Stephen, 'The Secrets of Paviland Cave', *Guardian*, 25 April 2011.

Mullan, John, *Anonymity: A Secret History of English Literature* (Princeton, NJ, 2007).

Mullin, Katherine, 'Poison More Deadly than Prussic Acid: Defining Obscenity after the 1857 Obscene Publications Act (1850–1885)', in David Bradshaw and Rachel Potter (eds.), *Prudes on the Prowl: Fiction and Obscenity in England* (Oxford, 2013), 11–29.

Murray-Wallace, Colin V., 'Understanding "Deep" Time – Advances since Archbishop Ussher?', *Archaeology in Oceania*, 31 (1996), 173–7.

Nash, David S., 'Unfettered Investigation: The Secularist Press and the Creation of Audience in Victorian England', *Victorian Periodicals Review*, 28 (1995), 123–35.

'National Trust Reference Creationism at the Giant's Causeway', *Guardian*, 5 July 2012.

Nieuwland, Ilja, *American Dinosaur Abroad: A Cultural History of Carnegie's Plaster Diplodocus* (Pittsburgh, 2019).

Numbers, Ronald L., 'Science and Religion', *Osiris*, 1 (1985), 59–90.

O'Connor, Bernard, *Digging for Dinosaurs: The Great Fenland Coprolite Rush* (Ely, 2011).

O'Connor, Ralph, 'Young-Earth Creationists in Early Nineteenth-Century Britain? Towards a Reassessment of "Scriptural Geology"', *History of Science*, 45 (2007), 357–403.

———, *The Earth on Show: Fossils and the Poetics of Popular Science, 1802–1856* (Chicago, 2008).

———, 'Victorian Saurians: The Linguistic Prehistory of the Modern Dinosaur', *Journal of Victorian Culture*, 17 (2012), 492–504.

Oxford Dictionary of National Biography (www.oxforddnb.com).

Palmer, Douglas, *Earth Time: Exploring the Deep Past from Victorian England to the Grand Canyon* (Chichester, 2005).

Pals, Daniel, 'The Reception of *Ecce Homo*', *Historical Magazine of the Protestant Episcopal Church*, 46 (1977), 63–84.

Park, Roberta J., 'Biological Thought, Athletics, and the Formation of "A Man of Character", 1830–1900', in J. A. Mangan and James Walvin (eds.), *Manliness and Morality: Middle-Class Masculinity in Britain and America, 1800–1940* (Manchester, 1987), 7–34.

Patterson, Elizabeth Chambers, *Mary Somerville and the Cultivation of Science, 1815–1840* (Hingham, MA, 1983).

Paul, Diane B., John Stenhouse, and Hamish G. Spencer, 'The two faces of Robert FitzRoy, captain of HMS *Beagle* and Governor of New Zealand', *Quarterly Review of Biology*, 88 (2013), 219–25.

Paul, Herbert W., *The Life of Froude* (New York, 1906).

Peck, Robert M., and Stephen M. Rowland, 'Benjamin Waterhouse Hawkins and the Early History of Three-Dimensional Palaeontological Art', in Renee M. Clary, Gary D. Rosenberg, and Dallas C. Evans (eds.), *The Evolution of Palaeontological Art* (Boulder, CO, 2021), 151–9.

Pettitt, Paul, and Mark White, 'John Lubbock, Caves, and the Development of Middle and Upper Palaeolithic Archaeology', *Notes and Records*, 68 (2014), 35–48.

Pierce, Patricia, *Jurassic Mary: Mary Anning and the Primeval Monsters* (Stroud, 2006).

Porter, Roy, 'Gentlemen and Geology: The Emergence of a Scientific Career, 1660–1920', *Historical Journal*, 21 (1978), 809–36.

Priest, Robert D., 'Reading, Writing, and Religion in Nineteenth-Century France: The Popular Reception of Renan's *Life of Jesus*', *Journal of Modern History*, 86 (2014), 258–94.

———, *The Gospel According to Renan: Reading, Writing, and Religion in Nineteenth-Century France* (Oxford, 2015).

Qureshi, Sadia, *Peoples on Parade: Exhibitions, Empire, and Anthropology in Nineteenth-Century Britain* (Chicago, 2011).

Raby, Peter, *Alfred Russel Wallace: A Life* (London, 2001).

Ralls, Walter, 'The Papal Aggression of 1850: A Study in Victorian Anti-Catholicism', *Church History*, 43 (1974), 242–56.

Randel, William Pierce, 'Huxley in America', *Proceedings of the American Philosophical Society*, 114 (1970), 73–99.

Rea, Tom, *Bone Wars: The Excavation and Celebrity of Andrew Carnegie's Dinosaur* (Pittsburgh, 2001).

Reardon, Bernard M. G., *Religious Thought in the Victorian Age: A Survey from Coleridge to Gore* (2nd edn, London, 2014).

Rectenwald, Michael, 'Secularism and the Cultures of Nineteenth-Century Scientific Naturalism', *British Journal for the History of Science*, 46 (2013), 231–54.

Rehbock, Philip F., 'Huxley, Haeckel, and the Oceanographers: The Case of Bathybius Haeckelii', *Isis*, 66 (1975), 504–33.

Rowlinson, John S., 'Chemistry Comes of Age: The 19th Century', in Robert J. P. Williams, John S. Rowlinson, and Allan Chapman (eds.), *Chemistry at Oxford: A History from 1600 to 2005* (London, 2009), 79–130.

Royle, Edward, *Radicals, Secularists, and Republicans: Popular Freethought in Britain, 1866–1915* (Manchester, 1980).

Rudwick, Martin J. S., 'The Strategy of Lyell's *Principles of Geology*', *Isis*, 61 (1970), 4–33.

——, *The Great Devonian Controversy: The Shaping of Scientific Knowledge among Gentlemanly Specialists* (Chicago, 1985).

——, *Scenes from Deep Time: Early Pictorial Representations of the Prehistoric World* (Chicago, 1992).

——, *Bursting the Limits of Time: The Reconstruction of Geohistory in the Age of Revolution* (London, 2005).

Rupke, Nicolaas A., '*Bathybius Haeckelii* and the Psychology of Scientific Discovery', *Studies in the History and Philosophy of Science*, 7 (1976), 53–62.

——, *The Great Chain of History: William Buckland and the English School of Geology* (Oxford, 1983).

——, *Richard Owen: Biology without Darwin* (2nd edn, Chicago, 2009).

Sarjeant, William A. S., and Justin B. Delair, 'An Irish naturalist in Cuvier's laboratory: the letters of Joseph Pentland, 1820–1832', *Bulletin of the British Museum (Natural History) Historical Series*, 6 (1980), 245–319.

Scott, Rosemary, 'The Sunday Periodical: *Sunday at Home*', *Victorian Periodicals Review*, 25 (1992), 158–62.

Secord, James A., *Victorian Sensation: The Extraordinary Publication, Reception, and Secret Authorship of Vestiges of the Natural History of Creation* (Chicago, 2000).

——, *Visions of Science: Books and Readers at the Dawn of the Victorian Age* (Oxford, 2014).

Sharpe, Tom, 'Henry de la Beche's 1829–1830 lithograph, *Duria Antiquior*', *Earth Sciences History*, 41 (2022), 47–63.

Shermer, Michael, *In Darwin's Shadow: The Life and Science of Alfred Russel Wallace: A Biographical Study on the Psychology of History* (Oxford, 2002).

Sinclair, David, *The Land That Never Was: Sir Gregor MacGregor and the Most Audacious Fraud in History* (Cambridge, MA, 2004).

Skinner, Simon A., *Tractarians and the 'Condition of England': The Social and Political Thought of the Oxford Movement* (Oxford, 2004).

Smiles, Samuel, *Self-Help: With Illustrations of Character and Conduct* (London, 1859).

Snell, K. D. M., and Paul S. Ell, *Rival Jerusalems: The Geography of Victorian Religion* (Cambridge, 2000).

Snobelen, Stephen D., 'To Discourse of God: Isaac Newton's Heterodox Theology and His Natural Philosophy', in *Science and Dissent in England, 1688–1945*, ed. Paul B. Wood (Aldershot, 2004), 39–65.

Sommer, Marianne, '"An Amusing Account of a Cave in Wales": William Buckland (1784–1856) and the Red Lady of Paviland', *British Journal for the History of Science*, 37 (2004), 53–74.

———, *Bones and Ochre: The Curious Afterlife of the Red Lady of Paviland* (Cambridge, MA, 2007).

Spanou, Petros, 'Soldiership, Christianity, and the Crimean War: The Reception of Catherine Marsh's *Memorials of Captain Hedley Vicars*', *Journal of Victorian Culture*, 27 (2022), 46–62.

Speller, John L., 'Alexander Nicoll and the Study of German Biblical Criticism in Early Nineteenth-Century Oxford', *Journal of Ecclesiastical History*, 30 (1979), 451–9.

Spokes, Sidney, *Gideon Algernon Mantell, Surgeon and Geologist* (London, 1927).

Stafford, Robert A., *Scientist of Empire: Sir Roderick Murchison, Scientific Exploration, and Victorian Imperialism* (Cambridge, 1990).

Stevens, Jennifer, *The Historical Jesus and the Literary Imagination, 1860–1920* (Liverpool, 2010).

Stothers, Richard B., 'Ancient Scientific Basis of the "Great Serpent" from Historical Evidence', *Isis*, 95 (2004), 220–38.

Strata: William Smith's Geological Maps (Chicago, 2020).

Styles, Phillip, 'James Ussher and His Times', *Hermathena*, 88 (1956), 12–33.

Tattersall, Iain, *The Last Neanderthal: The Rise, Success, and Mysterious Extinction of Our Closest Human Relatives* (Boulder, CO, 1999).

Taylor, Michael, 'British Conservatism, The Illuminati, and the Conspiracy Theory of the French Revolution, 1797–1802', *Eighteenth-Century Studies*, 47 (2014), 293–312.

———, *The Interest: How the British Establishment Resisted the Abolition of Slavery* (London, 2020).

Taylor, Michael A., 'The Plesiosaur's Birthplace: the Bristol Institution and its Contribution to Vertebrate Palaeontology', *Zoological Journal of the Linnean Society*, 112 (1994), 179–96.

———, 'Before the Dinosaur: The Historical Significance of the Fossil Marine Reptiles', in Jack M. Callaway and Elizabeth L. Nicholls (eds.), *Ancient Marine Reptiles* (San Diego, 1997), xix–xlvi.

——— and Hugh S. Torrens, 'Saleswoman to a New Science: Mary Anning and the Fossil Fish *Squaloraja* from the Lias of Lyme Regis', *Proceedings of the Dorset Natural History and Archaeological Society*, 108 (1986), 135–48.

Thompson, D., 'John Tyndall (1820–1893): A Study in Vocational Enterprise', *Vocational Aspects of Secondary and Further Education*, 9 (1957), 38–48.

Thompson, E. P., *The Making of the English Working Class* (New York, 1966).

Topham, Jonathan R., 'Beyond the "Common Context": The Production and Reading of the Bridgewater Treatises', *Isis*, 89 (1998), 233–62.

Torrens, Hugh S., 'Mary Anning (1799–1847) of Lyme; "The Greatest Fossilist the World Ever Knew"', *British Journal for the History of Science*, 28 (1995), 257–84.

Toynbee, Arnold J., *A Study of History* (12 vols, Oxford, 1934–61).

Turner, Frank M., *Between Science and Religion: Reaction to Scientific Naturalism in Late Victorian England* (New Haven, CT, 1974).

———, *Contesting Cultural Authority: Essays in Victorian Intellectual Life* (Cambridge, 1993).

Turner, S., C. V. Burek, and R. T. J. Moody, 'Forgotten Women in an Extinct Saurian (Man's) World', in R. T. J. Moody, E. Buffetaut, D. Naish, and D. M. Martill (eds.), *Dinosaurs and Other Extinct Saurians: A Historical Perspective* (London, 2010), 111–54.

Trevor-Roper, Hugh, 'James Ussher, Archbishop of Armagh' in Hugh Trevor-Roper, *Catholics, Anglicans, and Puritans: Seventeenth-Century Essays* (London, 1989), 120–65.

Tribe, David H., *President Charles Bradlaugh, M.P.* (London, 1971).

Ungureanu, James C., 'A Yankee at Oxford: John William Draper at the British Association for the Advancement of Science at Oxford, 30 June 1860', *Notes & Records of the Royal Society of London*, 70 (2016), 135–50.

Vance, Norman, *The Sinews of the Spirit: The Ideal of Christian Manliness in Victorian Literature and Religious Thought* (Cambridge, 1985).

von Arx, Jeffrey Paul, *Progress and Pessimism: Religion, Politics, and History in Late Nineteenth Century Britain* (Cambridge, MA, 1985).

Wallace, David Rains, *The Bonehunters' Revenge: Greed, and the Greatest Scientific Feud of the Gilded Age* (Boston, 1999).

Warren, Leonard, *Joseph Leidy: The Last Man Who Knew Everything* (New Haven, CT, 1998).

Webb, C. de B., 'Great Britain and "Die Republiek Natalia": An Early Case of U.D.I. and Sanctions', *Theoria: A Journal of Social and Political Theory*, 37 (1971), 15–30.

Wellnhofer, Peter, *Archaeopteryx: The Icon of Evolution*, trans. Frank Haase (Munich, 2009).

White, Andrew Dickson, *A History of the Warfare of Science with Theology in Christendom* (2 vols, New York, 1896).

White, George W., 'Announcement of Glaciation in Scotland: William Buckland (1784–1856)', *Journal of Glaciology*, 9 (1970), 143–5.

Williams, Paige, *The Dinosaur Artists: Obsession, Betrayal, and the Quest for Earth's Trophy* (London, 2018).

Wilson, A. N., *Charles Darwin: Victorian Mythmaker* (London, 2017).

Wilson, Ben, *Heyday: Britain and the Birth of the Modern World* (London, 2016).

Wilson, David, *Mr. Froude and Carlyle* [1898] (New York, 1970).

Wilson, Leonard J., 'Brixham Cave and Sir Charles Lyell's *The Antiquity of Man*: The Roots of Hugh Falconer's Attack on Lyell', *Archives of Natural History*, 23 (1996), 79–97.

Winn, William E., '*Tom Brown's Schooldays* and the Development of "Muscular Christianity"', *Church History*, 29 (1960), 64–73.

Wolffe, John, 'Elite and Popular Religion in the Religious Census of 30 March 1851', *Studies in Church History*, 42 (2006), 360–71.

Credits

INTEGRATED IMAGES

Skull of the '*Proteosaurus*': Biodiversity Heritage Library/Natural History Museum Library, London.

The fossilised skeleton of the Plesiosaurus: Biodiversity Heritage Library/California Academy of Sciences Library.

Thigh-bone misnamed as the '*scrotum humanum*': The Picture Art Collection/Alamy Stock Photo.

Jawbone of the *Megalosaurus*: Biodiversity Heritage Library/California Academy of Sciences Library.

Teeth used to identify the *Iguanodon*: Biodiversity Heritage Library/Natural History Museum Library, London.

The Temple of Serapis: Charles Lyell/Wellcome Collection.

Remains of the *Dimorphodon*: Biodiversity Heritage Library/California Academy of Sciences Library.

Fossilised remains of the *Hylaeosaurus*: Biodiversity Heritage Library/California Academy of Sciences Library.

Charles Darwin's first illustration of the 'tree of life': Reproduced with permission from John van Wyhe (ed., *The Complete Work of Charles Darwin Online*).

The *Iguanodon* Dinner at the Crystal Palace: Lithograph published by Day & Son, 1854/British Library, Add MS 50150, f. 225.

Darwin's Barnacles: Reproduced with permission from John van Wyhe (ed., *The Complete Work of Charles Darwin Online*).

Owen's illustrations of the human and chimpanzee brains: Reproduced with permission from John van Wyhe (ed., *The Complete Work of Charles Darwin Online*).

Neanderthal skull from the Neander Valley: John Lubbock, *Prehistoric Times* (1865)/public domain.

Frontispiece to *Evidence as to Man's Place in Nature*: Wellcome Collection.

Fossilised *Archaeopteryx* feather: Specimen MB.Av.100, reproduced by permission of the Museum für Naturkunde, Berlin.

Almost complete *Archaeopteryx* skeleton: The Natural History Museum/Alamy Stock Photos.

Representations of the *Hadrosaurus*: Keith Corrigan/Alamy Stock Photo.

Wallace's Bedford Level experiment: Reproduced with permission from John van Wyhe (ed., *The Complete Work of Charles Darwin Online*).

Remains of the *Hypsilophodon*: The History Collection/Alamy Stock Photo.

Hip-bones of birds, dinosaurs, and crocodiles: public domain.

G. W. Foote's *Comic Sketches*: public domain.

Edward Cope's Cretaceous landscape: public domain.

The *Dimorphodon* and the *Pterodactyl*: *Illustrated London News*/Mary Evans Library.

Comic News's mockery of Owen's proposal for a new natural history museum: public domain.

Charles Darwin's funeral: Wood engraving. Wellcome Collection.

Gertie the Dinosaur: Everett Collection Inc./Alamy Stock Photo

PLATE SECTION

William 'Strata' Smith's geological map of Britain: public domain.

James Ussher: Art Collection/Alamy Stock Photo.

Mary Anning: public domain.

Georges Cuvier: Coloured lithograph by N. E. Maurin/Wellcome Collection.

Gideon Mantell: Science History Images/Alamy Stock Photo.

William Buckland: Coloured lithograph printed by C. Hullmandel after N. Whittock/Wellcome Collection.

Awful Changes: Lithograph by Sir Henry De la Beche, 1830, after his drawing. Wellcome Collection.

Duria Antiquior: Art Collection 3/Alamy Stock Photo.

The Deluge: Yale Center for British Art, Paul Mellon Collection/public domain.

Country of the Iguanodon: Gift of Mrs Mantell-Harding, 1961/Te Papa Collection, Wellington.

Charles Lyell: Pictorial Press Ltd/Alamy Stock Photo.

Young Charles Darwin: Down House Museum, Kent/public domain.

The orangutans of the Regent's Park zoo: Coloured lithograph by Day & Haghe after W. Hawkins. Wellcome Collection.

Richard Owen: Engraving by D. J. Pound, 1861, after J. & C. Watkins/Wellcome Collection.

Thomas Huxley: Photograph by Cundall Downes & Co./Wellcome Collection.

The Crystal Palace dinosaurs: Colour and process print by Baxter, *c.* 1864/Wellcome Collection.

Alfred Russel Wallace: The Natural History Museum/Alamy Stock Photo.

Charles Kingsley: Photograph by Charles Watkins/Wellcome Collection.

Vanity Fair's depiction of Samuel Wilberforce and Thomas Huxley: Science History Images/Alamy Stock Photo.

John Tyndall: Photography by Lock & Whitfield/Wellcome Collection.

Charles Bradlaugh being arrested in Parliament: Colour lithograph by Tom Merry, 1888/Wellcome Collection.

Othniel Charles Marsh and the first Yale expedition to the American Badlands: Alpha Historica/Alamy Stock Photo.

Edward Drinker Cope: Chroma Collection/Alamy Stock Photo.

Charles Knight's *Leaping Laelaps*: public domain.

The Carnegie *Diplodocus* at the Natural History Museum: The Natural History Museum/Alamy Stock Photo

Natural History Museum: Photo-lithograph after M. B. Adams, 1879/Wellcome Collection.

Royal College of Science: Ilbusca/Getty Images.

Notes

ABBREVIATIONS

DCP = Darwin Correspondence Project
ODNB = *Oxford Dictionary of National Biography*

PREFACE

1. Steve Brusatte, *The Rise and Fall of the Dinosaurs: A New History of a Lost World* (London, 2018); Michael J. Benton, *The Dinosaurs Rediscovered: How a Scientific Revolution is Rewriting History* (London, 2019); Paige Williams, *The Dinosaur Artists: Obsession, Betrayal, and the Quest for Earth's Ultimate Trophy* (London, 2018).
2. My apologies to Professor Richard Huzzey of Durham University for stealing this joke.
3. 'National Trust reference Creationism at the Giant's Causeway', *Guardian*, 5 July 2012.
4. One of the best surveys of diverse Victorian attitudes to the Bible is: Timothy Larsen, *A People of One Book: The Bible and the Victorians* (Oxford, 2011).
5. Adrienne Mayor, *The First Fossil Hunters: Dinosaurs, Mammoths, and Myth in Greek and Roman Times* (Princeton, NJ, 2011); Richard B. Stothers, 'Ancient Scientific Basis of the "Great Serpent" from Historical Evidence', *Isis*, 95 (2004), 220–38.
6. For the importance of overseas and imperial exploration to British ideas about humankind, see: Peter Mandler, 'Looking around the World', in Adelene Buckland and Sadiah Qureshi (eds.), *Time Travellers: Victorian Perspectives on the Past* (Chicago, 2020), 24–41; and Sadia Qureshi, *Peoples on Parade: Exhibitions, Empire, and Anthropology in Nineteenth-Century Britain* (Chicago, 2011).

7. In *Darwin's Armada: Four Voyagers to the Southern Oceans and Their Battle for the Theory of Evolution* (London, 2009), Iain McCalman adds J. D. Hooker to this trinity of British explorers.
8. For the 1960s argument, see: Callum G. Brown, *The Death of Christian Britain: Understanding Secularisation, 1800–2000* (Abingdon, 2009). For the 1920s, see: S. J. D. Green, *The Passing of Protestant England: Secularisation and Social Change, c.1920–1960* (Cambridge, 2011). The classic examination of the subject is Owen Chadwick's *The Secularization of the European Mind in the Nineteenth Century* (Cambridge, 2011), while one of the better recent surveys of the literature is probably: Jeremy Morris, 'Secularization and Religious Experience: Arguments in the Historiography of Modern Religion', *Historical Journal*, 55 (2012), 195–219.
9. José Harris, *Private Lives, Public Spirits: A Social History of Britain, 1870–1914* (Oxford, 1993), p. 179.
10. Jeff Hardin, Ronald A. Numbers, and Ronald A. Binzley (eds.), *The Warfare between Science & Religion: The Idea that Wouldn't Die* (Baltimore, MD, 2018).
11. Major collections of essays on this tension include: Frank M. Turner, *Contesting Cultural Authority: Essays in Victorian Intellectual Life* (Cambridge, 1993); and Timothy Larsen, *Contested Christianity: The Political and Social Context of Victorian Theology* (Waco, TX, 2004).
12. For a counterpoint which presents a less belligerent view of even Huxley, see: Sheridan Gilley and Ann Loades, 'Thomas Henry Huxley: The War between Science and Religion', *The Journal of Religion*, 61 (1981), 285–308.

PROLOGUE: THE HOUSE OF USSHER

1. William Greenhill, 'To the Judicious Reader, and All Desirous to Understand the Chronology of Sacred Scriptures', in Thomas Allen, *A Chain of Scripture Chronology; from the Creation of the World to the Death of Jesus Christ* [1659] (London, 1668), [iii].
2. Arnold J. Toynbee, *A Study of History* (12 vols, Oxford, 1934–61), 9:180.
3. Richard Parr, *The Life of the Most Reverend Father in God, James Usher, Late Lord Arch-Bishop of Armagh, Primate and Metropolitan of All Ireland* (London, 1686), p. 58.
4. 'A Brief Relation of the Life and Memoirs of John, Lord Belasyse, Written and Collected by His Secretary, Joshua Moore', *Calendar of the Manuscripts of the Marquess of Ormonde, K.P., Preserved at Kilkenny Castle*, 2 (1903), 376–99, at p. 386.
5. Catharine Macaulay, *The History of England from the Accession of James I to that of the Brunswick Line* [1763–83] (8 vols, Cambridge, 2013), 4:191.

6. Samuel R. Gardiner, *History of the Great Civil War, 1642–1649* (4 vols, London, 1898), 2:290.
7. Parr, *Life of Usher*, p. 59.
8. Ibid.; Nicholas Bernard, *The Life and Death of the Most Reverend and Learned Father of Our Church, Dr James Ussher, Late Arch-Bishop of Armagh, and Primate of All-Ireland* (London, 1656), p. 100; *The Whole Works of the Most Rev. James Ussher, D.D., Lord Archbishop of Armagh, and Primate of All Ireland*, ed. Charles Richard Elrington (16 vols, Dublin, 1847), 1:244.
9. Parr, *Life of Usher*, p. 59.
10. Ibid.
11. Philip Styles, 'James Ussher and His Times', *Hermathena*, 88 (1956), 12–33, at p. 14.
12. James Ussher to Henry Fitzsimon, July 1600, in *The Correspondence of James Ussher, 1600–1656*, ed. Elizabethanne Boran (3 vols, Dublin, 2015), 1:2.
13. Gilbert Burnet, *The Life of William Bedell D.D., Lord Bishop of Killmore in Ireland* (London, 1692), p. 86.
14. Elrington, *Works of Ussher*, 1:283.
15. Parr, *Life of Usher*, p. 79.
16. J. A. Carr, *The Life and Times of Archbishop James Ussher, Archbishop of Armagh* (London, 1895), p. 43.
17. Bernard, *Life of Usher*, p. 9.
18. Hugh Trevor-Roper, 'James Ussher, Archbishop of Armagh' in *Catholics, Anglicans, and Puritans: Seventeenth-Century Essays* (London, 1989), 120–65, at p. 148; John Aikin, *The Lives of John Selden and Archbishop Ussher; with the Notices of the Principal English Men of Letters with Whom They Were Connected* (London, 1812), p. 263.
19. Elrington, *Works of Ussher*, 1:221.
20. Bernard, *Life of Ussher*, p. 94.
21. Parr, *Life of Usher*, pp. 44, 50.
22. Ibid., pp. 62–3.
23. Elrington, *Works of Ussher*, 1:261–2.
24. An English translation arrived in 1658: James Ussher, *The Annals of the World: Deduced from the Origin of Time, and Continued to the Beginning of the Emperour Vespasians Reign, and the Totall Destruction and Abolition of the Temple and Common-Wealth of the Jews* (London, 1658).
25. Trevor-Roper, 'James Ussher', p. 133.
26. Elrington, *Works of Ussher*, 1:89.
27. Patrick Wyse Jackson, *The Chronologers' Quest: Episodes in the Search for the Age of the Earth* (Cambridge, 2006), p. 19; Trevor-Roper, 'James Ussher',

p. 156; R. Buick Knox, *James Ussher: Archbishop of Armagh* (Cardiff, 1967), p. 105.

28. Bernard, *Life of Ussher*, p. 84.
29. J. G. C. M. Fuller, 'A Date to Remember: 4004 BC', *Earth Sciences History*, 24 (2005), 5–14, at pp. 7–9.
30. For an account of these problems and Ussher's methodology, see: James Barr, 'Why the World was Created in 4004 BC: Archbishop Ussher and Biblical Chronology', *John Rylands Bulletin*, 67 (1985), 575–608, esp. pp. 578–87.
31. Paul Barrett, 'Pre-Scientific Chronology: The Bible and the Origin of the World', *Proceedings of the American Philosophical Society*, 143 (1999), 379–87, at p. 381n8.
32. Colin V. Murray-Wallace, 'Understanding "Deep" Time – Advances since Archbishop Ussher?', *Archaeology in Oceania*, 31 (1996), 173–7, at p. 173.
33. Stephen Jay Gould in 'Fall in the House of Ussher', *Eight Little Piggies: Reflections in Natural History* (New York, 1994), 181–93, at pp. 190–91.
34. Elrington, *Works of Ussher*, 1:278–9; Parr, *Life of Usher*, p. 102.
35. John Milner, *A Defence of Arch-bishop Usher against Dr Cary and Dr Isaac Vossius* (Cambridge, 1694), p. 1.
36. James Boswell, *Life of Samuel Johnson, LL.D.* [1791] (Chicago, 1972), p. 183.
37. *The Memoirs of the Life of Edward Gibbon* [1796], ed. George Birkbeck Hill (London, 1900), p. 45.
38. Fuller, 'Date to Remember', p. 11.
39. Alan Ford, 'James Ussher', *ODNB*.
40. Gould, 'Fall in the House of Ussher', p. 185.
41. Jeffrey Hopes, 'Dating the World: The Science of Biblical Chronology', *XVII–XVIII*, 71 (2014), 65–83, at p. 80.
42. Martin J. S. Rudwick, *Bursting the Limits of Time: The Reconstruction of Geohistory in the Age of Revolution* (London, 2005), pp. 125, 127.
43. Patrick Brydone, *A Tour through Sicily and Malta in a Series of Letters to William Beckford, Esq., of Somerly in Suffolk* (London, 1806), p. 76.
44. Most information about Hutton comes from John Playfair, 'Biographical Account of the late James Hutton, M.D.', in *The Works of John Playfair, Esq.* (4 vols, Edinburgh, 1822), 4:33–120. The authoritative secondary account of his life is: Dennis R. Dean, *James Hutton and the History of Geology* (Ithaca, NY, 1992). A fine introduction is: Stephen Baxter, *Revolutions of the Earth: James Hutton and the True Age of the World* (London, 2003).
45. Jean Jones, 'James Hutton (1726–1797)', *ODNB*.
46. James Hutton, *Abstract of a Dissertation read in the Royal Society of Edinburgh ... Concerning the System of the Earth, its Duration, and Stability* (Edinburgh, 1785), p. 5.

47. Idem, 'Theory of the Earth; or an Investigation of the Laws Observable in the Composition, Dissolution, and Restoration of Land upon the Globe', *Transactions of the Royal Society of Edinburgh*, 1 (1788), 209–304, at p. 304.
48. The phrase 'deep time' was coined by American geologist John McPhee in *Basin and Range* (New York, 1981).

1. SHELLS BY THE SHORE

1. Jane Austen, *Northanger Abbey: and Persuasion* (4 vols, London, 1817), 3: 224–5.
2. Richard Russell, *A Dissertation on the Use of Sea Water in the Diseases of the Glands, particularly the Scurvy, King's-Evil, Leprosy, and the Glandular Consumption* (4th edn, London, 1760); George Roberts, *The History and Antiquities of the Borough of Lyme Regis and Charmouth* (London, 1834), p. 179.
3. Roger North, *The Life of the Right Honourable Francis North, Baron of Guildford, Lord Keeper of the Great Seal under King Charles II and King James II* (2nd edn, 3 vols, 1808), 1:228.
4. Roberts, *History of Lyme Regis*, p. 285.
5. W. D. Lang, 'Early Days of Natural History at Charmouth', *Proceedings of the Dorset Archaeological and Natural History*, 62 (1940), 97–113, at p. 98.
6. Deborah Cadbury, *The Dinosaur Hunters: A True Story of Scientific Rivalry & the Discovery of the Prehistoric World* (London, 2000), p. 7.
7. Douglas Palmer, *Earth Time: Exploring the Deep Past from Victorian England to the Grand Canyon* (Chichester, 2005), p. 122.
8. Patricia Pierce, *Jurassic Mary: Mary Anning and the Primeval Monsters* (Stroud, 2006), p. 14.
9. *Jane Austen's Letters to Her Sister Cassandra and Others*, ed. R. W. Chapman (2nd edn, London, 1952), p. 201.
10. W. D. Lang, 'Mary Anning (1799–1847) and the Pioneer Geologists of Lyme', *Proceedings of the Dorset Archaeological and Natural History Society*, 60 (1938), 142–64, at p. 144.
11. *Bath Chronicle and Weekly Gazette*, 27 December 1798.
12. Roberts, *History of Lyme*, p. 287.
13. W. D. Lang, 'Mary Anning's Escape from Lightning', *Proceedings of the Dorset Archaeological and Natural History Society*, 80 (1959), 91–3, at p. 91.
14. 'Mary Anning, the Fossil Finder', *All The Year Round*, 13 (1865), 60–63, at p. 61.

15. Roberts, *History of Lyme*, p. 286.
16. Jean-André Deluc, *Geological Travels* (3 vols, 1810–11), 2:83–4.
17. Roberts, *History of Lyme*, p. 288.
18. W. D. Lang, 'Mary Anning and Anna Maria Pinney', *Proceedings of the Dorset Archaeological and Natural History Society*, 76 (1955), 146–52, at p. 146.
19. Christopher McGowan, *The Dragon Seekers: How an Extraordinary Circle of Fossilists Discovered the Dinosaurs and Paved the Way for Darwin* (London, 2002), p. 24.
20. *Bath Chronicle and Weekly Gazette*, 12 November 1812.
21. *Statesman*, 13 November 1812; *General Evening Post*, 14 November 1812; *Bell's Weekly Messenger*, 15 November 1812.
22. *Madras Courier*, 21 December 1813.
23. 'Linnean Society', *Annals of Philosophy*, 5 (1815), p. 70.
24. Everard Home, 'An Account of the Fossil Skeleton of the Proteo-Saurus', *Philosophical Transactions of the Royal Society*, 109 (1819), 209–11, at p. 210.
25. Idem, 'Some Farther Account of the Fossils Remains of An Animal, of which a Description was Given to the Society in 1814', *Philosophical Transactions of the Royal Society*, 106 (1816), 318–21, at p. 320.
26. Idem, 'Reasons for Giving the Name Proteo-Saurus to the Fossil Skeleton which has been Described', *Philosophical Transactions of the Royal Society of London*, 109 (1819), 212–16, at p. 215.
27. Thomas Molyneux, 'A Discourse Concerning the Large Horns Frequently Found under the Ground in Ireland', *Philosophical Transactions of the Royal Society of London*, 19 (1697), 489–512, at p. 489.
28. Robert Hooke, 'Lectures and Discourses of Earthquakes, and Subterraneous Eruptions', in *The Posthumous Works of Robert Hooke . . . containing his Cutlerian Lectures and Other Discourses*, ed. Richard Waller (London, 1705), 277–450, at p. 434.
29. *Bath Chronicle*, 12 November 1812.
30. Everard Home, 'Additional Facts respecting the Fossil Remains of an Animal, on the Subject of which Two Papers have been Printed in the Philosophical Transactions', *Philosophical Transactions of the Royal Society of London*, 108 (1818), 24–32, at p. 28.
31. W. D. Lang, 'Mary Anning and a Very Small Boy', *Proceedings of the Dorset Archaeological and Natural History Society*, 84 (1962), 181.
32. Newton's exact words, translated from the Latin, were: 'And thus much concerning God; to discourse of whom from the appearances of things, does certainly belong to Natural Philosophy'. See Stephen D. Snobelen, 'To Discourse of God: Isaac Newton's Heterodox Theology and His Natural

Philosophy', in *Science and Dissent in England, 1688–1945*, ed. Paul B. Wood (Aldershot, 2004), 39–65, at p. 39.

33. William van Mildert, *An Historical View of the Rise and Progress of Infidelity, with a Refutation of its Principles and Reasonings* (2 vols, London, 1820), 2:99.
34. E. P. Thompson, *The Making of the English Working Class* (New York, 1966), p. 27.
35. Michael Taylor, 'British Conservatism, The Illuminati, and the Conspiracy Theory of the French Revolution, 1797–1802', *Eighteenth-Century Studies*, 47 (2014), 293–312, at pp. 305–6.
36. Boyd Hilton, *The Age of Atonement: The Influence of Evangelicalism on Social and Economic Thought, 1795–1865* (Oxford, 1988), p. 9.
37. David Brion Davis, *The Problem of Slavery in the Age of Revolution, 1770–1823* (Ithaca, NY, 1975), pp. 526–7.
38. Richard Kirwan, 'Hutton's Theory of the Earth', *British Critic*, 8 (1796), 468–80, at p. 476.
39. James Parkinson, *Organic Remains of a Former World: An Examination of the Mineralized Remains of the Vegetables and Animals of the Antediluvian World* (3 vols, London, 1804–11), 1:13.
40. Thomas W. Goodhue, 'The Faith of a Fossilist: Mary Anning', *Anglican and Episcopal History*, 70 (2001), 80–100, at p. 82.
41. Clive D. Field, 'Counting Religion in England and Wales: The Long Eighteenth Century, *c.*1680–*c.*1840', *Journal of Ecclesiastical History*, 63 (2012), 693–720, at pp. 711, 718.
42. Robert Currie, Alan Gilbert, and Lee Horsley, *Churches and Churchgoers: Patterns of Church Growth in the British Isles since 1700* (Oxford, 1977), pp. 65–6.
43. Hannah More to Zachary Macaulay, 6 January 1796, in *Memoirs of the Life and Correspondence of Mrs Hannah More*, ed. William Roberts (2 vols, New York, 1845), p. 473.
44. Hannah More, 'The Shepherd of Salisbury Plain', in *The Works of Hannah More* (7 vols, New York, 1835), 1:176–201, at p. 176.
45. Edmund Burke, *Reflections on the Revolution in France, and on the Proceedings in Certain Societies in London* (3rd edn, London, 1790), p. 114.
46. Austin Mitchell, 'The Association movement of 1792–3', *Historical Journal*, 4 (1961), 56–77, at pp. 58, 70.
47. Emmet Kennedy, *A Cultural History of the French Revolution* (New Haven, CT, 1989), p. 343.
48. 'Decree on the Worship of the Supreme Being', 7 May 1794, in J. F. Maclear (ed.), *Church and State in the Modern Age: A Documentary History* (Oxford, 1995), 88–90, at p. 88.

49. George Gleig (ed.), *Supplement to the Third Edition of the Encyclopaedia Britannica, or, a Dictionary of Arts, Sciences, and Miscellaneous Literature* (2 vols, Edinburgh, 1801), 1: preface.

2. UNDERGROUNDOLOGY

1. William Wordsworth, *The Excursion, Being a Portion of The Recluse* (London, 1814), p. 103.
2. *The Life and Correspondence of William Buckland, D.D., F.R.S.., Sometime Dean of Westminster*, ed. Elizabeth Oke Gordon (London, 1894), pp. 3, 2, 4.
3. John Jones, *Balliol College: A History* (2nd edn, 2005), p. 181.
4. *Oxford University Statutes*, trans. G. R. M. Ward and James Heywood (2 vols, London, 1851), 2:23–4.
5. Rev. W. Tuckwell, *Reminiscences of Oxford* (London, 1900), p. 61.
6. John S. Rowlinson, 'Chemistry Comes of Age: The 19th Century', in Robert J. P. Williams, John S. Rowlinson, and Allan Chapman (eds.), *Chemistry at Oxford: A History from 1600 to 2005* (London, 2009), 79–130, at pp. 79–81.
7. John Kidd, *An Answer to a Charge against the English Universities contained in the Supplement to the Edinburgh Encyclopaedia* (Oxford, 1818), pp. 7–8.
8. *Life of Buckland*, pp. 22, 8.
9. Ibid., pp. 14, 19, 36, 10.
10. Roy Porter, 'Gentlemen and Geology: The Emergence of a Scientific Career, 1660–1920', *Historical Journal*, 21 (1978), 809–36, at p. 821.
11. Simon J. Knell, 'The Road to Smith: How the Geological Society Came to Possess English Geology', in C. L. E. Lewis and S. J. Knell (eds.), *The Making of the Geological Society* (London, 2009), 1–48, at p. 8.
12. J. M. Edmonds, 'The Founding of the Oxford Readership in Geology, 1818', *Notes and Records of the Royal Society of London*, 34 (1979), 33–51, at pp. 37–8.
13. Ibid., pp. 35, 40, 39.
14. This description of the 1810s borrows from: Michael Taylor, *The Interest: How the British Establishment Resisted the Abolition of Slavery* (London, 2020), p. 30.
15. William Hone, *The Late John Wilkes's Catechism of a Ministerial Member* (London, 1817), p. 6.
16. Idem, *The Sinecurist's Creed, or Belief* (London, 1817), p. 7.
17. Idem, *The Political Litany, Diligently Revised* (London, 1817), p. 7.
18. Harriet Martineau, *The History of England during the Thirty Years' Peace, 1816–1846* (2 vols, London, 1849–50), 1:145.

19. Joss Marsh, *Word Crimes: Blasphemy, Culture, and Literature in Nineteenth-Century England* (Chicago, 1998), pp. 28, 24.
20. Thomas Paine, *The Age of Reason: Being an Investigation of True and Fabulous Theology* (3 vols, London, 1794–1807), 2:14.
21. John Martin, *A Letter to the Hon. Thomas Erskine, with a Postscript, to the Right Hon. Lord Kenyon, upon their Conduct at the Trial of Thomas Williams for Publishing Paine's Age of Reason* (3rd edn, London, 1797), p. 5.
22. Marsh, *Word Crimes*, pp. 20–21.
23. John Kidd, *A Geological Essay on the Imperfect Evidence in Support of a Theory of the Earth, Deducible either from its General Structure or from the Changes Produced on its Surface by the Operation of Existing Causes* (Oxford, 1815), pp. 1, 9, 1.
24. George Cumberland, 'On [the] Proper Objects of Geology', *Monthly Magazine*, 40 (1815), 130–33, at pp. 130–31.
25. Mrs R. Lee, *Memoirs of Baron Cuvier* (London, 1833), pp. 11–12.
26. *Life, Letters, and Journals of Sir Charles Lyell*, ed. Mrs Lyell (2 vols, London, 1881), 1:249–50.
27. Georges Cuvier, *Le Règne Animal, Distribué d'après son Organisation* (4 vols, Paris, 1816); idem, *Recherches sur les Ossemens Fossiles des Quadrupèdes* (5 vols, Paris, 1812–23).
28. Clifton C. Crais, *Sara Baartman and the Hottentot Venus: A Ghost Story and a Biography* (Princeton, NJ, 2009), esp. pp. 116–41.
29. Adrian Desmond, *The Hot Blooded Dinosaurs: A Revolution in Palaeontology* (London, 1975), p. 9.
30. Georges Cuvier, *Essay on the Theory of the Earth*, trans. Robert Jameson (2nd edn, Edinburgh, 1815), pp. 171–2, vi.
31. *Perthshire Courier*, 16 December 1813.
32. William Buckland, *Vindiciae Geologicae; or, the Connexion of Geology with Religion Explained, in an Inaugural Lecture Delivered before the University of Oxford, May 15, 1819, on the Endowment of a Readership in Geology by His Royal Highness, the Prince Regent* (Oxford, 1820), pp. 11–12, 18, 22–3.
33. Ibid., pp. 24–6, 31–2.
34. Martin J. S. Rudwick, *The Great Devonian Controversy: The Shaping of Scientific Knowledge among Gentlemanly Specialists* (Chicago, 1985), p. 70.
35. Shelley Emling, *The Fossil Hunter: Dinosaurs, Evolution, and the Woman Whose Discoveries Changed the World* (New York, 2009), p. 36.
36. Nick Davidson, *The Greywacke: How a Vicar, a Soldier, and a School Teacher Uncovered 300 Million Years of History* (London, 2021), p. 87.
37. Lang, 'Anning and Pinney', p. 148.
38. *Salisbury and Winchester Journal*, 3 April 1815.

39. Hugh Torrens, 'Mary Anning (1799–1847) of Lyme; "The Greatest Fossilist the World Ever Knew"', *British Journal for the History of Science*, 28 (1995), 257–84, at p. 261.
40. Ibid.
41. Emling, *Fossil Hunter*, pp. 71–2.
42. Torrens, 'Mary Anning', p. 262.

3. OF CAVES AND PADDLES

1. P. B. Duncan, 'On the Woman in Paviland Cavern', in C. G. B. Daubeny (ed.), *Fugitive Poems connected with Natural History and Physical Science* (Oxford, 1869), p. 122.
2. On getting to Paviland Cave, see: Stephen Moss, 'The Secrets of Paviland Cave', *Guardian*, 25 April 2011.
3. William Buckland, *Reliquiae Diluvianae: or, Observations on the Organic Remains contained in Caves, Fissures, and Diluvial Gravel, and on Other Geological Phenomena, Attesting the Action of an Universal Deluge* (London, 1823), pp. 83–8.
4. *Life of Buckland*, p. 31.
5. *Memories of Old Friends: Being Extracts from the Journals and Letters of Caroline Fox, of Penjerrick, Cornwall, from 1835 to 1871*, ed. Horace N. Pym (London, 1882), p. 5.
6. Nicolaas A. Rupke, *The Great Chain of History: William Buckland and the English School of Geology* (Oxford, 1983), p. 7; Charles Darwin to Edward Cresy, 5 July 1848, *The Correspondence of Charles Darwin*, ed. Frederick Burkhardt (30 vols, Cambridge, 1985–), 4:157.
7. Buckland, *Reliquiae Diluvianae*, p. 92.
8. Marianne Sommer, '"An Amusing Account of a Cave in Wales": William Buckland (1784–1856) and the Red Lady of Paviland', *British Journal for the History of Science*, 37 (2004), 53–74, at pp. 61–3. Sommer expanded her article into the most comprehensive study of the Paviland Cave: *Bones and Ochre: The Curious Afterlife of the Red Lady of Paviland* (Cambridge, MA, 2007).
9. Buckland, *Reliquiae Diluvianae*, pp. 6, 14.
10. William A. S. Sarjeant and Justin B. Delair, 'An Irish Naturalist in Cuvier's Laboratory: the Letters of Joseph Pentland, 1820–1832', *Bulletin of the British Museum (Natural History) Historical Series*, 6 (1980), 245–319, at p. 283.
11. Buckland, *Reliquiae*, p. 10.
12. Rupke, *Great Chain*, pp. 32–3.
13. Buckland, *Reliquiae*, p. 40.

14. *Life of Buckland*, p. 59.
15. Brenda Maddox, *Reading the Rocks: How Victorian Geologists Discovered the Secret of Life* (London, 2018), p. 69.
16. Rudwick, *Bursting the Limits of Time*, p. 629.
17. *Life of Buckland*, p. 77.
18. [Edward Copleston], 'Buckland–*Reliquiae Diluvianae*', *Quarterly Review*, 29 (1823), 138–65, at pp. 146, 157.
19. *Life of Buckland*, pp. 78, 77.
20. 'Autobiography of W. D. Conybeare' in *Letters and Exercises of the Elizabethan Schoolmaster John Conybeare, with Notes and a Fragment of Autobiography by the Very Rev. William Daniel Conybeare*, ed. Frederick Cornwallis Conybeare (London, 1905), 114–40, at pp. 121–2, 136.
21. Sarjeant and Delair, 'An Irish Naturalist', pp. 259, 274, 278.
22. Duncan C. L. Fitzwilliams, 'The Destruction of John Hunter's Papers', *Proceedings of the Royal Society of Medicine*, 42 (1949), 871–6, at p. 873.
23. 'Abstract of the Evidence of Members of the Royal College of Surgeons in London, Taken Before the Parliamentary Medical Committee in 1834', *Western Journal of the Medical and Physical Sciences*, 9 (1836), 353–61, at p. 353.
24. H. T. De la Beche and W. D. Conybeare, 'Notice of the Discovery of a New Fossil Animal, Forming a Link between the Ichthyosaurus and Crocodile, together with General Remarks on the Osteology of the Ichthyosaurus', *Transactions of the Geology Society*, 5 (1821), 559–94, at pp. 562–4, 572.
25. *Taunton Courier and Western Advertiser*, 24 December 1823.
26. De la Beche and Conybeare, 'New Fossil Animal', p. 559.
27. W. D. Conybeare, 'On the Discovery of an Almost Perfect Skeleton of the Plesiosaurus', *Transactions of the Geological Society of London*, 1 (2nd ser., 1824), 381–9, at p. 381; Emling, *Fossil Hunter*, p. 80.
28. Michael A. Taylor, 'The Plesiosaur's Birthplace: the Bristol Institution and its Contribution to Vertebrate Palaeontology', *Zoological Journal of the Linnean Society*, 112 (1994), 179–96, at p. 182.
29. Michael A. Taylor, 'Before the Dinosaur: The Historical Significance of the Fossil Marine Reptiles', in Jack M. Callaway and Elizabeth L. Nicholls (eds.), *Ancient Marine Reptiles* (San Diego, 1997), xix–xlvi, at xxiii–xxiv.
30. Cumberland, Konig, and Conybeare qu. in Cadbury, *Dinosaur Hunters*, pp. 104–05.
31. McGowan, *Dragon Seekers*, p. 131.
32. Taylor, 'Before the Dinosaur', xxiii.
33. W. D. Conybeare, *Outlines of the Geology of England and Wales, with an Introductory Compendium of the General Principles of that Science* (London, 1822).
34. Conybeare, 'Plesiosaurus', p. 382.

35. Gideon Mantell, *The Wonders of Geology* (2 vols, London, 1838), 2:435.
36. Taylor, 'Before the Dinosaur', xxiii.

4. A BONE OF PRODIGIOUS BIGNESS

1. George Byron, *Don Juan: Cantos IX, X, and XI* (London, 1823), p. 24.
2. Arthur McGregor, 'The Tradescants as Collectors of Rarities', in Arthur McGregor (ed.), *Tradescant's Rarities: Essays on the Foundation of the Ashmolean Museum, 1683, with a Catalogue of the Surviving Early Collections* (Oxford, 1983), 17–23, at p. 20.
3. A. J. Turner, 'Robert Plot', *ODNB*.
4. Robert Plot, *The Natural History of Oxford-Shire: Being an Essay toward the Natural History of England* (Oxford, 1677), pp. 131–6, index.
5. Richard Brookes, *The Natural History of Waters, Earths, Stones, Fossils, and Minerals, with their Virtues, Properties, and Medicinal Uses* (London, 1763), pp. 317–19.
6. E. A. Howlett, W. J. Kennedy, H. P. Powell and H. S. Torrens, 'New Light on the History of *Megalosaurus*, the Great Lizard of Stonesfield', *Archives of Natural History*, 44 (2017), 82–102, at pp. 84, 87, 89.
7. 'Donations to the Cabinet of Minerals', *Transactions of the Geological Society*, 5 (1819), 640–50, at p. 640.
8. Howlett et al., 'New Light', p. 94.
9. De la Beche and Conybeare, 'New Fossil Animal', p. 592.
10. James Parkinson, *Outlines of Oryctology: An Introduction to the Study of Fossil Organic Remains, Especially of Those Found in the British Strata* (London, 1822), p. 298.
11. William Buckland, 'Notice on the Megalosaurus or Great Fossil Lizard of Stonesfield', *Transactions of the Geological Society of London*, 1 (2nd ser., 1824), 390–96, at pp. 390, 393.
12. Francis T. Buckland, *Curiosities of Natural History* (4th edn, New York, 1859), p. 64.
13. Buckland, 'Notice on the Megalosaurus', p. 392.
14. *Life of Buckland*, p. 84.
15. The authoritative biography of Mantell is Dennis R. Dean, *Gideon Mantell and the Discovery of Dinosaurs* (Cambridge, 1999); others include E. M. R. Critchley, *Dinosaur Doctor: The Life of Gideon Mantell* (Stroud, 2010), and Sidney Spokes, *Gideon Algernon Mantell, Surgeon and Geologist* (London, 1927).
16. Dean, *Mantell*, p. 62.
17. Gideon Mantell, *A Pictorial Atlas of Fossil Remains, Consisting of Coloured Illustrations* (London, 1850), p. 14.

18. Idem, *The Fossils of the South Downs; or Illustrations of the Geology of Sussex* (London, 1822), vii.
19. Gordon Hake, *Memoirs of Eighty Years* (London, 1892), p. 60; *The Unpublished Journal of Gideon Mantell, 1819–1852*, ed. John A. Cooper (Brighton, 2010), p. 7.
20. *The Journal of Gideon Mantell, Surgeon and Geologist, Covering the Years 1818–1852*, ed. E. Cecil Curwen (Oxford, 1940), p. 6.
21. *Journal of Mantell*, pp. 9, 20, 8.
22. Ibid, p. 29.
23. Mantell, *South Downs*, viii.
24. *Journal of Mantell*, pp. 43–4, 47.
25. Ibid., pp. 43, 47.
26. Biographical information taken from: H. S. Torrens, 'William Smith [*called* Strata Smith] (1769–1839)', *ODNB*.
27. *Memoirs of William Smith, LL.D., Author of the 'Map of the Strata of England and Wales'*, ed. John Phillips (London, 1844), p. 34.
28. Smith's maps have been recently republished in *Strata: William Smith's Geological Maps* (Chicago, 2020).
29. *Journal of Mantell*, pp. 29, 43.
30. Mantell, *South Downs*, ix. For the general importance of Mrs Mantell to her husband's geology, see: Martina Kölbl-Ebert, 'Sketching Rocks and Landscapes: Drawing as a Female Accomplishment in the Service of Geology', *Earth Sciences History*, 31 (2012), 270–86.
31. Gideon Mantell, *The Geology of the South-East of England* (London, 1833), p. 268.
32. Idem, *Petrifactions and Their Teachings; or, a Hand-Book to the Gallery of Organic Remains of the British Museum* (London, 1851), p. 228.
33. Ibid., pp. 228–32.
34. Dean, *Gideon Mantell*, p. 75.
35. Mantell, *Petrifactions*, p. 230.
36. Dean, *Mantell*, p. 81.
37. Gideon Mantell, 'Notice on the Iguanodon, a Newly Discovered Fossil Reptile from the Sandstone of Tilgate Forest, in Sussex', *Philosophical Transactions of the Royal Society of London*, 115 (1825), 179–86, at p. 180.
38. Dean, *Gideon Mantell*, p. 88.
39. *Journal of Mantell*, p. 57.

5. FREE THE SCIENCE FROM MOSES

1. 'Transactions of the Geological Society of London' [review], *Quarterly Review*, 34 (1826), 507–39, at p. 524.

2. There are several classic studies of the tension between geology and religion in this period, including Charles Coulston Gillespie, *Genesis and Geology: A Study in the Relations of Scientific Thought, Natural Theology, and Social Opinion in Great Britain, 1790–1850* (Cambridge, MA, 1951), and Milton Millhauser, 'The Scriptural Geologists: An Episode in the History of Opinion', *Osiris*, 11 (1954), 65–86.
3. Mantell, *South Downs*, at pp. 1, ix, 3, 305.
4. George Bugg, *Scriptural Geology: or, Geological Phenomena, Consistent Only with the Literal Interpretation of the Sacred Scriptures* (2 vols, London, 1826–27), 2:351.
5. Bugg's biographical information is taken from Terry Mortenson, *The Great Turning Point: The Church's Catastrophic Mistake on Geology – Before Darwin* (Green Forest, AR, 2004), pp. 77–97.
6. Bugg, *Scriptural Geology*, 2:351, 1:xii, 1:vii, 2:39, 2:349.
7. Ralph O'Connor, 'Young-Earth Creationists in Early Nineteenth-Century Britain? Towards a Reassessment of "Scriptural Geology"', *History of Science*, 45 (2007), 357–403, at p. 368.
8. Granville Penn, *A Comparative Estimate of the Mineral and Mosaical Geologies: Revised, and Enlarged with Relation to the Latest Publications on Geology* (2nd edn, 2 vols, London, 1825), 1:xiv.
9. W. V. Farrar, 'Andrew Ure, F.R.S., and the Philosophy of Manufactures', *Notes and Records of the Royal Society of London*, 27 (1973), 299–324, at p. 299.
10. Andrew Ure, *A New System of Geology, in which the Great Revolutions of the Earth and Animated Nature, are Reconciled At Once to Modern Science and Sacred History* (London, 1829).
11. Henry Hart Milman, *The History of the Jews* (3 vols, London, 1829), 3:iv.
12. *Life of Lyell*, 1:263.
13. Boyd Hilton, *A Mad, Bad, and Dangerous People? England, 1783–1846* (Oxford, 2006), pp. 381–2.
14. This passage is based on the description of these tensions in Taylor, *The Interest*, pp. 165–9.
15. The most recent major study of Martin is: Max Adams, *The Prometheans: John Martin and the Generation that Stole the Future* (London, 2010); see also William Feaver, 'John Martin (1789–1854)', *ODNB*.
16. Thomas Balston, *John Martin, 1799–1854, His Life and Works* (London, 1947), p. 65.
17. *Life of Lyell*, 1:28, 1:68.
18. W. D. Lang, 'Three Letters by Mary Anning, Fossilist of Lyme', *Proceedings of the Dorset Archaeological and Natural History Society*, 66 (1945), 169–73, at p. 171. The peerless account of the dispute is Martin J. S. Rudwick,

The Great Devonian Controversy: The Shaping of Scientific Knowledge among Gentlemanly Specialists (Chicago, 1985).

19. *Queen of Science: Personal Recollections of Mary Somerville*, ed. Dorothy McMillan (Edinburgh, 2001), p. 100.
20. *Life of Lyell*, 1:185–9.
21. Ibid., 1:189–99. For their time in France, see also M. Kölbl-Ebert, 'The Geological Travels of Charles Lyell, Charlotte Murchison, and Roderick Impey Murchison in France and Northern Italy (1828)', *Geological Society of London: Special Publications*, 287 (2007), 109–17.
22. *Life of Lyell*, 1:209–10, 234.
23. Charles Lyell, *Principles of Geology: or, the Modern Changes of the Earth and Its Inhabitants, Considered as Illustrative of Geology* (3 vols, London, 1830–33), 1:454.
24. *Life of Lyell*, 1:268.
25. For a more detailed description of the printing process, see: Lucien Febvre and Henri-Jean Martin, *The Coming of the Book: The Impact of Printing, 1450–1800*, trans. David Gerard [1958] (London, 1976), pp. 61–8.
26. *Life of Lyell*, 1:258–62, 268, 271, 270.
27. Ibid., 1:270.
28. For Scrope's own geology, see: George Poulett Scrope, *Considerations on Volcanos, the Probable Causes of their Phenomena, the Laws which Determine their March, the Disposition of their Products, and their Connexion with the Present State and Past History of the Globe* (London, 1825).
29. *Life of Lyell*, 1:268, 270.
30. Lyell, *Principles*, 1:22.
31. *Life of Lyell*, 1:273.
32. [George Poulett Scrope], 'Principles of Geology', *Quarterly Review*, 43 (1830), 411–68, at pp. 414, 469.
33. *British Critic*, 15 (1834), p. 334; *New Monthly Magazine* (June 1832), p. 241.
34. Adam Sedgwick, 'Presidential Address', *Proceedings of the Geological Society of London*, 1 (1834), 270–316, at pp. 302–03.
35. W. D. Conybeare, 'Report on the Progress, Actual State, and Ulterior Prospects of Geological Science', *Report of the First and Second Meetings of the British Association for the Advancement of Science* (London, 1833), 365–414, at pp. 406–07.
36. *The Times*, 25 April 1831.
37. For the production and reception of the *Principles* generally, see: James A. Secord, *Visions of Science: Books and Readers at the Dawn of the Victorian Age* (Oxford, 2014), pp. 138–72. For Lyell's and Murray's strategy, see:

Martin J. S. Rudwick, 'The Strategy of Lyell's *Principles of Geology*', *Isis*, 61 (1970), 4–33.

38. Murray qu. in Secord, *Visions*, pp. 161–2.

6. THE PARLIAMENT OF SCIENCE

1. *Life of Lyell*, 1:234.
2. Pierce, *Jurassic Mary*, pp. 90–91.
3. Emling, *Fossil Hunter*, pp. 89–90.
4. *Exeter Flying Post*, 27 December 1827.
5. Lang, 'Three Letters', pp. 169–70.
6. Ibid., pp. 170, 169.
7. Idem, 'Mary Anning', pp. 160–61.
8. Idem, 'Anning and Pinney', p. 147.
9. *A Memoir of Miss Frances Augusta Bell, Who Died in Kentish Town*, ed. Johnson Grant (London, 1827), pp. 131–3.
10. *Salisbury and Winchester Journal*, 2 August 1824; *Taunton Courier*, 18 May 1825.
11. *Bath Chronicle and Weekly Gazette*, 27 October 1825.
12. Michael A. Taylor and Hugh S. Torrens, 'Saleswoman to a New Science: Mary Anning and the Fossil Fish *Squaloraja* from the Lias of Lyme Regis', *Proceedings of the Dorset Natural History and Archaeological Society*, 108 (1986), 135–48.
13. Pierce, *Jurassic Mary*, p. 98.
14. William Buckland, 'On the Discovery of a New Species of Pterodactyle in the Lias at Lyme Regis', *Transactions of the Geological Society of London*, 3 (2nd ser., 1829), 217–22, at pp. 217–18.
15. Henry De la Beche, *Remarks on the Geology of Jamaica* (London, 1827); idem, *Notes on the Present Condition of the Negroes in Jamaica* (London, 1825).
16. Anne Auriol, *Statement and Correspondence Consequent on the Ill-Treatment of Lady de la Beche by Major-General Henry Wyndham* (London, 1843), p. 95.
17. For De la Beche and *Awful Changes* (1831), see Ralph O'Connor, *The Earth on Show: Fossils and the Poetics of Popular Science, 1802–1856* (Chicago, 2008), pp. 77–80.
18. The most recent detailed study of the image is Tom Sharpe, 'Henry de la Beche's 1829–1830 lithograph, *Duria Antiquior*', *Earth Sciences History*, 41 (2022), 47–63.
19. Jack Morrell and Arnold Thackray (eds.), *Gentlemen of Science: Early Correspondence of the British Association for the Advancement of Science* (London, 1984), pp. 33, 26, 23, 24, 137.

20. Ibid., p. 42, referring to: [David Brewster], 'Reflexions on the Decline of Science in England, and on Some of its Causes' [review], *Quarterly Review*, 43 (1830), 305–42.
21. Morrell and Thackray, *Gentlemen*, pp. 34, 45, 48, 45, 36.
22. *Report of the First and Second Meetings of the British Association for the Advancement of Science; at York in 1831, and at Oxford in 1832* (2nd edn, London, 1835), pp. 47, 20.
23. *York Herald*, 1 October 1831; *Bury and Norwich Post*, 5 October 1831.
24. Morrell and Thackray, *Gentlemen*, p. 77.
25. Charles Dickens, *The Mudfog Papers and Other Sketches* [1837–38] (New York, 1880), which mimics the extensive annual reports of the BAAS at pp. 61 and 123.
26. *First and Second Meetings*, pp. 25, 41.
27. Morrell and Thackray, *Gentlemen*, p. 137.
28. *Life of Buckland*, p. 91.
29. G. V. Cox, *Recollections of Oxford* (2nd edn, London, 1870), p. 116.
30. Allan Chapman, *Caves, Coprolites, and Catastrophes: The Story of Pioneering Geologist and Fossil-Hunter William Buckland* (London, 2020), p. 84.
31. *Life of Buckland*, p. 90.
32. Ibid., pp. 92, 87.
33. Ibid., pp. 100, 102, 104–06.
34. E. T. Cook, *The Life of John Ruskin* (2 vols, London, 1911), 1:75–6. On Ruskin and Buckland, see: Van Akin Burd, 'Ruskin and his "Good Master", William Buckland', *Victorian Literature and Culture*, 36 (2008), 299–315.
35. For this and the Bucklands' eating habits, see: Richard Girling, *The Man Who Ate the Zoo: Frank Buckland, Forgotten Hero of Natural History* (London, 2016), p. 16.
36. William Buckland, 'On the Discovery of Coprolites, or Fossil Faeces, in the Lias at Lyme Regis, and in Other Formations', *Transactions of the Geological Society of London*, 3 (2nd ser., 1829), 223–36, at pp. 223–5.
37. *Life of Buckland*, p. 123.
38. Ibid., and for disagreement see: Elizabeth Chambers Patterson, *Mary Somerville and the Cultivation of Science, 1815–1840* (Hingham, MA, 1983), p. 94.
39. Morrell and Thackray, *Gentlemen*, p. 137.
40. For a detailed study of this discussion, see: Rebekah Higgitt and Charles W. J. Withers, 'Science and Sociability: Women as Audience at the British Association for the Advancement of Science, 1831–1901', *Isis*, 99 (2008), 1–27.
41. *First and Second Meetings*, pp. 301, 99, 402.
42. Ibid., pp. 104, 106, 107.

43. *Life of Lyell*, 1:388.
44. *First and Second Meetings*, p. 104.
45. Georges Cuvier, 'Extract from a Memoir on an Animal of Which the Bones are Found in the Plaster Stone around Paris, and Which Appears No Longer to Exist Alive Today', in Martin J. S. Rudwick, *Georges Cuvier, Fossil Bones, and Geological Catastrophes: New Translations & Interpretations of the Primary Texts* (Chicago, 1997), 35–41, at p. 36.
46. Jean-Baptiste Lamarck, *Philosophie Zoologique: ou, Exposition des Considérations Relatives à l'Histoire Naturelle des Animaux* (Paris, 1809).
47. The standard account of early developmental theories (and, of course, later ones) is: Peter J. Bowler, *Evolution, The History of an Idea* (London, 1983).
48. *Life of Lyell*, 1:168.
49. Georges Cuvier, 'Biographical Memoirs of M. de Lamarck', *Edinburgh New Philosophical Journal*, 20 (1836), 1–21, at pp. 1–2.
50. Stephen Jay Gould, 'Foreword', in Jean Chandler Smith (ed.), *Georges Cuvier: An Annotated Bibliography of His Published Works* (Washington, DC, 1993), vii–xii, at viii.
51. Etienne Geoffroy Saint-Hilaire, *Philosophie Anatomique: Des Organes Respiratoires sous le Rapport de la Détermination et de l'Identité de Leurs Pièces Osseuses* (Paris, 1818); Robert Grant, 'Observations on the Nature and Importance of Geology', *Edinburgh New Philosophical Journal*, 1 (1826), 293–302, at p. 300.
52. *Memoirs of Cuvier*, p. 72n.
53. Toby A. Appel, *The Cuvier–Geoffroy Debate: French Biology in the Decades before Darwin* (Oxford, 1997), p. 1.

7. A SHROPSHIRE LAD

1. Paul H. Barrett (ed.), 'A Transcription of Darwin's First Notebook on "Transmutation of Species"', *Bulletin of the Museum of Comparative Zoology*, 122 (1960), 245–96, at pp. 279–80.
2. *Charles Darwin's Beagle Diary*, ed. Richard Darwin Keynes (Cambridge, 1988), pp. 295–8.
3. Robert FitzRoy, 'Sketch of the Surveying Voyages of His Majesty's Ships Adventure and Beagle, 1825–1836', *Journal of the Royal Geographical Society of London*, 6 (1836), 311–43, at pp. 375–6.
4. *Charles Darwin: His Life Told in an Autobiographical Chapter, and in a Selected Series of His Published Letters*, ed. Francis Darwin (London, 1892), pp. 33–4.
5. [Charles Darwin], *Narrative of the Surveying Voyages of His Majesty's Ships Adventure and Beagle between the Years 1826 and 1836, Describing*

their Examination of the Southern Shores of South America, and the Beagle's Circumnavigation of the Globe, 1832–1836 (3 vols, London, 1839), 3:393.

6. Charles Darwin to W. D. Fox, 9–12 August 1835, *DCP*, Letter 282.
7. It is impossible, in the author's view, to write a better biography of Charles Darwin than Adrian Desmond and James Moore's *Darwin: The Life of a Tormented Evolutionist* [1991] (London, 2009), with Janet Browne's monumental *Charles Darwin* (2 vols, London, 1995–2002) not far behind.
8. *Darwin's Autobiography*, p. 8.
9. Ibid., pp. 17, 20.
10. Ibid., pp. 19, 18.
11. Alexander von Humboldt, *Personal Narrative of Travels to the Equinoctial Regions of the New Continent During the Years, 1799–1804*, trans. Helen Maria Williams (7 vols, London, 1829).
12. *Darwin's Autobiography*, p. 114.
13. Of the many accounts of *Beagle*'s voyage, the most recent is: Tom Chaffin, *Odyssey: Young Charles Darwin, The Beagle, and the Voyage that Changed the World* (London, 2022).
14. David Sinclair, *The Land That Never Was: Sir Gregor MacGregor and the Most Audacious Fraud in History* (Cambridge, MA, 2004).
15. Robert FitzRoy, *Narrative of the Surveying Voyages of His Majesty's Ships Adventure and Beagle between the Years 1826 and 1836, Describing Their Examination of the Southern Shores of South America, and the Beagle's Circumnavigation of the Globe. Proceedings of the Second Expedition, 1831–36, under the Command of Captain Robert Fitz-Roy, R.N.* (London, 1839), p. 18.
16. *Suffolk Chronicle*, 30 April 1831; Diane B. Paul, John Stenhouse, and Hamish G. Spencer, 'The Two Faces of Robert FitzRoy, Captain of HMS *Beagle* and Governor of New Zealand', *Quarterly Review of Biology*, 88 (2013), 219–25.
17. *Beagle Diary*, pp. 4, 11, 12.
18. Ibid., pp. 19, 22.
19. *Darwin's Autobiography*, p. 29.
20. *Beagle Diary*, p. 35.
21. FitzRoy, *Narrative of the Beagle*, p. 58.
22. Darwin to Susan Darwin, 14 July–7 August 1832, *DCP*, Letter 177.
23. Darwin to Caroline Darwin, 24 October–24 November 1832, *DCP*, Letter 188.
24. *Beagle Diary*, p. 122.
25. CD to Catherine Darwin, 8 November 1834, *DCP*, Letter 262.

26. *Beagle Diary*, p. 352.
27. FitzRoy, 'Sketch', p. 332.
28. Desmond and Moore, *Darwin*, p. 170.
29. Richard Darwin Keynes, *Fossils, Finches, and Fuegians: Darwin's Adventures on the Beagle* (Oxford, 2003), p. 328.
30. *Beagle Diary*, pp. 395–6, 434.
31. Darwin to W. D. Fox, 15 February 1836, *DCP*, Letter 299.
32. Mantell, *Journal*, pp. 85, 59, 62, 78.
33. Ibid., pp. 85–6; Mantell declined to mention to the king that he was only the co-author, with Rev. T. W. Horsfield, of *The History and Antiquities of Lewes and its Vicinity* (2 vols, Lewes, 1824).
34. Gideon Mantell, 'The Geological Age of Reptiles', *Edinburgh New Philosophical Journal*, 11 (1831), 181–5, at p. 181.
35. Mantell, *Journal*, p. 89.
36. Dean, *Mantell*, p. 111.
37. Mantell, *Geology of the South-East*, pp. 328, 317.
38. Gideon Mantell, 'Observations on the Remains of the Iguanodon, and Other Fossil Reptiles of the Strata of Tilgate Forest in Sussex', *Proceedings of the Geological Society of London*, 1 (1834), 410–11.
39. Mantell, *Journal*, pp. 110–11.
40. Ibid., pp. 109–10, 92.
41. Ibid., pp. 121–2.
42. W. E. Swinton, 'Gideon Mantell and the Maidstone *Iguanodon*', *Notes & Records of the Royal Society of London*, 8 (1951), 261–76, at pp. 264–6, referring to: Gideon Mantell, 'Notice of the Discovery of the Remains of the Iguanodon in the Lower Green Sand Formation of the South-East of England', *American Journal of Science and Arts*, 27 (1835), 355–60.
43. Mantell, *Journal*, pp. 123–4.
44. Ibid., pp. 125–6, 131.

8. THE GOODNESS OF GOD

1. Fox, *Memories of Old Friends*, p. 5.
2. John Campbell, *The Lives of the Lord Chancellors and Keepers of the Great Seal of England from the Earliest Times till the Reign of King George IV* (3 vols, London, 1845), 2:176n.
3. Jonathan R. Topham, 'Francis Henry Egerton, eighth Earl of Bridgewater (1756–1829)', *ODNB*.
4. W. H. Brock, 'The Selection of the Authors of the Bridgewater Treatises', *Notes and Records of the Royal Society of London*, 21 (1966), 162–79, at p. 166.

5. For the purpose and process behind the treatises, see: Jonathan R. Topham, 'Beyond the "Common Context": The Production and Reading of the Bridgewater Treatises', *Isis*, 89 (1998), 233–62.
6. William Buckland, *Geology and Mineralogy considered with Reference to Natural Theology* (2 vols, London, 1836).
7. *Life of Buckland*, pp. 193–4.
8. Sharon Turner, *The Sacred History of the World: As Displayed in the Creation and Subsequent Events to the Deluge; Attempted to be Philosophically Considered in a Series of Letters to a Son* (3 vols, New York, 1832), 1:iv–vi, 386–7.
9. Frederick Nolan, *The Analogy of Revelation and Science Established in a Series of Lectures delivered before the University of Oxford* (Oxford, 1833), pp. 96, 98, 48.
10. Mary Buckland qu. in Rupke, *Great Chain*, p. 49.
11. For the '*ancien régime*' thesis, see: J. C. D. Clark, *English Society, 1660–1832: Religion, Ideology, and Politics during the Ancien Regime* (2nd edn, Cambridge, 2000).
12. Thomas Elrington, *A Letter to the Clergy of the United Diocese, on the Church Temporalities Act* (Dublin, 1833), p. 7.
13. John Keble, *National Apostasy: Considered in a Sermon Preached in St Mary's Oxford, before His Majesty's Judges of Assize, on Sunday, July 14, 1833* (Oxford, 1833), pp. 14, 16.
14. *Letters and Correspondence of John Henry Newman during His Life in the English Church with a Brief Autobiography*, ed. Anne Mozley (2 vols, London, 1891), 1:109.
15. *Tracts for the Times* (90 vols, London, 1833–41).
16. For the thought of the Tractarians more generally, see: Simon A. Skinner, *Tractarians and the 'Condition of England': The Social and Political Thought of the Oxford Movement* (Oxford, 2004).
17. *Hurrell Froude: Memoranda and Comments*, ed. Louise Guiney (London, 1904), p. 132.
18. *Life of Buckland*, pp. 193–4.
19. Buckland, *Geology and Mineralogy*, vii–viii.
20. William Kirby, *On the Power and Goodness of God as Manifested in the Creation of Animals and in Their History, Habits, and Instincts* (2 vols, London, 1835), 1:33–4.
21. Buckland, *Geology and Mineralogy*, 1:212, 224, 233, 240.
22. 'Dr Buckland's *Bridgewater Treatise*', *Quarterly Review*, 46 (1836), 31–64, at p. 64.
23. 'Geology and Mineralogy', *Edinburgh Review*, 45 (1837), 1–39, at pp. 15, 39.

24. Charles Lyell, 'Presidential Address to the Geological Society', *Proceedings of the Geological Society of London*, 2 (1837), 479–523, at p. 517.
25. Darwin to Robert FitzRoy, 6 October 1836, *DCP*, Letter 310.
26. Charles Darwin, *Extracts from Letters Addressed to Professor Henslow* (Cambridge, 1835), p. 8.
27. Darwin to Caroline Darwin, 9 November 1836, *DCP*, Letter 321; Harriet Martineau, *Illustrations of Political Economy* (25 vols, London, 1832–4).
28. CD to Caroline Darwin, 7 December 1836, *DCP*, Letter 325.
29. Desmond and Moore set out Darwin's analysis, known as 'This is the Question Marry Not Marry [Memorandum on Marriage]' at *Darwin*, p. 257.
30. Robert FitzRoy to Darwin, 16 November 1837, *DCP*, Letter 387; Robert FitzRoy to Darwin, 26 February 1838, *DCP*, Letter 403.
31. Richard Owen to Darwin, 11 June 1839, *DCP*, Letter 519.
32. *Athenaeum* qu. in Desmond and Moore, *Darwin*, p. 284.
33. FitzRoy, *Voyage of the Beagle*, p. 679.
34. Erasmus Darwin, *Zoonomia; or, the Laws of Organic Life* [1794–6] (3rd edn, 2 vols, Boston, 1809), 1:397.
35. Sedgwick qu. in Desmond and Moore, *Darwin*, p. 236.
36. Robert Taylor, *The Diegesis: Being a Discovery of the Origin, Evidences, and Early History of Christianity* [1829] (Boston, MA, 1860).
37. *Lion*, 3 (1829), pp. 519, 776, 705–08.
38. Robert Grant, *An Essay on the Study of the Animal Kingdom: Being an Introductory Lecture delivered in the University of London* (London, 1829), at p. 6.
39. For the ambush of Grant, see Desmond and Moore, *Darwin*, pp. 274–6.
40. Darwin to Susan Darwin, 1 April 1838, *DCP*, Letter 407.
41. *Darwin on Evolution: The Development of the Theory of Natural Selection*, ed. Thomas F. Glick and David Kohn (Indianapolis, 1996), pp. 79, 102.
42. Qu. in Desmond and Moore, *Darwin*, p. 244.

9. TERRIBLE LIZARDS

1. *The Oracle of Reason; Or, Philosophy Vindicated*, 1 (1842), v.
2. On the importance of industrial activity to British palaeontology, see: Michael Freeman, *Tracks to a Lost World: Victorians and the Prehistoric* (New Haven, CT, 2004), esp. pp. 9–52.
3. Dennis R. Dean, 'Hitchcock's Dinosaur Tracks', *American Quarterly*, 21 (1969), 639–44, at pp. 639, 641;*The Complete Poetical Works of Henry Wadsworth Longfellow* (Cambridge, MA, 1893), p. 64.

4. Thomas Hawkins, *The Book of the Great Sea-Dragons, Ichthyosauri and Plesiosauri, Gedolim Taninim of Moses, Extinct Monsters of the Ancient Earth* (London, 1840), p. 25.
5. For an account of the vocabulary of fossil-hunting before 1841, see: Ralph O'Connor, 'Victorian Saurians: The Linguistic Prehistory of the Modern Dinosaur', *Journal of Victorian Culture*, 17 (2012), 492–504. For the relationship between dinosaurs and mythology in this period, see: John McGowan-Hartmann, 'Shadow of the Dragon: The Convergence of Myth and Science in Nineteenth Century Palaeontological Imagery', *Journal of Social History*, 47 (2013), 47–70.
6. Rev. Richard Owen (Jr), *The Life of Richard Owen by His Grandson, revised by C. Davies Sherborn* (2 vols, London, 1894), 1:8–9, 11.
7. Ibid., 1:19, 22–4.
8. Ibid., 1:27–8, 33, 35, 37.
9. Nicolaas A. Rupke, *Richard Owen: Biology without Darwin* (2nd edn, Chicago, 2009), pp. 4, 7.
10. *Life of Owen*, 1:61.
11. *Life of Owen*, 1:107, 104, 92.
12. Ibid., 1:117, 106.
13. Richard Owen, 'A Description of a Portion of the Skeleton of the Cetiosaurus, a Gigantic Extinct Saurian Reptile, Occurring in the Oolitic Formations of Different Portions of England', *Proceedings of the Geological Society of London*, 3 (1842), 457–62, at p. 462.
14. For the reconstruction of dinosaurs, see: Gowan Dawson, *Show Me the Bone: Reconstructing Prehistoric Monsters in Nineteenth-Century Britain and America* (Chicago, 2016).
15. For Owen's beliefs in this period, see: Adrian Desmond, 'Richard Owen's Reaction to Transmutation in the 1830s', *British Journal for the History of Science*, 18 (1985), 25–50.
16. Richard Owen, 'Report on British Fossil Reptiles', in *Report of the Eleventh Meeting of the British Association for the Advancement of Science; Held at Plymouth in July 1841* (London, 1842), 60–204, at pp. 60, 204, 200.
17. Adrian Desmond, 'Designing the Dinosaur: Richard Owen's Response to Robert Edmond Grant', *Isis*, 70 (1979), 224–34, at p. 230.
18. *Oracle of Reason*, 4 (1842), 25.
19. *Life and Letters of George Jacob Holyoake*, ed. Joseph McCabe (2 vols, London, 1908), 1:61.
20. [Henry Phillpotts], *Socialism. Second Speech of the Bishop of Exeter, in the House of Lords, February 4, 1840* (London, 1840), p. 2.
21. *Life of Holyoake*, 1:62–3.

22. *The Trial of George Jacob Holyoake, on an Indictment for Blasphemy, before Mr Justice Erskine, and a Common Jury, at Gloucester, August the 15th, 1842* (London, 1842), pp. 22–3.
23. *Cheltenham Chronicle*, 9 June 1842.
24. George Jacob Holyoake, *Sixty Years of an Agitator's Life* (2 vols, London, 1891), 1:148.
25. Idem, *The History of the Last Trial by Jury for Atheism in England: A Fragment of Autobiography* (London, 1850), p. 8.
26. *Trial of Holyoake*, pp. 1–2.
27. *Life of Holyoake*, 1:69–70.
28. *Hansard*, Commons, 18 July 1842, LXV, c243.
29. *Life of Holyoake*, 1:75.
30. *History of the Last Trial*, pp. 32–3.
31. *Trial of Holyoake*, pp. 4, 29.
32. *Life of Holyoake*, 1:71, 77.
33. *The Times*, 17 August 1842.
34. Darwin, *Autobiography*, p. 40.
35. Francis Darwin (ed.), *The Foundations of the Origin of Species: Two Sketches Written in 1842 and 1844* (Cambridge, 1909), xx, 2, 52, 7, 39, 41, 39n.
36. Ibid., pp. 6, 51.

10. ABOMINABLE BOOKS

1. Alfred Tennyson, *The Princess: A Medley* [1847] (London, 1860), pp. 37–9.
2. For Thackeray and the *Megatherium*, see: Gowan Dawson, 'Literary Megatheriums and Loose Baggy Monsters: Palaeontology and the Victorian Novel', *Victorian Studies*, 53 (2011), 203–30.
3. Dickens qu. in Gowan Dawson, 'Dickens, Dinosaurs, and Design', *Victorian Literature and Culture*, 44 (2016), 761–78, at p. 762; Charles Dickens, *Dealings with the Firm of Dombey and Son: Wholesale, Retail, and for Exportation* (2 vols, New York, 1848), 1:71.
4. Charles Lyell, *Elements of Geology* (2nd edn, 2 vols, London, 1841); Gideon Mantell, *The Medals of Creation; Or, First Lessons in Geology and in the Study of Organic Remains* (2 vols, London, 1844).
5. The best and most comprehensive study of phrenology is: Roger Cooter, *The Cultural Meaning of Popular Science: Phrenology and the Organization of Consent in Nineteenth-Century Britain* (Cambridge, 1984).
6. Franz Josef Gall, *On the Functions of the Brain and of Each of its Parts: On the Organ of the Moral Qualities and Intellectual Faculties, and the Plurality of the Cerebral Organs*, trans. Winslow Lewis (6 vols, Boston, 1835), 2:218, 70.

7. Charles Gibbon, *The Life of George Combe: Author of 'The Constitution of Man'* (2 vols, London, 1878), 1:73.
8. George Combe, *The Constitution of Man: Considered in Relation to External Objects* (Edinburgh, 1828), viii.
9. Cooter, *Cultural Meaning*, pp. 136, 154–5, 135, 120.
10. Roderick Murchison, Édouard de Verneuil, and Alexander von Keyserling, *The Geology of Russia in Europe and the Ural Mountains* (2 vols, London, 1845), vii.
11. Friedrich Engels, *The Condition of the Working Class in England in 1844*, trans. Florence Kelley Wischnewetzky [1845] (London, 1892), pp. 125–6.
12. Philip Jones, *Popular Phrenology: Tried by the Word of God, and Proved to be Antichrist, and Injurious to Individuals and Families* (London, 1845).
13. Secord, *Visions of Science*, p. 173.
14. [Robert Chambers], *Vestiges of the Natural History of Creation* (London, 1844), p. 156.
15. Ibid., pp. 154, 157.
16. Ibid., pp. 110, 98, 149.
17. James A. Secord, *Victorian Sensation: The Extraordinary Publication, Reception, and Secret Authorship of Vestiges of the Natural History of Creation* (Chicago, 2000), pp. 9–10, 168–9.
18. Ibid., pp. 229, 222, 227.
19. [Adam Sedgwick], 'Natural History of Creation', *Edinburgh Review*, 82 (1845), 1–85, at p. 3.
20. Secord, *Victorian Sensation*, p. 223.
21. [Robert Chambers], *Vestiges of the Natural History of Creation* (10th edn, London, 1853), ix.
22. John Mullan, *Anonymity: A Secret History of English Literature* (Princeton, NJ, 2007), p. 98.
23. Secord, *Victorian Sensation*, p. 18.
24. Darwin to W. D. Fox, 24 April 1845, *DCP*, Letter 859.
25. Ibid.; Darwin to Charles Lyell, 8 October 1845, *DCP*, Letter 919.
26. *Memoirs of Sir Andrew Crombie Ramsay*, ed. Archibald Geikie (London, 1895), p. 103.
27. Rudolf Chambers Lehmann, *Memories of Half a Century* (London, 1908), p. 8.

11. NEMESIS

1. Alfred Tennyson, *In Memoriam, A.H.* (London, 1850), cantos lv–lvi.
2. The best biographies of Froude are Ciaran Brady's *James Anthony Froude: An Intellectual Biography of a Victorian Prophet* (Oxford, 2013) and Julia

Markus's *J. Anthony Froude: The Last Undiscovered Great Victorian* (New York, 2007), which have both influenced this section.

3. Ludwig Feuerbach, *The Essence of Christianity*, trans. Mary Ann Evans [1841] (2nd edn, New York, 1855), pp. 10, 13.
4. David Friedrich Strauss, *The Life of Jesus, Critically Examined* [trans. Mary Ann Evans] (4th edn, 3 vols, London, 1846), 1:33–5.
5. Henry Parry Liddon, *Life of Edward Bouverie Pusey*, ed. J. O. Johnston and Robert J. Wilson (4 vols, London, 1894), 1:72. The career of Alexander Nicoll suggests that the claim about 'only two persons' knowing German is apocryphal: John L. Speller, 'Alexander Nicoll and the Study of German Biblical Criticism in Early Nineteenth-Century Oxford', *Journal of Ecclesiastical History*, 30 (1979), 451–9.
6. Liddon, *Life of Pusey*, 1:77.
7. Jeffrey Paul von Arx, *Progress and Pessimism: Religion, Politics, and History in Late Nineteenth Century Britain* (Cambridge, MA, 1985), p. 179.
8. [J. A. Froude], *Shadow of the Clouds* (London, 1847), described in Michael Bentley, 'Victorian Historians and the Larger Hope', in Michael Bentley (ed.), *Public and Private Doctrine: Essays in British History presented to Maurice Cowling* (Cambridge, 1993), 127–49, at p. 130.
9. J. A. Froude, *The Nemesis of Faith* (2nd edn, London, 1849), pp. 7, 26, 24, 121, 27.
10. Ibid., pp. 11, 17, 12.
11. David Wilson, *Mr. Froude and Carlyle* [1898] (New York, 1970), p. 50.
12. *Athenaeum* qu. in Waldo Hilary Dunn, *James Anthony Froude: A Biography* (Oxford, 1961), p. 138.
13. *Morning Herald*, 10 March 1849.
14. *Wiltshire and Gloucestershire Standard*, 14 May 1892.
15. Froude to Clough qu. in Dunn, *Froude*, pp. 132, 134.
16. Simon Heffer, *High Minds: The Victorians and the Birth of Modern Britain* (London, 2013), p. 180.
17. Herbert W. Paul, *The Life of Froude* (New York, 1906), p. 46.
18. *London Evening Standard*, 12 March 1849.
19. Dunn, *Froude*, p. 233.
20. Markus, *Froude*, p. 43.
21. Darwin to Emma Darwin, 22 May 1848, *DCP*, Letter 1177.
22. Darwin to Richard Owen, n.d. but 1849, *DCP*, Letter 1089.
23. Darwin to Hugh Cuming, 4 November 1849, *DCP*, Letter 1265.
24. Darwin to J. D. Hooker, 28 March 1849, *DCP*, Letter 1236.
25. Desmond and Moore, *Darwin*, pp. 366, 396–7.
26. Charles Darwin, *The Structure and Distribution of Coral Reefs* (London, 1842); idem, *Geological Observations on the Volcanic Islands* (London, 1844).

27. Darwin to J. D. Hooker, 11 January 1844, *DCP*, Letter 729.
28. Desmond and Moore, *Darwin*, pp. 344–5.
29. Francis William Newman, *A History of the Hebrew Monarch from the Administration of Samuel to the Babylonish Captivity* (London, 1847), iv; idem, *The Soul, Her Sorrows and Her Aspirations: An Essay towards the Natural History of the Soul, as the True Basis of Theology* (London, 1849), pp. 189, 185; idem, *Phases of Faith; or, Passages from the History of My Creed* (London, 1850), pp. 122, 115.
30. Desmond and Moore, *Darwin*, pp. 377–8.
31. Ibid., p. 375.
32. Darwin to W. D. Fox, 29 April 1851, *DCP*, Letter 1425.
33. Annie's death and illness are narrated in Desmond and Moore, *Darwin*, pp. 375–87, qu. at pp. 375, 379, 384.
34. George Richardson Porter, *The Progress of the Nation, in its Various Social and Economical Relations, from the Beginning of the Nineteenth Century* (3rd edn, London, 1851), xviii.
35. Charles Dickens, *Hard Times: For These Times* (Leipzig, 1854), p. 5.
36. William Cobbett, *Rural Rides* (London, 1830), p. 83.
37. For the papal bull and the British response, see: Walter Ralls, 'The Papal Aggression of 1850: A Study in Victorian Anti-Catholicism', *Church History*, 43 (1974), 242–56.
38. *Census of Great Britain, 1851: Religious Worship: England and Wales: Report and Tables* (London, 1853), viii.
39. Horace Mann, *Letters on the Existing Law of Registration of Births, Deaths, and Marriages, with Suggestions for its Improvement* (London, 1849).
40. *Census of Religious Worship*, cxix.
41. *Census of Religious Worship*, cxix, clxix.
42. The most authoritative study of the 1851 religious census is K. D. M. Snell and Paul S. Ell, *Rival Jerusalems: The Geography of Victorian Religion* (Cambridge, 2000), where these numbers are collated at pp. 423–4.
43. *Census of Religious Worship*, clviii.
44. *The Times*, 9 January 1854.
45. 'The Census of 1851: Religious Worship', *Hansard*, Lords, 11 July 1854, CXXXV, cc24, 29.
46. *The Truth about Church Extension: An Exposure of Certain Fallacies and Misstatements contained in the Census Reports on Religious Worship and Education* (London, 1857), p. 1.
47. K. S. Inglis, 'Patterns of Religious Worship in 1851', *Journal of Ecclesiastical History*, 11 (1960), 74–86, at pp. 80, 78.
48. George Eliot, *Silas Marner: The Weaver of Raveloe* (Leipzig, 1861), pp. 127–8.

49. John Wolffe, 'Elite and Popular Religion in the Religious Census of 30 March 1851', *Studies in Church History*, 42 (2006), 360–71, at p. 361.
50. Emling, *Fossil Hunter*, p. 189.
51. Lang, 'Three Letters', p. 171.
52. Goodhue, 'Faith of a Fossilist', p. 90.
53. *Journal of Mantell*, p. 108.
54. *Life of Owen*, 1:164.
55. 'Mary Anning', *All the Year Round*, p. 62.
56. M. Aurousseau (ed.), *The Letters of F. W. Ludwig Leichhardt* (3 vols, Cambridge, 1968), 1:232.
57. C. G. Carus, *The King of Saxony's Journey through England and Scotland in the Year 1844* (London, 1846), pp. 197–8.
58. Thomas W. Goodhue, 'Mary Anning's Commonplace Book' (Lyme Regis Museum, undated) [www.lymeregismuseum.co.uk], p. 3.
59. 'Mary Anning', *All the Year Round*, p. 62.
60. Pierce, *Jurassic Mary*, pp. 50–51.
61. Henry De la Beche, 'Anniversary Address of the President', *Quarterly Journal of the Geological Society of London*, 4 (1848), xxi–xxxi, at p. xxiv.

12. THE PALACE

1. Charles Dickens, *Bleak House* (London, 1853), p. 1.
2. *Punch: or, the London Charivari*, 20 (1851), p. 21.
3. *Hansard*, Commons, 30 June 1851, CXVII, c1439.
4. *Official Descriptive and Illustrated Catalogue of the Great Exhibition of the Works of Industry of All Nations, 1851* (London, 1851), p. 18.
5. Theodore Martin, *The Life of His Royal Highness the Prince Consort* (2 vols, New York, 1879), 2:303.
6. Henry Mayhew and George Cruikshank, *1851: or, The Adventures of Mr and Mrs Sandboys and Family, Who Came Up to London to 'Enjoy Themselves' and to See the Great Exhibition* (London, 1851), pp. 127–8.
7. *The Life and Letters of Thomas Henry Huxley*, ed. Leonard Huxley (2 vols, London, 1903), 1:5.
8. Adrian Desmond, *Huxley* (London, 1997), pp. 164–5. For a view of the Great Exhibition as a religious rather than secular phenomenon, see: Geoffrey Cantor, *Religion and the Great Exhibition of 1851* (Oxford, 2011).
9. Alison Bashford, *An Intimate History of Evolution: The Story of the Huxley Family* (London, 2022), p. 5.
10. Desmond, *Huxley*, pp. 6–15.
11. *Life and Letters of Huxley*, 1:21–3. Adrian Desmond has recently written of Huxley's apprenticeship in 'T. H. Huxley's Turbulent Apprenticeship Years:

John Charles Cooke and the John Salt Scandal', *Archives of Natural History*, 48 (2021), 215–26.

12. Desmond, *Huxley*, p. 15. For the broader relationship between radical science and radical politics, see Adrian Desmond's *The Politics of Evolution: Morphology, Medicine, and Reform in Radical London* (Chicago, 1992).
13. Desmond, *Huxley*, pp. 23, 26.
14. Thomas H. Huxley, 'On a Hitherto Undescribed Structure in the Human Hair Sheath', *Medical Gazette*, 1 (1845), 1340–41.
15. Desmond, *Huxley*, p. 37.
16. *Life of Huxley*, 1:27.
17. Desmond, *Huxley*, pp. 55, 58.
18. The paper was published as 'On the Anatomy and the Affinities of the Family of the *Medusae*', *Philosophical Transactions of the Royal Society*, 139 (1849), 413–37.
19. Desmond, *Huxley*, pp. 68, 70, 73.
20. Ibid., p. 79.
21. William Carron, *Narrative of an Expedition Undertaken Under the Direction of the Late Mr. Assistant Surveyor E. B. Kennedy for the Exploration of the Country Lying Between Rockingham Bay and Cape York* (Sydney, 1849), p. 78.
22. Desmond, *Huxley*, pp. 144, 117, 112, 118.
23. Ibid., pp. 133, 137.
24. Ibid., p. 141.
25. The fullest account of the *Rattlesnake*'s voyage is probably: Jordan Goodman, *The Rattlesnake: A Voyage of Discovery to the Coral Sea* (London, 2005).
26. 'Professor Owen, F.R.S., &c.', *The Year-Book of Facts in Science and Art* (London, 1852), 1–4, at p. 4.
27. *Life of Huxley*, 1:64.
28. *Life of Holyoake*, p. 88.
29. *Oracle of Reason*, 2:73–5.
30. *The Movement; and Anti-Persecution Gazette*, 1 (1843), 1.
31. For a good summary of these developments, see: David S. Nash, 'Unfettered Investigation: The Secularist Press and the Creation of Audience in Victorian England', *Victorian Periodicals Review*, 28 (1995), 123–35.
32. George Jacob Holyoake, *English Secularism: A Confession of Belief* (London, 1896), p. 34.
33. *Reasoner*, 19 November 1851.
34. Ibid., 3 December 1851.
35. *Reasoner*, 4 February 1852.
36. Martineau qu. in George Jacob Holyoake, *The Principles of Secularism Illustrated* (3rd edn, London, 1871), p. 8.

37. On the relationship of Holyoake and the early secularists to science, see: Michael Rectenwald, 'Secularism and the Cultures of Nineteenth-Century Scientific Naturalism', *British Journal for the History of Science*, 46 (2013), 231–54.
38. *Journal of Mantell*, pp. 140, 140n4.
39. S. Turner, C. V. Burek, and R. T. J. Moody, 'Forgotten Women in an Extinct Saurian (Man's) World', in R. T. J. Moody, E. Buffetaut, D. Naish and D. M. Martill (eds.), *Dinosaurs and Other Extinct Saurians: A Historical Perspective* (London, 2010), 111–54, at p. 121.
40. *Journal of Mantell*, pp. 140n5, 141n3.
41. Ibid., pp. 140–41.
42. Gideon Mantell, 'Memoir on a Portion of the Lower Jaw of the Iguanodon, and on the Remains of the Hylaeosaurus and Other Saurians, Discovered in the Strata of Tilgate Forest, in Sussex', *Philosophical Transactions of the Royal Society of London*, 131 (1841), 131–51; idem, 'Case of Abscess of the Prostate Gland', *Lancet*, 2 (1841), 301; idem, 'Cultivation of the Canine Brain', *Lancet*, 2 (1841), 907–08.
43. *Journal of Mantell*, pp. 142, 148–9, 152.
44. Gideon Mantell, *Thoughts on a Pebble; or, A First Lesson in Geology* (6th edn, London, 1842), pp. 19–21.
45. Idem, *Medals of Creation*, xi, ix.
46. *Sussex Advertiser*, 16 July 1844; Thomas Hood, 'A Geological Excursion to Tilgate Forest, A.D. 2000', in *Relics from the Wreck of a Former World* (New York, 1847), 53–6, at p. 53.
47. 'A Member of the Council of the Clapham Athenaeum', *A Reminiscence of Gideon Algernon Mantell* (London, 1853), pp. 3–4.
48. [Richard Owen], 'Dr Mantell', *Literary Gazette and Journal of Belles Lettres, Science, and Art for the Year 1852* (London, 1852), 842.
49. Benjamin Silliman, 'Obituary: Gideon Algernon Mantell', *American Journal of Science and Arts*, 15 (1853), 147–50, at p. 149.
50. *Lady's Newspaper and Pictorial Times*, 27 November 1852.
51. 'Member', *Reminiscence*, p. 5.
52. Gideon Mantell, *Thoughts on a Pebble: or, a First Lesson in Geology* (8th edn, London, 1849), pp. 28–9.

13. DEATH AND THE DINOSAURS

1. Elizabeth Gaskell, *North and South* (Leipzig, 1855), p. 221.
2. For the technological revolutions of the 1850s, see: Ben Wilson, *Heyday: Britain and the Birth of the Modern World* (London, 2016).

3. *Hansard*, Lords, 31 January 1854, CXXX, cc2–4.
4. The best recent history of the war is: Orlando Figes, *Crimea: The Last Crusade* (London, 2010).
5. 'One of the Royal Fusiliers' [Timothy Gowing], *A Soldier's Experience, or A Voice from the Ranks: Showing the Cost of War in Blood and Treasure* (Nottingham 1892), pp. 10–11.
6. *The Times*, 12 October 1854.
7. Stefanie Markowitz, *The Crimean War in the British Imagination* (Cambridge, 2009), p. 31.
8. Petros Spanou, 'Soldiership, Christianity, and the Crimean War: The Reception of Catherine Marsh's *Memorials of Captain Hedley Vicars*', *Journal of Victorian Culture*, 27 (2022), 46–62, at p. 52.
9. William Robertson, *Sermon preached on the Day of Humiliation on Account of the War, ordered by Her Majesty, on Wednesday, 26 April, 1854, in New Greyfriars' Church, Edinburgh* (Edinburgh, 1854), pp. 7, 9.
10. [Catherine Marsh], *Memorials of Captain Hedley Vicars, Ninety-Seventh Regiment* (London, 1856), pp. 132, 286.
11. *The Sunday at Home: A Family Magazine for Sabbath Reading*, 1 (1854), p. 6.
12. John Ruskin to Henry Acland, 24 May 1851, in *The Works of John Ruskin*, ed. E. T. Cook and Alexander Wedderburn (39 vols, London, 1903–12), 36:115.
13. *The Life of Frederick Denison Maurice, Chiefly Told in His Own Letters*, ed. Frederick Maurice (2 vols, London, 1884), 1:14.
14. Richard Deacon, *The Cambridge Apostles: A History of Cambridge University's Elite Intellectual Secret Society* (New York, 1986), p. 8.
15. *Life of Maurice*, 2:44, 1:516–17.
16. Frederick Denison Maurice, *The Epistle to the Hebrews; Being the Substance of Three Lectures Delivered in the Chapel of the Honourable Society of Lincoln Inn* (London, 1846); idem, *The Lord's Prayer: Nine Sermons preached in the Chapel of Lincoln's Inn* (2nd edn, London, 1849); idem, *The Religions of the World and their Relations to Christianity* [1847] (3rd edn, London, 1854).
17. Idem, *Theological Essays* (Cambridge, 1853), vi.
18. Bernard M. G. Reardon, *Religious Thought in the Victorian Age: A Survey from Coleridge to Gore* (2nd edn, London, 2014), p. 136.
19. R. W. Jelf, *Grounds for Laying before the Council of King's College London, Certain Statements contained in a Recent Publication, entitled, 'Theological Essays'* (2nd edn, London, 1853), p. 6.
20. Maurice, *Theological Essays*, p. 477.
21. *Morning Advertiser*, 21 July 1853.

22. F. D. Maurice, *The Word 'Eternal' and the Punishment of the Wicked: A Letter to the Rev. Dr. Jelf, Canon of Christ Church, and Principal of King's College* (5th edn, Cambridge, 1854), v–vi.
23. Bernard M. G. Reardon, '(John) Frederick Denison Maurice', *ODNB*.
24. Joseph Paxton, *What is to Become of the Crystal Palace?* (London, 1851), p. 1.
25. 'Denarius' [Henry Cole], *Shall We Keep the Crystal Palace, and Have Riding and Walking in all Weathers among Flowers, Fountains and Sculptures?* (London, 1851), p. 8.
26. *The Palace and Park: Its Natural History, and its Portrait Gallery, together with a Description of the Pompeian Court* (London, 1855), p. 15.
27. Jeffrey A. Auerbach, *The Great Exhibition of 1851: A Nation on Display* (New Haven, CT, 1999), pp. 194–5.
28. *Palace and Park*, pp. 16–17.
29. Samuel Phillips, *Crystal Palace: A Guide to the Palace & Park* (London, 1858), p. 158.
30. For the exhibition's sense of moving through time, see: Nancy Rose Marshall, '"A Dim World, Where Monsters Dwell": The Spatial Time of the Sydenham Crystal Palace Dinosaur Park', *Victorian Studies*, 49 (2007), 286–301.
31. *Palace and Park*, p. 159.
32. 'The Crystal Palace', *Illustrated Magazine*, 1 (January 1854), 33–7, at p. 35; H. G. Wells, *Kipps: The Story of a Simple Soul* (New York, 1906), p. 391.
33. 'Crystal Palace', pp. 34, 35.
34. *Routledge's Guide to the Crystal Palace and Park at Sydenham: with Descriptions of the Principal Works of Science and Art, and of the Terraces, Fountains, Geological Formations, and Restoration of Extinct Animals, therein Contained* (London, 1854), pp. 185, 193.
35. Martin J. S. Rudwick, *Scenes from Deep Time: Early Pictorial Representations of the Prehistoric World* (Chicago, 1992), p. 140.
36. Richard Owen, *Geology and Inhabitants of the Ancient World* (London, 1854), pp. 5–6.
37. Benjamin Waterhouse Hawkins, *On Visual Education as Applied to Geology, Illustrated by Diagrams and Models of the Geological Restorations at the Crystal Palace* (London, 1854), pp. 3–4.
38. McCarthy and Gilbert, *Palace Dinosaurs*, pp. 19–22.
39. *Illustrated London News*, 24 June 1854.
40. Samuel Leigh Sotheby, *A Letter Addressed to the Directors of the Crystal Palace Company* (London, 1855), pp. 32–3.
41. Buckland, *Curiosities*, p. 40.
42. [Harriet Martineau], 'The Crystal Palace', *Westminster Review*, 62 (1854), 534–50, at p. 540. My thanks to Petros Spanou for this reference.

43. Charles Waterton, *Essays on Natural History: Third Series* (London, 1857), p. 61.
44. Samuel Leigh Sotheby, *A Few Words by Way of a Letter Addressed to the Shareholders of the Crystal Palace Company: With Remarks on the Various Subjects Necessarily Brought under the Notice of the Committee, Appointed August 9, 1855, to Investigate the Affairs of the Company* (3rd edn, London, 1856).
45. 'The Crystal Palace, Sydenham: The Geological Restorations', *Observer*, 23 September 1855.
46. George W. White, 'Announcement of Glaciation in Scotland: William Buckland (1784–1856)', *Journal of Glaciology*, 9 (1970), 143–5, at p. 145.
47. *Life of Buckland*, pp. 220, 219.
48. Ibid., pp. 220, 222, 223, 224.
49. For Buckland and sanitary reform, see Chapman, *Caves*, pp. 201–06.
50. *Life of Buckland*, p. 268.
51. Ibid., p. 269; *The Leisure Hour: A Family Journal of Instruction and Recreation* (1852–1905).
52. *Life of Buckland*, pp. 269, 271.

14. THE BRANCHING TREE

1. William Makepeace Thackeray, *The Newcomes: Memoirs of a Most Respectable Family* [1854–5] (Boston, 1869), p. 336.
2. The biographies of Wallace which have informed the relevant sections of this book are: Peter Raby, *Alfred Russel Wallace: A Life* (London, 2001), and Michael Shermer, *In Darwin's Shadow: The Life and Science of Alfred Russel Wallace: A Biographical Study on the Psychology of History* (Oxford, 2002).
3. Alfred Russel Wallace, *My Life: A Record of Events and Opinions* (2 vols, London, 1905), 1:269.
4. Ibid., 1:272, 1:277, 1:281, 1:283.
5. Ibid., 1:287, 1:303, 1:305.
6. Ibid., 1:306, 1:308, 1:320, 1:323.
7. Ibid., 1:326.
8. For Murchison's commitment to scientific imperialism, see: Robert A. Stafford, *Scientist of Empire: Sir Roderick Murchison, Scientific Exploration, and Victorian Imperialism* (Cambridge, 1990).
9. Alfred Russel Wallace, *Letters from the Malay Archipelago*, ed. John van Wyhe and Kees Rookmaaker (Oxford, 2013), pp. 6–8, 12.
10. Idem, *My Life*, 1:334–6, 1:338
11. Idem, *The Malay Archipelago: The Land of the Orangutan, and the Bird of Paradise: A Narrative of Travel, with Studies of Man and Nature* (New York, 1869), pp. 51–7.

12. Idem, *My Life*, 1:342.
13. Idem, 'On the Law which has Regulated the Introduction of New Species', *Annals and Magazine of Natural History*, 16 (1855), 184–96, at pp. 186, 195, 191.
14. Desmond and Moore, *Darwin*, p. 407.
15. Darwin to J. D. Hooker, 12 October 1849, *DCP*, Letter 1260; Desmond and Moore, *Darwin*, p. 414.
16. Charles Darwin, *A Monograph of the Sub-Class Cirripedia, with Figures of All the Species: The Lepadidae; or Pedunculated Cirripedes* (London, 1851); idem, *A Monograph on the Fossil Lepadidae, or Pedunculated Cirripedes of Great Britain* (London, 1851); idem, *A Monograph of the Sub-Class Cirripedia, with Figures of all the Species. The Balanidae (or Sessile Cirripedes); the Verrucidae, etc* (London, 1854); idem, *A Monograph on the Fossil Balanidae and Verrucidae of Great Britain* (London, 1854), pp. 2, 24.
17. Thomas H. Huxley, 'Lectures on General Natural History: Lecture XII', *Medical Times & Gazette*, 15 (1857), 238–41, at p. 241n(a).
18. Desmond and Moore, *Darwin*, p. 444.
19. Darwin, *Autobiography*, p. 164.
20. Charles Darwin, 'Memorandum', December 1855, *DCP*, Letter 1812.
21. Darwin to W. E. Darwin, 29 November 1855, *DCP*, Letter 1689; Desmond and Moore, *Darwin*, p. 429.
22. Desmond and Moore, *Darwin*, pp. 425, 427.
23. Edward Blyth to Darwin, 8 December 1855, *DCP*, Letter 1792.
24. Desmond and Moore, *Darwin*, p. 438.
25. Darwin to Alfred Russel Wallace, 1 May 1857, *DCP*, Letter 2086.
26. Desmond, *Huxley*, pp. 172, 182, 230.
27. Herbert Spencer, *An Autobiography* (2 vols, London, 1904), 1:101, 1:105.
28. *The Economist*, preliminary number (1843), p. 4.
29. Herbert Spencer, *Social Statics: or, The Conditions Essential to Human Happiness Specified, and the First of Them Developed* (London, 1851), p. 282.
30. Idem, 'The Development Hypothesis' [1852], in Herbert Spencer, *Essays Scientific, Political & Speculative* (London, 1858), 389–95, at p. 394.
31. John Stuart Mill, *A System of Logic, Ratiocinative and Inductive* (2 vols, London, 1843), 1:454.
32. Herbert Spencer, 'Progress: Its Law and Causes', *Westminster Review*, 67 (1857), 445–85, at pp. 465, 483–4.
33. Desmond, *Huxley*, pp. 183, 175, 204, 176.
34. *Morning Post*, 26 April 1852.
35. T. H. Huxley, 'The Vestiges of Creation', *British and Foreign Medico-Chirurgical Review*, 13 (1854), 332–43, at pp. 333, 332, 334, 333.
36. *Life of Huxley*, 1:180.

37. Desmond, *Huxley*, pp. 201, 208, 205, 203, 202, 203.
38. Ibid., pp. 189, 207, 211, 213.

15. THE DEVIL'S GOSPEL

1. George Eliot, 'The Lifted Veil', *Blackwood's Edinburgh Magazine*, 86 (1859), 24–47, at p. 26.
2. Wallace, *Malay Archipelago*, pp. 313, 314.
3. Idem, *My Life*, 1:361.
4. Ibid., 1:361–2.
5. Ibid., 1:363.
6. Charles Darwin and Alfred Russel Wallace, *Journal of the Proceedings of the Linnean Society*, 3 (1859), 'On the Tendency of Species to Form Varieties; and on the Perpetuation of Varieties and Species by Natural Means of Selection', 45–62, at pp. 54, 57, 58.
7. Darwin to Charles Lyell, 18 June 1858, *DCP*, Letter 2285.
8. *Correspondence of Darwin*, 7:504.
9. Darwin to J. D. Hooker, 29 June 1858, *DCP*, Letter 2297.
10. *Journal of the Proceedings of the Linnean Society*, 3 (1859), liv.
11. Thomas Bell, 'Presidential Address', *Journal of the Proceedings of the Linnean Society*, 4 (1860), viii–xx, at p. viii.
12. Darwin to A. R. Wallace, 1 May 1857, *DCP*, Letter 2086; Wallace, *Letters from Malaya*, p. 65.
13. Darwin to A. R. Wallace, 25 January 1859, *DCP*, Letter 2405.
14. Wallace, *Letters from Malaya*, pp. 180, 181.
15. Desmond and Moore, *Darwin*, p. 459; Thomas H. Huxley, 'On the Persistent Types of Animal Life', *Notices of the Proceedings at the Meetings of the Members of the Royal Institution*, 3 (1858–60), 151–3, at p. 153.
16. Desmond and Moore, *Darwin*, p. 475.
17. J. D. Hooker, 'Introductory Essay', in *Flora Tasmaniae* (2 vols, London, 1860), 1:i–cxxviii, at p. ii; J. D. Hooker to Darwin, 21 November 1859, *DCP*, Letter 2539.
18. Charles Darwin, *On the Origin of Species by Means of Natural Selection, or the Preservation of Favoured Races in the Struggle for Life* (1st edn, London, 1859), pp. 130, 490.
19. Ibid., p. 280; Thomas H. Huxley, 'The Darwinian Hypothesis' [1859], in *Collected Essays* (2 vols, New York, 1894), 2:1–21, at 2:1.
20. Darwin to J. D. Hooker, 27 October or 3 November 1859, *DCP*, Letter 2512.
21. John Mayhall, *The Annals and History of Leeds and Other Places in the County of York* (London, 1860), p. 680.

22. Darwin to A. R. Wallace, 13 November 1859, *DCP*, 2529.
23. Darwin, *Origin* (1st edn), p. 167.
24. Desmond and Moore, *Darwin*, p. 489.
25. *Hansard*, Lords, 10 July 1857, vol. 146, cc1209–78, at cc1236, 1222.
26. Katherine Mullin, 'Poison More Deadly than Prussic Acid: Defining Obscenity after the 1857 Obscene Publications Act (1850–1885)', in David Bradshaw and Rachel Potter (eds.), *Prudes on the Prowl: Fiction and Obscenity in England* (Oxford, 2013), 11–29.
27. 'The Ballad of Roaring Hanna', *Punch*, 33 (1857), p. 131; *Belfast Mercury*, 15 September 1857.
28. [W. J. Conybeare], *Church Parties: An Essay, reprinted from 'The Edinburgh Review'* (London, 1854), pp. 1, 80.
29. Samuel Smiles, *Self-Help: With Illustrations of Character and Conduct* (London, 1859).
30. Norman Vance, *The Sinews of the Spirit: The Ideal of Christian Manliness in Victorian Literature and Religious Thought* (Cambridge, 1985), p. 4.
31. Donald E. Hall, 'Muscular Christianity: Reading and Writing the Male Social Body' in Donald E. Hall (ed.), *Muscular Christianity: Embodying the Victorian Age* (Cambridge, 1994), 3–14, at p. 8.
32. Charles Kingsley, *Westward Ho! Or the Voyages and Adventures of Sir Amyas Leigh* (London, 1855).
33. Idem, *Two Years Ago* (3 vols, Cambridge, 1857); [A. P. Stanley], 'Two Years Ago' [review], *Saturday Review of Politics, Literature, Science, and Art*, 3 (1857), 176–7, at pp. 176–7.
34. Stanley, 'Two Years Ago', p. 176.
35. Thomas Hughes, *Tom Brown's School Days* (London, 1857). See also: William E. Winn, '*Tom Brown's Schooldays* and the Development of "Muscular Christianity"', *Church History*, 29 (1960), 64–73.
36. Idem, *Memoir of a Brother* [1859] (2nd edn, London, 1874), p. 96.
37. Idem, *Tom Brown at Oxford* [1859] (3 vols, Boston, MA, 1868), 1: 168–70.
38. Roberta J. Park, 'Biological Thought, Athletics, and the Formation of "A Man of Character", 1830–1900', in J. A. Mangan and James Walvin (eds.), *Manliness and Morality: Middle-Class Masculinity in Britain and America, 1800–1940* (Manchester, 1987), pp. 7–34, at p. 7.
39. My thanks to Professor Martin Hewitt for permitting me to read an early draft of his manuscript 'Darwin's Generations: The Reception of Darwinian Evolution in Britain, 1859–1909' (unpublished manuscript, 2023).
40. Charles Kingsley to Darwin, 18 November 1859, *DCP*, Letter 2534.
41. Jonathan Conlin, *Evolution and the Victorians: Science, Culture, and Politics* (London, 2013), p. 93.

42. Wallace, *Letters from Malaya*, p. 231.
43. *The Times*, 26 December 1859.
44. *The Gardeners' Chronicle and Agricultural Gazette*, 31 December 1859.
45. *The Life and Letters of the Reverend Adam Sedgwick*, ed. John Willis Clark and Thomas McKenny Hughes (London, 1890), p. 357.
46. *Spectator*, 7 April 1860.
47. Richard Owen to Darwin, 12 November 1859, *DCP*, Letter 2526; Darwin to Charles Lyell, 10 December 1859, *DCP*, Letter 2575.
48. [Richard Owen], 'Darwin on the Origin of Species' [review], *Edinburgh Review*, 111 (1860), 487–532, at pp. 494, 501, 521.
49. Darwin to A. R. Wallace, 18 May 1860, *DCP*, Letter 2807.
50. Richard Owen, 'Address of the President', in *Report of the Twenty-Eighth Meeting of the British Association for the Advancement of Science; Held at Leeds in September 1858* (London, 1859), xlix–cx, at p. li.

16. MONKEYANA

1. George Eliot, *The Mill on the Floss* (2 vols, Leipzig, 1860), 2:351.
2. Andrew Dickson White, *A History of the Warfare of Science with Theology in Christendom* (2 vols, New York, 1896), 1:71.
3. James R. Moore, *The Post-Darwinian Controversies: A Study of the Protestant Struggle to Come to Terms with Darwin in Great Britain and America, 1870–1900* (Cambridge, 1979), p. 60.
4. Debate has raged since the 1970s, with contributions of the last few years including: Richard England, 'Censoring Huxley and Wilberforce: A New Source for the Meeting that the *Athenaeum* "Wisely Softened Down"', *Notes & Records of the Royal Society of London*, 71 (2017), 371–84; James C. Ungureanu, 'A Yankee at Oxford: John William Draper at the British Association for the Advancement of Science at Oxford, 30 June 1860', *Notes & Records of the Royal Society of London*, 70 (2016), 135–50; and Nanna Katrine Lurers Kaalund, 'Oxford Serialized: Revisiting the Huxley–Wilberforce Debate through the Periodical Press', *History of Science*, 52 (2014), 429–53. The most comprehensive study of the debate is: Ian Hesketh, *Of Apes and Ancestors: Evolution, Christianity, and the Oxford Debate* (Toronto, 2009).
5. J. Vernon Jensen, *Thomas Henry Huxley: Communicating for Science* (Newark, NJ, 1991), p. 70.
6. John William Draper, 'On the Intellectual Development of Europe, Considered with Reference to the Views of Mr. Darwin and Others', *Report of the Thirtieth Meeting of the British Association for the Advancement of*

Science; Held at Oxford in June and July 1860 (London, 1861), 115–16, at p. 115.

7. Ungureanu, 'Yankee at Oxford', pp. 141, 135.
8. Ambrose Bierce, *The Devil's Dictionary* [1906] (Cleveland, 1911), p. 235.
9. [Samuel Wilberforce], 'Darwin's Origin of Species', *Quarterly Review*, 108 (1860), 118–38, at pp. 118, 136, 137.
10. Frank A. J. L. James, 'An "Open Clash Between Science and the Church"? Wilberforce, Huxley, and Hooker on Darwin at the British Association, Oxford 1860', in David M. Knight and Matthew D. Eddy (eds.), *Science and Beliefs: From Natural Philosophy to Natural Science, 1700–1900* (Aldershot, 2005), 171–93.
11. 'British Association: Section D – Zoology and Botany', *Athenaeum*, 1707 (1860), 64–5, at p. 65.
12. J. Vernon Jensen, 'Return to the Wilberforce–Huxley Debate', *British Journal for the History of Science*, 21 (1988), 161–79, at p. 167n55.
13. Desmond, *Huxley*, p. 141; *Life of Huxley*, 1:193.
14. Desmond, *Huxley*, p. 230; *Life of Huxley*, 1:162.
15. *Life of Huxley*, 1:202.
16. Ungureanu, 'Yankee at Oxford', p. 136.
17. J. D. Hooker to Darwin, 2 July 1860, *DCP*, Letter 2852.
18. Frederick Temple, *The Present Relations of Science to Religion: A Sermon, Preached on Act Sunday, July 1, 1860, before the University of Oxford* (Oxford, 1860), pp. 14, 15, 18.
19. Robert Cooper (ed.), *The London Investigator: A Monthly Journal of Secularism*, 1 (London, 1854–55), pp. 2, 8–12, 57.
20. The classic account of evolutionary theory and British society remains: J. W. Burrow, *Evolution and Society: A Study in Victorian Social Theory* (Cambridge, 1966).
21. *Life of Lyell*, 2:213.
22. Richard Owen, 'On the Characters, Principles of Division and Primary Groups of the Class Mammalia', *Journal of the Proceedings of the Linnean Society*, 2 (1858), 1–37, at pp. 14, 19, 20.
23. Darwin to J. D. Hooker, 5 July 1857, *DCP*, Letter 2117.
24. *Life of Huxley*, 1:101.
25. Owen, 'Presidential Address', li.
26. T. H. Huxley, 'The Croonian Lecture: On the Theory of the Vertebrate Skull', *Proceedings of the Royal Society of London*, 9 (1859), 381–457, at p. 417.
27. Desmond, *Huxley*, p. 253.
28. Ibid., p. 230; *Life of Huxley*, 1:163.
29. *Life of Huxley*, 1:163; Desmond, *Huxley*, p. 286.
30. Desmond, *Huxley*, pp. 286–7.

31. J. M. I. Klaver, *The Apostle of the Flesh: A Critical Life of Charles Kingsley* (Leiden, 2006), p. 480; *Life of Huxley*, 1:233.
32. *Life of Huxley*, 1:233–9.
33. Desmond, *Huxley*, p. 290.
34. Darwin to Huxley, 3 January 1861, *DCP*, Letter 3041; T. H. Huxley, 'On the Zoological Relations of Man with the Lower Remains', *Natural History Review*, 1 (1861), 67–84, at p. 67.
35. Huxley, 'Zoological Relations', pp. 67, 68, 84.
36. Desmond, *Huxley*, p. 292.
37. Paul B. Du Chaillu, 'The Geographical Features and Natural History of a Hitherto Unexplored Region of Western Africa', *Proceedings of the Royal Geographical Society of London*, 5 (1860–61), 108–12, at pp. 108, 110.
38. Ibid., p. 111.
39. Richard Owen, 'The Gorilla and the Negro', *Athenaeum*, 1743 (1861), 395–6, at pp. 396, 395.
40. T. H. Huxley, 'Man and the Apes', *Athenaeum*, 1744 (1861), 433; Richard Owen, 'The Gorilla and the Negro' (II), *Athenaeum*, 1745 (1861), 467.
41. T. H. Huxley, 'Man and the Apes' (II), *Athenaeum*, 1746 (1861), 498.
42. Darwin to Huxley, 1 April 1861, *DCP*, Letter 3107.
43. *Punch*, 40 (1861), p. 206.
44. T. H. Huxley, 'On Species and Races, and Their Origin', *Notices of the Proceedings at the Meetings of the Members of the Royal Institution*, 3 (1862), 195–200, at p. 199.
45. *The Life and Letters of Benjamin Jowett, M.A., Master of Balliol College, Oxford*, ed. Evelyn Abbott and Lewis Campbell (2nd edn, 2 vols, London, 1897), 1:297.
46. *Essays and Reviews* [1860] (4th edn, London, 1861), i, pp. 1, 48, 52, 129.
47. Ibid., pp. 209, 330, 337.
48. Ieuan Ellis, *Seven Against Christ: A Study of 'Essays and Reviews'* (Leiden, 1980), ix.
49. 'Essays and Reviews' [review], *Edinburgh Review*, 113 (1861), 461–500, at p. 465.
50. *The Times*, 28 February 1861.
51. Desmond and Moore, *Darwin*, p. 501; Darwin to Frederick Temple, 28 February 1861, *DCP*, Letter 2628.
52. [Samuel Wilberforce], 'Essays and Reviews' [review], *Quarterly Review*, 248–305, at pp. 248, 249.
53. Ellis, *Seven Against Christ*, ix.
54. White, *History of the Warfare*, 1:346.
55. 'Convocation', *Contemporary Review*, 1 (1866), 250–69, at p. 265.
56. *Life of Pusey*, p. 70.

17. GOING THE WHOLE ORANG

1. Charles Kingsley, *The Water-Babies: A Fairy Tale for a Land Baby* [1862–3] (London, 1880), p. 79.
2. Rev. James I. Good, *History of the Reformed Church of Germany* (Reading, PA, 1894), pp. 351–2.
3. George Busk, 'On the Crania of the Most Ancient Races of Man. By Professor D. Schaaffhausen, of Bonn', *Natural History Review*, 1 (1861), 155–76, at p. 166.
4. Ian Tattersall, *The Last Neanderthal: The Rise, Success, and Mysterious Extinction of Our Closest Human Relatives* (Boulder, CO, 1999), pp. 81, 76.
5. See Jacob W. Gruber, 'Brixham Cave and the Antiquity of Man', in Tim Murray (ed.), *Histories of Archaeology: A Reader in the History of Archaeology* (Oxford, 2008), 13–45; and Leonard J. Wilson, 'Brixham Cave and Sir Charles Lyell's *The Antiquity of Man*: The Roots of Hugh Falconer's Attack on Lyell', *Archives of Natural History*, 23 (1996), 79–97.
6. Busk, 'On the Crania', pp. 155, 172–3.
7. William Henry Flower, 'On the Posterior Lobes of the Cerebrum of the Quadrumana', *Philosophical Transactions of the Royal Society of London*, 152 (1862), 185–203, at pp. 187, 199.
8. Richard Owen, 'On the Characters of the Aye-Aye, as a Test of the Lamarckian and Darwinian Hypothesis of the Transmutation and Origin of Species', *Report of the Thirty-Second Meeting of the British Association for the Advancement of Science; Held at Cambridge in October 1862* (London, 1863), 114–16, at pp. 114, 116.
9. Huxley to Darwin, 9 October 1862, *DCP*, Letter 3755.
10. Richard Owen, 'On the Zoological Significance of the Cerebral and Pedial Characters of Man', *Report of the Thirty-Second Meeting of the British Association for the Advancement of Science; Held at Cambridge in October 1862* (London, 1863), 116–18, at p. 117.
11. Charles Kingsley, 'Speech of Lord Dundreary in Section D, on Friday Last, on the Great Hippocampus Question', in *Charles Kingsley: His Letters and Memories of His Life*, ed. Frances Kingsley (2 vols, London, 1877), 2: 140–43.
12. 'Men or Monkeys!', *British Medical Journal*, 18 October 1862.
13. 'The British Association at Cambridge', *Quarterly Review*, 113 (1863), 362–92, at p. 365.
14. Charles Lyell, *Travels in North America; with Geological Observations of the United States, Canada, and Nova Scotia* (2 vols, London, 1845); idem, *A Second Visit to the United States of North America* (2 vols, London, 1849), 1:176, 1:209.

15. Lyell to Darwin, 3 October 1859, *DCP*, Letter 2501.
16. *Life of Lyell*, 2:319.
17. Lyell to Darwin, 3 October 1859, *DCP*, Letter 2501; Lyell to Huxley, 17 June 1859, *DCP*, Letter 2469A.
18. Charles Lyell, *The Geological Evidences of the Antiquity of Man with Remarks on Theories on the Origin of Species by Variation* (London, 1863), pp. 9, 194.
19. Ibid., pp. 373, 366.
20. Darwin to Lyell, 6 March 1863, *DCP*, Letter 4028.
21. *Life of Lyell*, 2:361.
22. T. H. Huxley, *Evidence as to Man's Place in Nature* (London, 1863), pp. 1, 55, 57, 60, 70, 103, 106, 105, 108.
23. Darwin to Huxley, 18 February 1863, *DCP*, Letter 3996; Desmond, *Huxley*, pp. 312–13.
24. *Leeds Times*, 7 March 1863.
25. *Hereford Times*, 24 October 1863.
26. Desmond, *Huxley*, p. 314; [Charles Carter Blake], 'Professor Huxley on Man's Place in Nature', *Edinburgh Review*, 117 (1863), 278–93, at pp. 283, 282.
27. Charles Darwin, *On the Origin of Species by Means of Natural Selection, or the Preservation of Favoured Races in the Struggle for Life* (2nd edn, London, 1860), p. 481.
28. *Letters of Kingsley*, 2:137.
29. Kingsley to Darwin, 18 November 1859, *DCP*, Letter 2534.
30. *Letters of Kingsley*, 2:171, 2:137.
31. Charles Kingsley, *The Water-Babies: A Fairy Tale for a Land Baby* (London, 1863), pp. 74, 70, 75.

18. X

1. Jules Verne, *Journey to the Centre of the Earth* [1864] (London, 1876), p. 189.
2. Desmond and Moore, *Darwin*, p. 502.
3. Darwin, *Autobiography*, p. 245.
4. Desmond and Moore, *Darwin*, p. 509; Darwin to Hooker, 6 July 1861, *DCP*, Letter 3200.
5. Darwin to Hooker, 13 July 1861, *DCP*, Letter 3207.
6. Desmond and Moore, *Darwin*, p. 509.
7. Darwin to John Murray III, 24 September 1861, *DCP*, Letter 3264.
8. Charles Darwin, *On the Various Contrivances by which British and Foreign Orchids are Fertilised by Insects and on the Good Effects of Intercrossing* (London, 1862), p. 1.

9. Darwin to Hooker, 14 March 1862, *DCP*, Letter 3472.
10. For Darwin and botany, see: Steve Jones, *Darwin's Island: The Galapagos in the Garden of England* (London, 2009).
11. Desmond and Moore, *Darwin*, p. 517.
12. Charles Darwin, *On the Movements and Habits of Climbing Plants* (London, 1865), pp. 117–18.
13. Alex Menez, 'Custodian of the Gibraltar Skull: The History of the Gibraltar Scientific Society', *Earth Sciences History*, 37 (2018), 34–62.
14. George Busk, 'On a Very Ancient Human Cranium from Gibraltar', *Report of the Thirty-Fourth Meeting of the British Association for the Advancement of Science; Held at Bath in September 1864* (London, 1865), 91–2, at p. 92.
15. David A. Wells (ed.), *Annual of Scientific Discovery: Or, Year-Book of Facts in Science and Art for 1865* (London, 1865), p. 288.
16. Roy M. MacLeod, 'The X-Club: A Social Network of Science in Late-Victorian England', *Notes and Records of the Royal Society of London*, 24 (1970), 305–22, at pp. 306–07.
17. Herbert Spencer, *The Principles of Biology* (2 vols, London, 1864), passim, first at 1:444.
18. The best recent biography of Tyndall is: Roland Jackson, *The Ascent of John Tyndall: Victorian Scientist, Mountaineer, and Public Intellectual* (Oxford, 2018).
19. John Tyndall, *New Fragments* (New York, 1892), p. 231.
20. D. Thompson, 'John Tyndall (1820–1893): A Study in Vocational Enterprise', *Vocational Aspects of Secondary and Further Education*, 9 (1957), 38–48, at p. 47.
21. J. Vernon Jensen, 'The X Club: Fraternity of Victorian Scientists', *British Journal for the History of Science*, 5 (1970), 63–72, at p. 63.
22. Ruth Barton, *The X Club: Power and Authority in Victorian Science* (Chicago, 2018), p. 6.
23. Sedgwick to Darwin, 24 November 1859, *DCP*, Letter 2548.
24. Hugh Falconer to William Sharpey, 25 October 1864, *DCP*, Letter 4644.
25. Desmond and Moore, *Darwin*, p. 526.
26. Hooker to Darwin, 23 November 1864, *DCP*, Letter 4667.
27. Darwin to Hooker, 26 November 1864, *DCP*, Letter 4682.
28. *The Works of Charles Darwin*, ed. Paul H. Barrett and R. B. Freeman (29 vols, New York, 1986–1992), 29:123.
29. C. de B. Webb, 'Great Britain and "Die Republiek Natalia": An Early Case of U.D.I. and Sanctions', *Theoria: A Journal of Social and Political Theory*, 37 (1971), 15–30, at pp. 17–18.
30. *The Life of John William Colenso, D.D., Bishop of Natal* (2 vols, London, 1888), 1:47.

31. John William Colenso, *Ten Weeks in Natal: A Journal of a First Tour of Visitation among the Colonists and Zulu Kafirs of Natal* (London, 1855), p. 5.
32. Idem, *First Steps of the Zulu Mission (Oct. 1859)* (London, 1860), pp. 9, 53, 1.
33. Idem, *Zulu–English Dictionary* (Pietermaritzburg, 1861), viii.
34. Jeff Guy, *The Heretic: A Study of the Life of John William Colenso, 1814–1883* (Pietermaritzburg, 1983), p. 80.
35. For an authoritative collection of essays on Colenso's 'crisis', see: Jonathan A. Draper (ed.), *The Eye of the Storm: Bishop John William Colenso and the Crisis of Biblical Inspiration* (London, 2003).
36. John William Colenso, *The Pentateuch and Book of Joshua Critically Examined: Part I* [1862] (London, 1865), p. 8.
37. *The Life of Robert Gray, Bishop of Cape Town*, ed. Charles Gray (2 vols, London, 1876), 2:20.
38. John William Colenso, *St Paul's Epistle to the Romans: Newly Translated, and Explained from a Missionary Point of View* (Cambridge, 1861), p. 215.
39. Idem, 'On Missions to the Zulus in Natal and Zululand', *Social Science Review and Journal of the Sciences*, 18 (1865), 482–510, at p. 509.
40. Idem, *Pentateuch*, pp. 8, 11.
41. *Life of Colenso*, 1:vii.
42. Colenso, *Pentateuch*, p. 15.
43. Pusey qu. in Larsen, *Contested Christianity*, p. 69.
44. *What is Bishop Colenso's New Book? And Who is Bishop Colenso?* (Manchester, 1865), p. 2.
45. *Life of Gray*, 2:21, 2:30.
46. *Trial of the Bishop of Natal for Erroneous Teaching* (Cape Town, 1863), p. 4.
47. Desmond, *Huxley*, pp. 316, 685n9.
48. *Life of Colenso*, 1:260, 1:262.
49. *The Lambeth Conferences of 1867, 1878, and 1888: with the Official Reports and Resolutions, together with the Sermons Preached at the Conferences* (London, 1889), pp. 13, 16.

19. FEATHERS

1. Peter Wellnhofer, *Archaeopteryx: The Icon of Evolution*, trans. Frank Haase (Munich, 2009), p. 47.
2. Ibid., pp. 44, 51, 49.
3. Gavin de Beer, *Archaeopteryx Lithographica* (London, 2009), p. 2.
4. *Life of Owen*, 2:131.

5. *The Times*, 12 November 1862. The scoop is narrated in Paul Chambers, *Bones of Contention: The Fossil that Shook Science* (London, 2002), at pp. 83–6.
6. Richard Owen, 'On the Archaeopteryx of von Meyer, with a Description of the Fossil Remains of a Long-Tailed Species, from the Lithographic Stone of Solenhofen', *Philosophical Transactions of the Royal Society of London*, 153 (1863), 33–47, at pp. 33–4, 46.
7. Owen's position on the *Archaeopteryx* is covered in: Chambers, *Bones of Contention*, at pp. 88–9; and in Thor Hanson, *Feathers: The Evolution of a Natural Miracle* (New York, 2011), pp. 21–2.
8. Charles Darwin, *On the Origin of Species by Means of Natural Selection, or the Preservation of Favoured Races in the Struggle for Life* (4th edn, London, 1866), p. 346.
9. 'The Feathered Reptile of Solenhofen', *Intellectual Observer*, 1 (1862), 367–8, at p. 368.
10. *Loughborough Monitor*, 4 December 1862.
11. *Cardiff Times*, 5 December 1862.
12. Adrian Desmond, 'Redefining the X Axis: "Professionals", "Amateurs", and the Making of Mid-Victorian Biology: A Progress Report', *Journal of the History of Biology*, 34 (2001), 3–50, at p. 21.
13. The most comprehensive analysis of Darwin's views on race and slavery is: Adrian Desmond and James Moore, *Darwin's Sacred Cause: How a Hatred of Slavery Shaped Darwin's Views on Human Evolution* (New York, 2009).
14. Desmond, *Huxley*, p. 352.
15. Ibid., p. 358.
16. Ibid., p. 352.
17. Ernst Haeckel, *Generelle Morphologie der Organismen: Allgemeine Grundzüge der Organischen Formen-Wissenschaft, Mechanisch Begrundet Durch Die von Charles Darwin* (Berlin, 1866), first at p. 166.
18. 'Interesting Fossil Bones in Philadelphia', *Friends' Intelligencer*, 15 (1859), 715–17, at p. 715.
19. J. P. Lesley, 'Obituary Notice of Wm. Parker Foulke', *Proceedings of the American Philosophical Society*, 10 (1868), 481–510, at p. 492.
20. Ibid., at p. 493; ['Foulke and Leidy on the *Hadrosaurus*'], *Proceedings of the Academy of Natural Sciences of Philadelphia*, 10 (1859), 213–18, at p. 215.
21. 'Foulke and Leidy on the *Hadrosaurus*', p. 217.
22. Leonard Warren, *Joseph Leidy: The Last Man Who Knew Everything* (New Haven, CT, 1998).
23. 'Foulke and Leidy on the *Hadrosaurus*', p. 215.
24. Joseph Leidy, *Cretaceous Reptiles of the United States* (Philadelphia, 1864), p. 97.

25. E. D. Cope, '[On the *Archaeopteryx* and *Compsognathus*]', *Proceedings of the Academy of Natural Sciences of Philadelphia*, 19 (1867), 234–5.
26. For the influence of German zoology and dinosaurs on Huxley, see: Mario A. Di Gregorio, 'The Dinosaur Connection: A Reinterpretation of T. H. Huxley's Evolutionary View', *Journal of the History of Biology*, 15 (1982), 397–418.
27. T. H. Huxley, 'On the Animals which are Most Nearly Intermediate between Birds and Reptiles', *Annals and Magazine of Natural History*, 2 (1868), 66–75, at pp. 66–9.
28. Ibid., pp. 70–75.
29. Mario A. di Gregorio, *From Here to Eternity: Ernst Haeckel and Scientific Faith* (Gottingen, 2005), pp. 432–3.
30. T. H. Huxley, 'On Some Organisms Living at Great Depths in the North Atlantic Ocean', *Quarterly Journal of Microscopical Science*, 8 (1868), 203–12, at p. 205.
31. 'Discussion on Captain S. Osborn's Paper', *Proceedings of the Royal Geographical Society*, 15 (1871), 37–9, at p. 38.
32. T. H. Huxley, *On the Physical Basis of Life* [1868] (2nd edn, New Haven, CT, 1870), p. 1.
33. Philip F. Rehbock, 'Huxley, Haeckel, and the Oceanographers: The Case of Bathybius Haeckelii', *Isis*, 66 (1975), 504–33, at pp. 508–09. See also: Nicolaas A. Rupke, '*Bathybius Haeckelii* and the Psychology of Scientific Discovery', *Studies in the History and Philosophy of Science*, 7 (1976), 53–62.
34. T. H. Huxley, 'Triassic Dinosauria', *Nature*, 1 (1869), 23–4, at p. 23.
35. Idem, 'Geological Reform' [1869], in Huxley, *Discourses, Biological and Geological: Essays* (New York, 1896), 305–39, at pp. 316, 307.
36. Idem, 'On the Ethnology of Britain', *Journal of the Ethnological Society of London*, 2 (1870), 382–4.
37. Desmond, *Huxley*, p. 375.
38. *The Times*, 25 August 1869.
39. T. H. Huxley, *Lay Sermons, Essays, and Reviews* (London, 1870).
40. Darwin to Huxley, 19 March 1869, *DCP*, Letter 6670.
41. *Spectator*, 29 January 1870.
42. Desmond, *Huxley*, p. 374.
43. T. H. Huxley, 'Agnosticism', *Eclectic Magazine*, 49 (1889), 433–50, at p. 443. For 'agnosticism' as merely a denomination of Dissenting Protestantism, see: Bernard V. Lightman, 'Interpreting Agnosticism as a Nonconformist Sect: T. H. Huxley's "New Reformation"', in Paul Wood (ed.), *Science, Technology and Culture, 1700–1945* (Aldershot, 2004), pp. 197–214.
44. Idem, 'A Liberal Education; and Where to Find It', *Macmillan's Magazine*, 17 (1868), 367–78, at p. 369.

45. John Lubbock, *Pre-Historic Times as Illustrated by Ancient Remains* (London, 1865) at pp. 2, 3; Paul Pettitt and Mark White, 'John Lubbock, Caves, and the Development of Middle and Upper Palaeolithic Archaeology', *Notes and Records*, 68 (2014), 35–48.
46. Bernard V. Lightman, 'Huxley and the Devonshire Commission', in Gowan Dawson (ed.), *Victorian Scientific Naturalism: Community, Identity, Continuity* (Chicago, 2014), pp. 101–30.

20. TIDES OF FAITH

1. Matthew Arnold, 'Dover Beach' in *New Poems* (2nd edn, London, 1868), 112–14, at p. 113.
2. Olive Anderson, 'Gladstone's Abolition of Compulsory Church Rates: A Minor Political Myth and its Historiographical Career', *Journal of Ecclesiastical History*, 25 (1974), 185–98.
3. Padraic C. Kennedy, '"Underhand Dealings with the Papal Authorities": Disraeli and the Liberal Conspiracy to Disestablish the Irish Church', *Parliamentary History*, 27 (2008), 19–29.
4. Jennifer Stevens, *The Historical Jesus and the Literary Imagination, 1860–1920* (Liverpool, 2010), p. 43.
5. Ernest Renan, *The Life of Jesus*, trans. Charles Edwin Wilbour [1863] (New York, 1864), pp. 331, 339, 341, 347, 346, 348.
6. Robert D. Priest, 'Reading, Writing, and Religion in Nineteenth-Century France: The Popular Reception of Renan's *Life of Jesus*', *Journal of Modern History*, 86 (2014), 258–94, at pp. 261, 268. The definitive study of the book is: Robert D. Priest, *The Gospel According to Renan: Reading, Writing, and Religion in Nineteenth-Century France* (Oxford, 2015).
7. The book's denunciation by the Cardinal Archbishop of Reims was carried, inter alia, in: *Berkshire Chronicle*, 8 August 1863; *Preston Chronicle*, 8 August 1863; *Scotsman*, 9 August 1863.
8. *Dunfermline Saturday Press*, 12 March 1864; *Leeds Intelligencer*, 2 January 1864; *Belfast Newsletter*, 11 November 1864.
9. Ian Hesketh, 'Behold the (Anonymous) Man: J. R. Seeley and the Publishing of "Ecce Homo"', *Victorian Review*, 38 (2012), 93–112, at p. 97.
10. [J. R. Seeley], *Ecce Homo: A Survey of the Life and Work of Jesus Christ* [1865] (London, 1866), v–vi, p. 42.
11. Ian Hesketh, *Victorian Jesus: J. R. Seeley, Religion, and the Cultural Significance of Anonymity* (Toronto, 2017), p. 3.
12. Daniel Pals, 'The Reception of *Ecce Homo*', *Historical Magazine of the Protestant Episcopal Church*, 46 (1977), 63–84, at pp. 73, 74, 75, 77.

13. Hesketh, *Victorian Jesus*, p. 3.
14. R. T. Shannon, 'John Ronald Seeley', *ODNB*.
15. J. R. Seeley, *The Expansion of England: Two Courses of Lectures* (London, 1891), p. 8.
16. Martin Hewitt, *The Dawn of the Cheap Press in Victorian Britain: The End of the 'Taxes on Knowledge', 1849–1869* (London, 2014).
17. Holyoake, *Life and Letters*, 2:267.
18. George Jacob Holyoake, *Self-Help by the People: Thirty-Three Years of Co-Operation in Rochdale* [1858] (9th edn, 2 vols, London, 1882), 1:86.
19. *Charles Bradlaugh: A Record of His Life and Work*, ed. Hypatia Bradlaugh Bonner (2 vols, London, 1895), 1:8, 1:18.
20. 'Secular Progress', *London Investigator*, 36 (1857), 360–61, at p. 360; Annie Besant, *Autobiographical Sketches* (London, 1885), p. 116.
21. G. W. Foote, *Secularism: The True Philosophy of Life: An Exposition and a Defence* (London, 1879), p. 32.
22. J. Ewing Ritchie, *Days and Nights in London: or, Studies in Black and Gray* (London, 1880), p. 106.
23. Walter L. Arnstein, *The Bradlaugh Case: A Study in Late Victorian Opinion and Politics* (Oxford, 1965), p. 14; Besant, *Sketches*, p. 90.
24. Arnstein, *Bradlaugh Case*, pp. 14–15.
25. Edward Royle, *Radicals, Secularists, and Republicans: Popular Freethought in Britain, 1866–1915* (Manchester, 1980), p. 7.
26. Arnstein, *Bradlaugh Case*, p. 15.
27. *National Reformer*, 3 May 1868.
28. G. W. Foote, *Reminiscences of Charles Bradlaugh* (London, 1891), p. 24.
29. *Record of Bradlaugh*, 1:145, 1:137, 1:139, 1:149.
30. George Campbell, 'On Tertiary Leaf-Beds in the Isle of Mull', *Quarterly Journal of the Geological Society of London*, 7 (1851), 89–103; Charles Lyell to Charles Darwin, 25 September 1874, *DCP*, Letter 9658.
31. George Campbell to Charles Lyell, 1 May 1867, in *George Douglas Campbell, Eighth Duke of Argyll, K.G., K.T. (1823–1900): Autobiography and Memoirs*, ed. Dowager Duchess of Argyll (2 vols, London, 1906), 2:486.
32. George Campbell, *The Reign of Law* [1867] (5th edn, London, 1871), pp. 208, 262; Peter J. Bowler, 'Darwinism and the Argument from Design: Suggestions for a Reevaluation', *Journal of the History of Biology*, 10 (1977), 29–43, at pp. 37–8.
33. George Campbell, *Primeval Man: An Examination of Some Recent Speculations* [1869] (3rd edn, London, 1870), pp. 73–4.
34. James R. Moore, *The Post-Darwinian Controversies: A Study of the Protestant Struggle to Come to Terms with Darwin in Great Britain and America* (Cambridge, 1979), p. 117.

35. Jacob W. Gruber, *A Conscience in Conflict: The Life of St George Jackson Mivart* (New York, 1960), p. 22.
36. [St George Mivart], 'Difficulties of the Theory of Natural Selection', *The Month*, 11 (1869), 35–53, at p. 37.
37. Darwin, *Origin* (1st edn), p. 282.
38. William Thomson, 'Geological Dynamics', *Geological Magazine*, 6 (1869), 472–6, at p. 473.
39. Idem, 'On Geological Time', *Transactions of the Geological Society of Glasgow*, 3 (1868), 1–29, at pp. 16, 1.
40. T. H. Huxley, 'The Anniversary Address of the President', *Quarterly Journal of the Geological Society of London*, 25 (1869), xxviii–liii, at p. xxxviii, liii.
41. Joe D. Burchfield, *Lord Kelvin and the Age of the Earth* (Chicago, 1990), p. 2.
42. Hesba Stretton, *Jessica's First Prayer* (London, 1866).
43. On the book's success, see: J. S. Bratton, 'Hesba Stretton's Journalism', *Victorian Periodicals Review*, 12 (1979), 60–70; Rosemary Scott, 'The Sunday Periodical: *Sunday at Home*', *Victorian Periodicals Review*, 25 (1992), 158–62; and Patricia Demers, 'Sarah Smith [*pseud.* Hesba Stretton]', *ODNB*.
44. Christine Garwood, *Flat Earth: The History of an Infamous Idea* (London, 2007), p. 73.
45. 'Parallax' [Samuel Rowbotham], *Zetetic Astronomy: Earth not a Globe! An Experimental Inquiry into the True Figure of the Earth: Proving it a Plane, without Axial or Orbital Motion* (London, 1865), pp. 7, 109.
46. *Scientific Opinion*, 12 January 1870.
47. Shermer, *In Darwin's Shadow*, pp. 221–2.
48. Raby, *Wallace*, p. 187. See also: Malcolm Jay Kottler, 'Alfred Russel Wallace, the Origin of Man, and Spiritualism', *Isis*, 65 (1974), 144–92.
49. Raby, Wallace, p. 203.
50. Desmond and Moore, *Darwin*, p. 569.
51. Shermer, *In Darwin's Shadow*, p. 259; John Hampden, *Is Water Level or Convex After All? The Bedford Canal Swindle Detected & Exposed* (Swindon, 1870), p. 14.
52. Wallace, *My Life*, 2:365, 2:368.
53. Hampden, *Is Water Level?*, pp. 3, 5.
54. Wallace, *My Life*, 2:371, 2:374, 2:375, 2:364.

21. THE ANGEL AND THE APE

1. Samuel Butler, *Erewhon: or, Over the Range* (London, 1872), p. 190.
2. Desmond and Moore, *Darwin*, pp. 529, 553.
3. Campbell, *Reign of Law*, p. 188.
4. Darwin to John Murray, 3 January 1867, *DCP*, Letter 5346.

5. Charles Darwin, *The Variation of Animals and Plants under Domestication* (2 vols, London, 1868), 1:1, 1:45.
6. Darwin to Huxley, 27 May 1865, *DCP*, Letter 4837.
7. Huxley to Darwin, 16 July 1865, *DCP*, Letter 4875.
8. Darwin, *Variation under Domestication*, 2:374.
9. John Price to Darwin, 5 March 1868, *DCP*, Letter 5982.
10. David Galton, 'Did Darwin Read Mendel?', *QJM*, 102 (2009), 587–9.
11. Huxley to Darwin, 22 June 1870, *DCP*, Letter 7239.
12. John Murray to Darwin, 6 August 1870, *DCP*, Letter 7295.
13. Charles Darwin, *The Descent of Man, and Selection in Relation to Sex* (2 vols, London, 1871), 1:2–3.
14. Ibid., 1:10, 1:14, 1:33, 1:106, 1:199, 1:204, 1:226, 2:7.
15. Desmond and Moore, *Darwin*, pp. 573, 581.
16. 'Darwin on *The Descent of Man*', *Edinburgh Review*, 134 (1871), 195–235, at p. 196.
17. *The Times*, 8 April 1871.
18. St George Mivart, *On the Genesis of Species* (2nd edn, London, 1871), p. 14.
19. Idem, 'Descent of Man', *Quarterly Review*, 131 (1871), 47–90, at pp. 89–90.
20. Darwin to Huxley, 30 September 1871, *DCP*, Letter 7976; Desmond, *Huxley*, p. 407.
21. *New York Times*, 23 and 28 April 1871.
22. 'The Latest Form of Infidelity', *Raleigh Christian Advocate*, 6 September 1871.
23. 'Darwin Demolished', *Chicago Tribune*, 26 February 1871.
24. Alfred Russel Wallace, 'The Descent of Man', *Academy*, 2 (1871), 177–83, at pp. 177, 183.
25. William Preyer to Darwin, 27 April 1871, *DCP*, Letter 772; Laszlo Dapsy to Darwin, 17 June 1871, *DCP*, Letter 7817.
26. *Spectator*, 11 March 1871.
27. For what Darwin got wrong, see: Jeremy M. Desilva (ed.), *A Most Interesting Problem: What Darwin's* Descent of Man *Got Right and Wrong about Human Evolution* (Princeton, 2021).
28. Charles Bell, *The Hand, Its Mechanism, and Vital Endowments as Evincing Design* (London, 1833), paraphrased in Darwin, *Descent of Man*, p. 5.
29. Charles Darwin, *The Expression of the Emotions in Man and Animals* (London, 1872), pp. 141, 59.
30. Ibid., pp. 28, 132, 352.
31. Bashford, *Intimate History*, p. 38.
32. Desmond, *Huxley*, p. 444.
33. Ibid., pp. 449, 450, 451, 454, 457.

34. Rev. W. Fox, 'On the Skull and Bones of an Iguanodon', *Report of the Thirty-Eighth Meeting of the British Association for the Advancement of Science; Held at Norwich in August 1868* (London, 1869), 64–5.
35. T. H. Huxley, 'On *Hypsilophodon*, a new Genus of *Dinosauria*', *The Geological Magazine, or Monthly Journal of Geology*, 6 (1869), 573–4; idem, 'On *Hypsilophodon Foxii*, a new *Dinosaurian* from the Wealden of the Isle of Wight', *Quarterly Journal of the Geological Society of London*, 26 (1870), 3–12, at pp. 5, 7, 6, 12.
36. Idem, 'Further Evidence of the Affinity between the Dinosaurian Reptiles and Birds', *Quarterly Journal of the Geological Society of London*, 26 (1870), 12–34.
37. Charles Darwin, *On the Origin of Species by Means of Natural Selection, or the Preservation of Favoured Races in the Struggle for Life* (3rd edn, London, 1861), pp. 345–6.
38. Idem, *The Origin of Species by Means of Natural Selection, or the Preservation of Favoured Races in the Struggle for Life* (6th edn, London, 1872), pp. 302–03.
39. John Phillips, *Life on the Earth: Its Origin and Succession* (London, 1860), pp. 3, 217.
40. Idem, *Geology of Oxford and the Valley of the Thames* (Oxford, 1871), p. 196.
41. John Phillips to Darwin, 14 March 1874, *DCP*, Letter 9360; John Morrell, 'John Phillips', *ODNB*.
42. 'Men of the Day, No. 57', *Vanity Fair*, 1 March 1873.
43. William Henry Davenport Adams, *Life in the Primeval World: Founded on Meunier's 'Les Animaux D'Autrefois'* (London, 1872), pp. 198, 199, 201, 215.
44. J. A. Secord, 'Harry Govier Seeley', *ODNB*.
45. David E. Fastovsky and David B. Weishampel, *Dinosaurs: A Concise Natural History* (4th edn, Cambridge, 2021), pp. 95–7.
46. Bernard O'Connor, *Digging for Dinosaurs: The Great Fenland Coprolite Rush* (Ely, 2011).
47. T. H. Huxley, 'On *Acanthopolis Horridus*, a New Reptile from the Chalk-Marl', *Geological Magazine*, 4 (1867), 65–7.
48. H. G. Seeley, *Index to the Fossil Remains of Aves, Ornithosauria, and Reptilia, from the Secondary System of Strata, Arranged in the Woodwardian Museum of the University of Cambridge* (Cambridge, 1869), pp. 94–5.

22. LEARNED BABOONS

1. Thomas Hardy, *A Pair of Blue Eyes* (3 vols, London, 1873), 2:185.
2. John William Draper, *History of the Conflict Between Science and Religion* (New York, 1874).

3. Ibid., pp. 11, vi–vii, 363.
4. Ronald L. Numbers, 'Science and Religion', *Osiris*, 1 (1985), 59–90, at p. 59.
5. Andrew Dickson White, 'Scientific and Industrial Education in the United States', *Popular Science Monthly*, 5 (1874), 170–91, at p. 190.
6. Idem, *The Warfare of Science* (New York, 1876), p. 7.
7. John Tyndall, 'Introduction', in Andrew Dickson White, *The Warfare of Science* (London, 1876), iii–iv, at pp. iii, iv.
8. Ibid., iv.
9. *Life of Huxley*, 1:440.
10. Ruth Barton, 'John Tyndall, Pantheist: A Rereading of the Belfast Address', *Osiris*, 3 (1987), 111–34, at p. 115.
11. *Belfast News-Letter*, 14 May 1862; John Tyndall, *Address Delivered before the British Association assembled at Belfast* (London, 1874), p. 64.
12. Tyndall, *Belfast Address*, pp. 4, 19, 38, 54.
13. Ibid., pp. 61, 64, 60, 61.
14. Among the many studies of the address, see: Matthew Brown, 'Darwin at Church: John Tyndall's Belfast Address', in James H. Murphy (ed.), *Evangelicals and Catholics in Nineteenth-Century Ireland* (Dublin, 2005), pp. 235–46.
15. Andrew R. Holmes, 'Presbyterians and Science in the North of Ireland before 1874', *British Journal of the History of Science*, 41 (2008), 541–65, at pp. 542–3.
16. Diarmid A. Finnegan and Jonathan Jeffrey Wright, 'Catholics, Science, and Civic Culture in Victorian Belfast', *British Journal for the History of Science*, 48 (2015), 261–87, at p. 280.
17. *Religion and Science; the Letters of 'Alpha' on the Influence of Spirit upon Imponderable Actienic Molecular Substances, and the Life-Forces of Mind and Matter* (Boston, 1875), p. 25.
18. Tyndall, *Belfast Address*, viii.
19. Darwin to Tyndall, 12 August 1874, *DCP*, Letter 9599.
20. Lyell to Darwin, 1 September 1874, *DCP*, Letter 9619.
21. Charles Bradlaugh, *The Bible: What It Is* (London, 1870), v, pp. 105, 203.
22. 'Iconoclast', *The Bible: What It Is* (London, 1861), p. 99.
23. Bradlaugh, *The Bible* (1870), dedication, vi.
24. Idem, *The Land, The People, and the Coming Struggle* [1871] (3rd edn, London, 1880), p. 16.
25. Idem, *The Impeachment of the House of Brunswick* [1872] (2nd edn, London, 1873), p. 5; Arnstein, *Bradlaugh Case*, p. 15.
26. Adolphe S. Headingley, *The Biography of Charles Bradlaugh* (London, 1883), p. 144.
27. David H. Tribe, *President Charles Bradlaugh, M.P.* (London, 1971), p. 145.

28. Besant, *Sketches*, pp. 74, 90, 91.
29. Ibid., pp. 92, 95, 96.
30. Anne Taylor, 'Annie Besant', *ODNB*.
31. Besant, *Sketches*, pp. 99, 117.
32. Charles Knowlton, *Fruits of Philosophy: An Essay on the Population Question* [1832] (Rotterdam, 1877), pp. 12, 48.
33. Besant, *Sketches*, pp. 120, 123.
34. *Edinburgh Evening News*, 30 May 1877; *Dundee Courier*, 25 May 1877; *Leicester Daily Post*, 18 May 1877.
35. Bradlaugh to Darwin, 5 June 1877, *DCP*, Letter 10984.
36. Besant, *Sketches*, p. 136.
37. Darwin to Bradlaugh, 6 June 1877, *DCP*, Letter 10988.
38. Besant, *Sketches*, pp. 141, 159.
39. Jim Herrick, *Humanism: An Introduction* (Amherst, NY, 2005), p. 68.
40. G. W. Foote, *Comic Bible Sketches reprinted from 'The Freethinker'* (London, 1885), vii.
41. Foote would publish an account of his prosecution and imprisonment in *Prisoner for Blasphemy* (London, 1886).

23. BADLANDS

1. Thomas Hardy, *The Return of the Native* (3 vols, London, 1878), 2:74.
2. Mark Jaffe, *The Gilded Dinosaur: The Fossil War between E. D. Cope and O. C. Marsh and the Rise of American Science* (New York, 2000), pp. 27, 22.
3. Charles Kingsley, 'The Wonders of the Shore', *North British Review*, 22 (1855), 1–56, at pp. 9–10.
4. Othniel Charles Marsh, 'Description of the Remains of a new Enaliosaurian', *American Journal of Science and Arts*, 34 (1862), 1–16.
5. Idem, 'Thomas Henry Huxley', *American Journal of Science*, 50 (1895), 177–83, at p. 187.
6. 'The Cardiff Giant A Humbug', *Proceedings of the Massachusetts Historical Society*, 11 (1871), 161–2, at p. 161.
7. *New York Herald*, 1 December 1869.
8. Url Lanham, *The Bone Hunters* (New York, 1975), p. 18.
9. Othniel Charles Marsh, 'Notice of a New and Diminutive Species of Fossil Horse (*Equus Parvulus*), from the Tertiary of Nebraska', *Annals and Magazine of Natural History*, 3 (1869), 95–6, at p. 96.
10. Lanham, *Bone Hunters*, p. 52.
11. Charles E. Beecher, 'Memoir of O. C. Marsh', *American Journal of Science*, 7 (1899), 403–28, at pp. 404, 405.

12. Timothy Dwight, *Memories of Yale Life and Men, 1854–1899* (New York, 1903), p. 410.
13. Jaffe, *Gilded Dinosaur*, pp. 27, 26.
14. 'The Yale College Expedition of 1870', *Harper's New Monthly Magazine*, 43 (1871), 663–71, at p. 663.
15. Ibid., p. 664.
16. Jaffe, *Gilded Dinosaur*, p. 38.
17. Lanham, *Bone Hunters*, pp. 194–5; O. C. Marsh, 'On a New Subclass of Fossil Birds (Odontornithes)', *American Journal of Science*, 5 (1873), 233–4, at p. 234.
18. Jaffe, *Gilded Dinosaur*, p. 87.
19. Darwin to Othniel Charles Marsh, 25 January 1873, *DCP*, Letter 9829.
20. *Life of Huxley*, 1:497.
21. Jane P. Davidson, *The Bone Sharp: The Life of Edward Drinker Cope* (Philadelphia, 1997), p. 118.
22. Lanham, *Bone Hunter*, pp. 117–19.
23. Ibid., pp. 121–3.
24. Jill Hunting, *For Want of Wings: A Bird with Teeth and a Dinosaur in the Family* (Norman, OK, 2002), p. 97.
25. David Rains Wallace, *The Bonehunters' Revenge: Greed, and the Greatest Scientific Feud of the Gilded Age* (Boston, 1999), pp. 100–08.
26. Ibid., pp. 142, 184.
27. Charles H. Sternberg, *The Life of A Fossil Hunter* (New York, 1909), p. 33.
28. Edward Drinker Cope, *On the Origin of Genera* (Philadelphia, 1869), pp. 4–5.
29. Othniel Charles Marsh, *Introduction and Succession of Vertebrate Life in America* (New Haven, CT, 1877), pp. 3, 22.
30. Edward Drinker Cope, *On the Origin of the Fittest: Essays on Evolution* (New York, 1887), pp. 225, 168.
31. William Pierce Randel, 'Huxley in America', *Proceedings of the American Philosophical Society*, 114 (1970), 73–99, at p. 73.
32. *Life of Huxley*, 1:495.
33. Wallace, *Bonehunters' Revenge*, p. 135.
34. T. H. Huxley, *American Addresses: With a Lecture on the Study of Biology* (London, 1886), pp. 16, 18, 66–7.
35. For the popularity of British scientists in the United States at this time, see: Diarmid A. Finnegan, *The Voice of Science: British Scientists on the Lecture Circuit in Gilded Age America* (Pittsburgh, 2021).
36. Randy Moore, *Dinosaurs by the Decades: A Chronology of the Dinosaur in Science and Popular Culture* (Oxford, 2014), p. 78.
37. Darwin to Wallace, 5 June 1876, *DCP*, Letter 10531.

38. T. H. Huxley, 'Evolution in Biology' [1878], in *Science and Culture, and Other Essays* (New York, 1882), 281–316, at p. 316.
39. Idem, 'The Coming of Age of "The Origin of Species"', in *Science and Culture*, 317–32, at p. 330.
40. Richard Lydekker, 'Notices of New and Other Vertebrata from Indian Territory and Secondary Rocks', *Records of the Geological Society of India*, 9 (1876), 30–49, at p. 40.
41. William Davies, 'On the Exhumation and Development of a Large Reptile (*Omosaurus Armatus*, Owen), from the Kimmeridge Clay, Swindon, Wilts', *Geological Magazine*, 3 (1876), 193–7, at pp. 193, 195.
42. Richard Owen, *A History of British Fossil Reptiles* (4 vols, London, 1849–84), 1:557.
43. Secord, 'Seeley', *ODNB*.
44. *Buxton Herald*, 22 April 1875; H. G. Seeley, *Dragons of the Air: An Account of Extinct Flying Reptiles* (New York, 1901), vi–vii.
45. *Morning Post*, 3 May 1875.
46. *Evening Mail*, 28 November 1877.
47. Louis Figuier, *La Terre avant le Déluge* (Paris, 1863).
48. *Illustrated London News*, 8 May 1875.

24. THE CITADELS FALL

1. Samuel Butler, *Unconscious Memory: A Comparison between the Theory of Dr. Ewald Hering . . . and the 'Philosophy of the Unconscious' of Dr. Edward von Hartmann* (London, 1880), p. 124.
2. *The Times*, 18 April 1881.
3. *The Times*, 16 April 1881.
4. J. B. Bullen, 'Alfred Waterhouse's Romanesque "Temple of Nature": The Natural History Museum, London', *Architectural History*, 49 (2006), 257–85, at p. 271.
5. *The Times*, 16 April 1881.
6. *Life of Owen*, 2:29, 2:31.
7. *Hansard*, Commons, 24 April 1860, vol. 158, cc37–53, at c43.
8. *Report from the Select Committee on the British Museum, together with the Proceedings of the Committee, Minutes of Evidence, and Appendix* (London, 1860), p. 4.
9. Ibid., pp. 89, ix.
10. *Life of Owen*, 2:40, 2:36, 2:43–4.
11. *Comic News*, 18 July 1863.
12. *Life and Letters of Sir Joseph Dalton Hooker, O.M., G.C.S.I.* (2 vols, London, 1918), 2:172.

13. *Life of Owen*, 2:28
14. Ibid., 2:55, 2:240.
15. Sophie Forgan and Graeme Gooday, 'Constructing South Kensington: The Buildings and Politics of T. H. Huxley's Working Environments', *British Journal for the History of Science*, 29 (1996), 435–68, at pp. 453–4.
16. Tom Rea, *Bone Wars: The Excavation and Celebrity of Andrew Carnegie's Dinosaur* (Pittsburgh, 2001), p. 15.
17. Wallace, *Bonehunters' Revenge*, p. 149.
18. Lanham, *Bone Hunters*, p. 175.
19. Jaffe, *Gilded Dinosaur*, p. 232.
20. Lanham, *Bone Hunters*, p. 175.
21. Jaffe, *Gilded Dinosaur*, p. 232.
22. Ibid., pp. 234, 238.
23. Lanham, *Bone Hunters*, pp. 249, 252.
24. E. D. Cope, *Theology of Evolution: A Lecture* (Philadelphia, 1887), pp. 29, 39.
25. Henry Fairfield Osborn, '*Tyrannosaurus* and other Cretaceous Carnivorous Dinosaurs', *Bulletin of the American Museum of Natural History*, 21 (1905), 259–65.
26. John Morley, *The Life of William Gladstone* (3 vols, London, 1903), 2:596, 2:593, 2:594. See also: David Brooks, 'Gladstone and Midlothian: The Background to the First Campaign', *Scottish Historical Review*, 64 (1985), 42–67.
27. Annie Besant, *The Law of Population: Its Consequences and Its Bearing upon Human Conduct and Morals* (New York, 1878), p. 31.
28. *Is It Reasonable to Worship God? Verbatim Report of Debate between R. A. Armstrong and C. Bradlaugh* (London, 1878).
29. Charles Bradlaugh, *When Were Our Gospels Written?* (4th edn, London, 1881), p. 12.
30. *Secularism: Unphilosophical, Immoral, and Anti-Social: Verbatim Report of a Three Nights' Debate between the Rev. Dr. McCann and Charles Bradlaugh, in the Hall of Science, London* (London, 1881), pp. 20–21.
31. Arnstein, *Bradlaugh Case*, pp. 28, 31.
32. See: Michael Clark, 'Jewish Identity in British Politics: the Case of the First Jewish MPs, 1858–87', *Jewish Social Studies*, 13 (2007), 93–126.
33. *Hansard*, Commons, 3 May 1880, vol. 252, cc20–29.
34. Arnstein, *Bradlaugh Case*, pp. 43–4.
35. Bradlaugh, *House of Brunswick*, p. 102.
36. Paula Bartley, *Queen Victoria* (London, 2016), p. 234; Queen Victoria's Journal, 13 January 1881 (http://www.queenvictoriasjournals.org).
37. Arnstein, *Bradlaugh Case*, p. 43.

38. *British Medical Journal*, 26 November 1881.
39. Arnstein, *Bradlaugh Case*, pp. 54, 58.
40. Ibid., p. 97.
41. *The Times*, 27 April 1882.
42. T. H. Huxley, 'Darwin', *Nature*, 25 (1882), 597.
43. *The Times*, 27 April 1882.
44. Desmond and Moore, *Darwin*, pp. 634–5, 636, 645.
45. Ibid., p. 650.
46. Charles Darwin, *The Effects of Cross and Self Fertilisation in the Vegetable Kingdom* (London, 1876); idem, *The Power of Movement in Plants* (London, 1880), p. 559.
47. Janet Browne, Adrian Desmond, and James Moore, 'Charles Darwin', *ODNB*.

EPILOGUE: REX

1. Qu. in Richard Fallon, *Reimagining Dinosaurs in Late Victorian and Edwardian Literature: How the 'Terrible Lizard' Became a Transatlantic Cultural Icon* (Cambridge, 2021), p. 29.
2. For the enduring tension between science and religion, see: Frank M. Turner, *Between Science and Religion: Reaction to Scientific Naturalism in Late Victorian England* (New Haven, CT, 1974), and Peter J. Bowler, *Reconciling Science and Religion: The Debate in Early Twentieth-Century Britain* (Chicago, 2001). For the prominence of alternative theories of evolution in the early twentieth century, see: Peter J. Bowler, *The Eclipse of Darwinism: Anti-Darwinian Evolution Theories in the Decades around 1900* (Baltimore, MD, 1983).
3. Andrew Lycett, *The Man Who Created Sherlock Holmes: The Life and Times of Sir Arthur Conan Doyle* (London, 2007), p. 347.
4. Fallon, *Reimagining Dinosaurs*, pp. 136, 154, 155.
5. Percy Harrison Fawcett, *Lost Trails, Lost Cities: From His Manuscripts, Letters, and Other Records*, ed. Brian Fawcett (New York, 1953), p. 131.
6. Arthur Conan Doyle, *The Lost World* (New York, 1912), pp. 136, 161, 299.
7. Frank Mackenzie Savile, *Beyond the Great Wall: The Secret of the Antarctic* (New York, 1903), pp. 184, 214, 184.
8. Wardon Allan Curtis, 'The Monster of Lake LaMetrie', *Pearson's Magazine*, 7 (1899), 341–7, at p. 347.
9. Fallon, *Reimagining Dinosaurs*, pp. 38, 70.
10. Bernard V. Lightman, *Victorian Popularizers of Science: Designing Nature for New Audiences* (Chicago, 2009), p. 452.

11. Fallon, *Reimagining Dinosaurs*, p. 44.
12. *St James's Gazette*, 9 May 1899.
13. *Daily Telegraph & Courier*, 21 February 1905.
14. *Abergavenny Chronicle*, 3 May 1901.
15. Ulrich Merkl, *Dinomania: The Lost Art of Winsor McCay, the Secret Origins of King Kong, and the Urge to Destroy New York* (New York, 2015), pp. 278, 97.
16. Donald Crafton, *Before Mickey: The Animated Film, 1898–1928* (Chicago, 2015), p. 112.
17. Charles R. Knight, *Autobiography of an Artist*, ed. Jim Ottaviani (New York, 2005), pp. 6, 42.
18. Rea, *Bone Wars*, pp. 29, 41, 58.
19. O. C. Marsh, 'The Principal Characters of American Jurassic Dinosaurs', *American Journal of Science and Arts*, 16 (1878), 411–16, at p. 414.
20. Rea, *Bone Wars*, pp. 55, 104, 55, 82.
21. Ibid., p. 3.
22. *The Times*, 13 May 1905; Rea, *Bone Wars*, pp. 1–5.
23. Ilja Nieuwland, *American Dinosaur Abroad: A Cultural History of Carnegie's Plaster Diplodocus* (Pittsburgh, 2019), ch. 6 for the general international competition for copies of the *Diplodocus*.
24. Rea, *Bone Wars*, p. 201.
25. The passage on zircon dating relies on Hannah Fry and Adam Rutherford, *Rutherford & Fry's Complete Guide to Absolutely Everything* (New York, 2021), pp. 108–10.

Index

Aberdeen, George Hamilton-Gordon, 4th Earl, 142, 166
Abergavenny Chronicle, 328
abolitionism, xviii, 90, 102, 160, 212, 215, 245, 323
Academy of Natural Sciences, Philadelphia, 246, 247, 248–9
Acanthopholis, 280
Acland, Henry, 38
Adam and Eve, 7, 274
Adams, William Henry Davenport, 280
Addington, Henry, 29
Adelaide, Queen consort of the United Kingdom, 94
Age of Reason, The (Paine), xiv, 29–30, 107–8, 135
Ahaz, King of Judah, 7
Albert, Prince consort, 129, 151, 169, 185, 196, 219, 254
Alcott House, Richmond, 158
Aleppo, Syria, 6
Alexander disaster (1815), 35
Alexander I, Emperor of Russia, 41
Alexander II, Emperor of Russia, 264
Alexander the Great, 7
Alhambra, Granada, 170
All Souls College, Oxford, 279
All the Year Round (Dickens), 145
alleles, 271
Allosaurus, 305, 314, 327
Alps, 79, 105, 287
Altmühl Valley, Bavaria, 240–41, 248, 250, 281, 325
Amazon region, 178, 326
American Association, 303
American Civil War (1861–5), 247, 285, 297, 299, 330
American Journal of Science, 97
American Museum of Natural History, 313, 328, 329
American Naturalist, 300
American Philosophy Society, 302
Amherst College, Massachusetts, 112
ammonites, 14, 24, 26
anatomy, 113–14, 115
Andes, 88, 107
Anglo-Saxon period (*c.* 410–1066), 53, 100
Anglo-Teutonic fraternity, 196
Ankistrodon, 252
ankylosaurs, 280
Annales Veteris Testamenti (Ussher), 5
Annals and Magazine of Natural History, 181, 184–5
Annet, Peter, 20
Anning, Joseph, 15, 16, 17–20, 34, 42, 73

Anning, Mary, xiii, xiv, xvi, 11, 15, 16–20, 34–6, 71–5, 76, 145
Alexander disaster (1815), 35
bezoar stone discoveries, 80
death (1847), 146–7
Dimorphodon discovery, 74–5, 111, 145
'Green Sand Mania', 72
Ichthyosaurus discoveries, xiii, xv, 17–20, 33, 36, 42, 46, 73, 111, 141, 147, 323
London visit (1829), 72–3
Murchisons, relationship with, 63, 72
Plesiosaurus discovery, 43–4, 46, 111, 147
religious views, 21, 59, 145
Squaloraja discoveries, 73
Anning, Molly, 14–16, 34, 36, 73, 146
Anning, Richard, 14–16, 18, 34, 51
Anning Fossil Depot, Lyme Regis, 73–4
anno domini, 6
Antarctica, 138, 182
Anthropological Society, 245
anti-Catholicism, xix, 3, 23, 62, 101, 141, 195
Anti-Persecution Union, 159
anti-Semitism, 195
Antiquity of Man (Lyell), 222–3
Apatosaurus, 314
apostasy, 107–8
Archaeopteryx, xvii, xviii, 241–6, 242, 243, 277, 304, 305, 325
Argentina, 92, 96
Argyll, George Campbell, 8th Duke, 261–2, 270, 273, 281, 321
Armagh, James Ussher, Archbishop, xii, xiii, xv, xvii, 1–9, 10, 101, 254, 263, 321, 334
Arnold, Matthew, 136, 255, 257, 261, 276
Arnold, Thomas, 196, 206
arsenic, 94
Aru Islands, 189
Ashfield House, Hertfordshire, 99
Ashmole, Elias, 47
Ashmolean Museum, Oxford, 25, 47, 49, 308, 309
Athanasian Creed, 167
atheism, xiv, 23, 33, 103, 118, 153, 158, 159–60, 215
Bradlaugh, 259, 290–91, 316–20
Athenaeum Club, London, 238
Athenaeum, 107, 135
Athens, ancient, 170
Atlantic mud, 251
Atlantosaurus, 305
Augustus, Roman emperor, 6–7
Austen, Jane, 13, 14
Australia, xvi, 93, 152, 154, 155–6, 157, 334
Austria, 26, 36, 313, 332
Auvergne, France, 64
Aveling, Edward, 319, 321
avian-hipped dinosaurs, xvii, 245–51, 277–8, 278, 300, 303
Awful Changes (De la Beche), 75
Axminster, Devon, 24
aye-ayes, 220
Azores, 93

Baartman, Sarah, 31–2
Babbage, Charles, 57, 76, 115
Babylonia (1894–539 BC), 7, 62
Bacon, Francis, 20, 194, 197
Baden Powell, Robert, 1st Baron, 213
Bagehot, Walter, 253
Bahia, Brazil, 91, 93
Bainbridge, John, 6
Bali, 189

Balliol College, Oxford, 25
Bampton Lectures, 101
Banbury, Oxfordshire, 125
Baptists, 21, 167
Barbados, 57
barnacles, 139, 140, 177, 181–3, *183*
Barnum, Phineas Taylor, 298
Basevi, George, 98
Bates, Henry, 177
Bateson, William, 271
Bath Chronicle, 73
Bathybius haeckelii, 251
Bathygnathus, 252
Battle of Actium (31 BC), 7
Battle of Alma (1854), 165
Battle of Balaclava (1854), 165
Battle of Blood River (1838), 234
Battle of Congella (1842), 234
Battle of Edgehill (1642), 4
Battle of Gaugamela (331 BC), 7
Battle of Inkerman (1854), 165
Battle of the Little Bighorn (1867), 303
Battle of Naseby (1645), 2
Bavaria, Kingdom of (1806–1918), xvii, 75, 111, 240–41, 248, 250, 281
Beagle, HMS, xviii, 84, 87–94, 100, 105, 106, 115, 137, 138–9
Becquerel, Henri, 333
Beddoes, Thomas, 25
Bedford river, 265
bees, xvi
behemoth, xi
Belém, Brazil, 177–8
belemnites, 14, 26
Belfast Newsletter, 256
Belfast, Ulster, 186–7, 195, 262, 288
Bell, Alexander Graham, 304
Bell, Charles, 274
Bell, Fanny, 73
Bell's Weekly Messenger, 18
Belshazzar, 62
Ben Nevis, Scotland, 174
Bensted, William Harding, 97
Benton, Michael, xii
Berkshire, England, 26
Berlin, Germany, 313, 332
Bermuda, 179
Besant, Annie, 292–5, 317
Besant, Frank and Mabel, 292, 294
Beyond the Great South Wall (Savile), 327
bezoar stones, 80
Bible, xii–xiii, 5–10, 21, 62–3, 135, 158–9, 213
  Book of Esther, 7
  Book of Ezra, 7
  Book of Genesis, *see* Book of Genesis
  Book of Job, xi, 267
  Book of Joshua, 7, 62, 236
  Book of Nehemiah, 7
  Book of Zechariah, 236
  Books of Kings, 6, 176
  Colenso's study of, 235–9
  Corinthians, 126–7, 139
  Epistle to the Hebrews, 167
  Gospel of John, 168, 256
  Gospel of Matthew, 7
  Pentateuch, 236–8
bipedalism, 51, 245–51, 277–8
Birch, Thomas, 35–6
birds, xvi, 107
  see also avian-hipped dinosaurs
birds of paradise, 189
Birmingham, West Midlands, 22, 114, 118, 141, 158, 324
birth control, 293–4
Black Sea, 165, 180
Black Ven cliff, Dorset, 16, 35, 43

Blackfriars, London, 253
Blackpool Tower, 324
Blake, Charles Carter, 225
Blake, William, 31
Blasphemous and Seditious Libels Act (1819), 28
blasphemy, 28–30, 159, 260–61
  Carlile, 30, 107–8, 239
  extinction and, 19
  Froude, 144
  Holyoake, 118–20, 158
  Home, 28, 239
  Paine, 29–30
  Renan, 256
  Taylor, 107–8
Bleak House (Dickens), 151
Blomfield, Arthur, 135
Blyth, Edward, 184
Board of Trade, 140
Boer people, 234
Bolivia, 90, 326
Boltwood, Bertram, 333
Bone Wars (*c.* 1872–92), xii, 300–306, 313–16, 333
Book of Common Prayer, 238
Book of Esther, 7
Book of Ezra, 7
Book of Genesis, xii, viii, 6, 9, 19, 20, 101, 240
  Adam and Eve, 7, 274
  Bradlaugh on, 290
  Buckland on, 31, 34, 99
  Chambers on, 127
  Darwin on, 88
  dating of, xii, 6, 9
  days of creation, xiv, 9, 32–3, 34, 101, 325
  Foote on, 295
  Hutchinson on, 328
  Lyell on, 67, 69
  mankind, creation of, xv, 83, 207, 325
  Mantell on, 163
  Noachian flood myth, *see* Noachian flood myth
  Paine on, 29–30
  Temple on, 206
Book of Job, xi, 267
Book of Joshua, 7, 62, 236
Book of Nehemiah, 7
Book of Zechariah, 236
Books of Kings, 6, 176
Borneo, 181, 184, 189
Boston, Massachusetts, 222
Boswell, James, 8
Bourbon Restoration (1814–30), 33
Boyle, Robert, 20
brachiosaurs, xvi
Bradbury, Ray, 330
Bradlaugh, Charles, xii, xviii, 258–61, 290–95, 317
brain, 207–8, 210–12, 218–21
Bramble, HMS, 155
Brazil, xvi, 91, 93, 177–8, 312, 326, 327
'Break, Break, Break' (Tennyson), 123
Brewster, David, 76–7, 81
Bridgewater, Francis Egerton, 8th Earl, 99
Bridgewater Treatises, 99–100, 103, 112, 129, 174, 274
Bright, John, 317
Brighton, East Sussex, 53, 96, 98, 160, 170
Bristol, England, 11, 21, 42, 44, 118
  British Association meeting (1836), 103
  Southwell blasphemy conviction (1841), 118
Bristol Channel, 4
British Association, 76–81, 199
  Belfast meeting (1853), 186–7
  Belfast meeting (1874), 288

Bristol meeting (1836), 103
Busk's report (1862), 230
Cambridge meeting (1862), 219–20
Ipswich meeting (1851), 185
Leeds meeting (1858), 208
Liverpool meeting (1870), 252, 253, 288
Norwich meeting (1868), 251
Oxford meeting (1832), 80–81
Oxford meeting (1847), 131
Oxford meeting (1860), 203–6, 285
Plymouth meeting (1841), 112
presidential election (1870), 252
York meeting (1831), 77
British Critic, 69
British Empire, xvi, 78, 126, 140, 164, 180, 234, 323
British Library, London, 99
British Medical Journal, 221
British Museum, London, 308, 309, 310, 312
Anning and, 72
Archaeopteryx fossil, 241
Carlyle at, 124
Darwin and, 105, 182
Davies at, 307
Du Chaillu's collection, 211
Ichthyosaurus fossils, 36, 43
Huxley at, 185
Hypsilophodon fossil, 277
Mantell's collection, 160
Owen and, 199, 219, 241, 277, 310
Seeley at, 280
Brixham Cave, Devon, 217
Brodie, Benjamin, 205
Brontë, Charlotte, 123, 129
Brontë, Emily, 123
Brontosaurus, 314, 327
Brookes, Richard, 48–9
Brougham, Henry, 76
Brown, Barnum, 316
Browning, Robert, 123
Brunswick Clubs, 61
Brusatte, Steve, xvi
Brute Force (1914 film), 328
Brydone, Patrick, 9
Buck, Peter, xvi
Buckingham, Richard Temple, 2nd Duke, 44
Buckland, Frank, 103
Buckland, Mary, 78–9, 101, 103
Buckland, William, xiv, xv, 11, 23, 24–8, 31–4, 37–41, 54, 59, 60, 78–80, 174–6
Bridgewater Treatise (1836), 100, 103–5, 174
British Association establishment (1831), 76
Chambers' speech (1847), 131
Conybeare, relationship with, 42
Copley Medal, 41, 232
coprolites, study of, 80
Crystal Palace sculptures, 173
death (1856), 176
Dimorphodon, 74–5
European tour (1816), 26, 42
Grant's opossum speech (1838), 108–9
Hylaeosaurus, 96
Iguanodon, 56, 57
Kirkdale cave, 39–41, 50, 56, 78, 232
Mantell, sponsorship of, 53
Megalosaurus, xvi, 46, 49–51, 82, 111, 245
Morland, wedding to (1825), 78–9
Oxford conference (1832), 81
Paviland cave, 37–9, 41, 50
Plesiosaurus, 44
religious views, 31, 33–4, 59, 100, 116, 328

Buckland, William – *cont.*
  Reliquiae Diluvianae (1823), 41
  transmutation, views on, 103, 116
  Vindiciae Geologicae (1819), 28, 33–4, 42
  Westminster dean (1845–56), 175–6
  zoophagy, 79
Buenos Aires, Argentina, 92
'Buffalo Bill', 299
Bugg, George, 60
Bulaq Museum, Cairo, 276
Bulgaria, 316
Bullock, William, 18
Bulwer-Lytton, Edward, 161
Bunsen, Robert, 231
Burchell, William John, 40
Burke, Edmund, 22
Burnett, William, 154
Burton, James, 78
Busk, George, 217–18, 230, 231, 232
Butler, Samuel, 269, 309
butterflies, xvi
Byron, George Gordon, 6th Baron, 47, 160

Caesar, Julius, 252, 287
Cairngorms, Scotland, 10
Caius, John, 216
Calvinism, 6, 124, 143, 167, 253, 262
Cambridge, Cambridgeshire, 78, 219–20
  University, *see* University of Cambridge
Canada, 252, 297, 306–7
Canary Islands, 89, 91, 115, 261
cannibalism, 224, 327
Canterbury, Archbishop of
  Howley, William, 100
  Longley, Charles, 215, 238
  Sumner, John Bird, 214
Cape Colony (1806–1910), 40, 93, 154, 234, 237–8
Cape Horn, 157
Cape Verde, 91, 93
Cardiff, Wales, 1–2, 4
Cardiff Giant, 298
Cardiff Times, The, 244
Carlile, Richard, 30, 107–8, 239, 259
Carlin, William Edwards, 313–14
Carlisle, George Howard, 6th Earl, 77
Carlyle, Thomas, 114, 124, 135, 153, 289, 312
Carnegie, Andrew, 330–33
Carpenter, William, 186, 310
Carus, Carl Gustav, 146
catastrophism, 32
Catholicism, xii, xiii, xiv, xvii, xix, 3, 23, 62, 101, 141, 286
  anti-Catholicism, xix, 3, 23, 141, 195
  census (1851), 143
  Ecclesiastical Titles Act (1851), 141
  Oxford Movement, 102, 132, 133, 139, 168, 271
  Papacy, 23, 62, 141, 255, 286
  Roman Catholic Relief Act (1829), 62, 101, 323
  Tyndall and, 287
  Universalis Ecclesiae papal bull (1850), 141
Cato Street Conspiracy (1820), 28
cats, 270
cell theory, 270
Celtic Christianity, 3
Celts, 151, 217, 252
census (1851), 140–45
Central Asia, xiii

Cetiosaurus, xvi, 116, 279
Ceylon, 180, 333
Chalmers, Thomas, 100
Chamberlain, Joseph, 316–17
Chambers, Robert, 122, 127–31, 144, 186, 190, 193–4, 198
  Darwin and, 139, 177
  Huxley and, 187, 205, 208
Chandos, Richard Temple, Marquess, 44
Chapman, John, 186, 187
Charge of the Light Brigade (1854), 165
Charing Cross, London, 5, 153
Charles I, King of England, Scotland and Ireland, 2, 4, 5
Charles II, King of England, Scotland and Ireland, 61
Charles X, King of France, 84
Charles Island, Galapagos, 93
Charmouth, Dorset, 14, 16, 36
Chartism, xviii, 117, 118, 119, 126, 159, 167, 187
Cheap Repository Tracts (More), 22
Cheltenham, Gloucestershire, 118
Cheltenham Chronicle, 119
Cheyenne, Wyoming, 328
Chicago Tribune, 274
Chichester, Thomas Pelham, 2nd Earl, 53
Chickering Hall, New York, 304
Chile, 87–8, 92
chimera, xiii
chimpanzees, 208, 209, 219, 224, 272
Chipping Norton, Oxfordshire, 116
Christ Church College, Oxford, 49, 78, 103, 133, 168
Christ in the House of His Parents (Millais), 166
Christian Socialism, 167, 169, 253
Christianity, iv, xix, 1–10, 166
  Anglicanism, *see* Church of England
  Athanasian Creed, 167
  blasphemy, 19, 28–30, 107–8, 118–20, 144, 159, 239
  Calvinism, 6, 124, 143, 167, 253, 262
  Catholicism, *see* Catholicism
  Celtic Christianity, 3
  census (1851), 140–45
  Christian Socialism, 167, 169, 253
  Creation myth, *see* Creation myth
  creationism, *see* creationism
  Dissenting Protestants, xiv, 21, 51, 62, 101, 142, 143, 144, 158, 323
  Enlightenment and, xv, 20
  eschatology, 168
  extinction and, xv, 19
  miracles, 213
  muscular Christianity, 195–7, 297, 324
  Nativity, 6
  Noachian flood myth, *see* Noachian flood myth
  Protestantism, xiv, 9, 21, 51, 61, 62, 101, 142–4, 158, 264, 323
Christmas Carol, A (Dickens), 123
Church of England, 21, 23, 31, 59, 88, 101, 102, 116, 135, 139, 213
  census (1851), 142–4
  Colenso's excommunication (1863), 238
  Darwin and, 198
  Essays and Reviews (1860), 213–15, 236
  Huxley and, 199
  Irish Church Act (1869), 255
  Jelf and, 168

Church of England – *cont.*
  Lambeth Conference (1867), 238
  muscular Christianity, 195–7, 297, 324
  Natal bishopric, 234–9
  Owen and, 187
  Oxford Movement, 102, 132, 133, 139, 168, 271
  tax reform (1868), 255
  Thirty-Nine Articles (1571), 136, 167, 213, 238, 259
Church Temporalities Act (1833), 101–2
Churchill, John, 128, 130
Churchill, Randolph, 319
Civil List, 118
Clapham Athenaeum, 161, 162, 163
Clapham Common, London, 160, 161
Clarendon Building, Oxford, 80
Claridge's, London, 262
Clift, Caroline, 114–15
Clift, William, 56, 114, 163
Clough, Arthur Hugh, 136
Clowes, William, 67
Cobb, Lyme Regis, 13, 44
Cobbett, William, 141
Cobden, Richard, 307
Cody, William Frederick, 299
Cole, Henry, 169
Cole, Mary, 37–8
Colenso, John William, 234–9, 245, 257, 290
Collège de France, Paris, 31
Colombia, 178
Colorado, United States, 305, 316, 331
Colt, Samuel, 152
Colway Manor, Dorset, 18
Colyton, Devon, 14
Combe, George, 124–7, 130
Comic News, 311–12
Commonwealth, 238
Como Bluff, Wyoming, 306, 313–15, 316
Compsognathus, 248–9, 250, 278, 300, 304, 305, 325
Concepción, Chile, 87, 92
Congregationalists, 143
conservatism, xv, 20
Constitution of Man, The (Combe), 124–7
Convocation of Canterbury, 215
Conybeare, William Dean, xv, 11, 25, 26, 41–6, 49, 50, 53, 54, 75, 79
  British Association establishment (1831), 76
  enaliosauria, 112
  European tour (1816), 26, 42
  Iguanodon, 56, 57, 163
  Lyell, views on, 69
  Oxford conference (1832), 81
  Plesiosaurus, xiv, xv, 44–6, 50, 56
  religious views, 59, 69
Conybeare, William John, 195
Cook, Thomas, 152
Cooper, Robert, 207
Cope, Edward Drinker, 248, 277, 297, 300–306, 313–16
Copernicus, Nicolaus, 286, 288
Copley Medal, 41, 232–3, 269
coprolites, 80
Corinthians, 126–7, 139
Cornell University, 286
Cornwell, Oxfordshire, 47–9
Corpus Christi College, Oxford, 24–8, 78
correlation, principle of, 82, 244
Cottenham, Cambridgeshire, 280
Countess of Huntingdon's Connexion, 143

Coventry, West Midlands, 153
craniology, 113, 124–7
Creation myth, xv, xvi–xiii, xvii–xviii, 3, 5–10, 19–20, 21, 31, 32–4, 59
  Adam and Eve, 7, 274
  Chambers on, 127
  Darwin on, 121–2, 127
  mankind, creation of, 83, 207, 325
  Noachian flood, *see* Noachian flood myth
  six days, xiv, 9, 32–3, 34, 101, 325
creationism, xii, xiii, xv, xvii, 5–10, 31, 60, 100–101
Cretaceous period, 280, 300
Crimean War (1853–6), 164–6, 180, 196, 291
crocodiles, 14, 17, 49, 56, 117, 306
  see also *Proteosaurus*
Crowborough, Sussex, 326
Crystal Palace, London, xii, 151–2, 164, 169–74, 224, 247, 250, 308
Cuckfield, West Sussex, 53, 95
Cult of Reason, 23
Cumberland, George, 30–31, 45
Curie, Marie and Pierre, 333
Curtis, Wardon Allan, 327
Custer, George, 303
Cuvier, Georges, 11, 31–4, 36, 38, 41, 45, 49, 50, 60, 82–4, 127
  correlation, principle of, 82, 244
  Crystal Palace sculptures, 172
  death (1832), 82
  fixity of species, 82–4, 205
  Iguanodon, 56, 57, 63, 162–3
  Morland, relationship with, 78, 79
  Teleosaurus, 171
Dacentrurus, 307
Daily Telegraph, The, 328
Dalton, John, 81
Dante Alighieri, 155
Dartmoor, England, 26
Darwin, Annie, 121, 139–40, 321
Darwin, Charles, xii–xx, 84, 87–94, 105–110, 136–40, 181–99, 228–33, 269–76, 320–22
  Annie's death (1851), 140
  Archaeopteryx and, 244
  barnacles, study of, 139, 140, 177, 181–3, *183*
  Beagle voyage (1831–1836), xvi, 84, 87–94, 100, 105, 106, 115, 137, 138–9
  Besant, relationship with, 293–4
  Charles Junior's death (1858), 191
  Copley Medal, 232–3, 269
  death (1882), 320–22
  Descent of Man, The (1871), 271–5, 325
  dinosaurs, views on, 278–9, 306
  Essays and Reviews, views on, 214
  evolution, 272, 281
  Expression of the Emotions, The (1872), 275–6
  Fertilisation of Orchids, The (1862), 229
  fossil record, views on, 193, 279, 306
  Grant's opossum speech (1838), 109
  Great Exhibition (1851), 152
  illness, 137–8, 229, 269
  Jamaica Committee (1865–9), 245
  Kingsley on, 216
  Lay Sermons, views on, 253

Darwin, Charles – *cont.*
Linnean Society readings (1858), 191–2
Mivart, relationship with, 273–4
natural selection, *see* natural selection
On the Origin of Species (1859), xvi, 122, 192–9, 204–6, 222, 228, 261, 278–9, 306
orangutans, study of, 109–10, 325
pangenesis, 270–71
plants, study of, 183, 228–30, 269, 270, 321
Pour le Mérite, 271
religious views, 139–40, 194, 321
tree analogy, *109*, 110, 193
Tyndall, relationship with, 289
Variation of Plants and Animals, The (1868), 270
Vestiges, views on, 130
Wallace, relationship with, 191–2, 197, 228, 274
Darwin, Charles Jr, 191
Darwin, Emma, 90, 106, 138, 152, 269
Darwin, Eras, 88, 105
Darwin, Erasmus, 107
Darwin, Etty, 191
Darwin, William, 121
Davies, William, 307
Davy, Humphry, 41, 77, 232
Dawson, George Mercer, 306
De la Beche, Henry, 11, 34–5, 36, 42, 43, 44, 49, 64, 75, 146, 308
De la Beche, Laetitia, 75
Deal, Kent, 179
Dee river, 265
deep-sea detritus, 251
Deluc, Jean-André, 16
Deluge, *see* Noachian flood myth
Democritus, 288
Denmark, 222
Denver, Colorado, 305
Derby, Edward Stanley, 14th Earl, 195
Derby, Edward Stanley, 15th Earl, 252, 321
Dern, Laura, xii
Descartes, René, 259, 288
Descent of Man, The (Darwin), 271–5, 325
Desmond, Adrian, 140, 156, 245, 253, 273
Deuterosaurus, 252
Devonshire, William Cavendish, 7th Duke, 321
Dickens, Charles, 77, 115, 123, 140, 145, 151, 170, 242
Dickinson, Emily, 112
Diegesis (Taylor), 108
Dilke, Charles, 317
Dimorphodon, xvi, 74–5, *74*, 111, 145, 307, *308*, 325
dinosaurs
avian-hipped dinosaurs, xvii, 245–51, 277–8, *278*, 300, 303
coining of term, 117, 141
Ornithischia and Saurischia, 280
sauropods and theropods, 315
see also fossils
diphtheria, 191
Diplodocus, 305, 330–33
Disraeli, Benjamin, 129, 151, 204, 311, 316
dissections, 113–14, 153
Dissenting Protestants, xiv, 21, 51, 62, 101, 142, 143, 144, 158, 323
Dombey and Son (Dickens), 123
Don Giovanni (Mozart), 115
Don Juan (Byron), 47

Douglas, James, 52
'Dover Beach' (Arnold), 255, 257, 261
Down House, Orpington, 137–8, 177, 182–4, 191, 194, 228, 231, 270
Doyle, Arthur Conan, 46, 326–7
dragons, x, xiii
Dragons of the Air (Seeley), 328
Dramatic Lyrics (Browning), 123
Draper, John William, 204–5, 285–6
Drayton Manor, Staffordshire, 175
Dresden, Saxony, 146
Du Chaillu, Paul, 210–11
Dublin, Ireland, 3
Dublin, Richard Whately, Archbishop, 198
dugongs, 42
Duncan, Philip, 37
Dunfermline, Fife, 256
Durban, Natal, 234
Durham, William van Mildert, Bishop, 20
Duria Antiquior (De la Beche), 75, 308
Dutch East Indies (1800–1949), xvi, 179–81, 184, 189–90, 197, 230, 264–5
Dutch Republic (1579–1795), 216
Dwight, Timothy, 299
Dyster, Frederick, 208

Ealing, Middlesex, 153
earthquakes, 87–8, 92, 189
East End, London, 153
Ecce Homo (Seeley), 256–8, 280, 324
Ecclesiastical Titles Act (1851), 141
Economist, The, 169, 185, 186
Edinburgh, Scotland, 10, 76, 88–9, 108, 124, 130–31
Edinburgh Evening News, 293
Edinburgh Journal, 130
Edinburgh Review, 104, 129, 199, 208, 214, 225, 273
Edinburgh University, 88, 108, 113
Education Act (1870), 264
Edward I, King of England, 13
Edward IV, King of England, 175
Edward VII, King of the United Kingdom, 331, 332
Egerton, Philip, 212
Egerton Papers, British Library, 99
Egremont, George Wyndham, 3rd Earl, 75, 96
Egypt, 62, 83, 170, 196, 222, 259, 276
Egyptian Hall museum, Piccadilly, 18, 36
Elasmosaurus, 300, 327
electric eels, 89
Elements of Geology (Lyell), 221
elephants, 39
Elgin Marbles, 72
Eliot, George, xvi, 133–4, 144, 186, 189, 203, 237, 257
Elotherium, 329
Elrington, Thomas, 102
Ely, Cambridgeshire, 280
emigration, 118, 190, 323
enaliosauria, 112
Encyclopaedia Britannica, 23
Engels, Friedrich, 126, 225
English Civil War (1642–51), 1–2, 4, 61
Enlightenment (*c.* 1685 – c. 1815), xv, 9, 10, 20, 173
Eobasileus, 302
Eosaurus, 297
Epicurus, 288
Epistle to the Hebrews, 167
Epistle to the Romans (Colenso), 236
Equus parvulus, 298

Erewhon (Butler), 269
Erie Canal, 297
eschatology, 168
Essays and Reviews, 213–15, 236
Essence of Christianity, *The* (Feuerbach), 133, 186
Established Church Act (1836), 141
Esther, Book of, 7
Ethics (Spinoza), 186
Ethnological Society, 252
Etna, Sicily, 9, 261
Eton College, Berkshire, 168
eugenics, 210
evangelicals, 143
Evans, Mary Ann, xvi, 133–4, 144, 186, 189, 203, 237, 257
Evening Mail, 308
Everton FC, 324
Evidence Amendment Act (1853), 319
Evidence as to Man's Place in Nature (Huxley), 223–5
evolution, 272, 281, 325, 328
Excursion, *The* (Wordsworth), 24
Exeter College, Oxford, 132, 135–6
Exeter, Henry Phillpotts, Bishop, 118, 136
Expression of the Emotions, *The* (Darwin), 275–6
extinction, xv, 19, 32, 127
Eyre, Edward, 245
Ezra, Book of, 7

Falconer, Hugh, 232
Falkland Islands, 92, 157
Falmouth, Cornwall, 93, 105
Faraday, Michael, 76, 81, 175, 216, 231, 287
Farrar, Frederick, 321
Fawcett, Percy Harrison, 326
feathers, xvii, 227, 240–44, 325
Featherstonehaugh, George, 71, 129
Feldhofer Grotto, Neander Valley, 217
Fell, John, 9
feminism, xiii
Fenian Rising (1867), 255, 291
Fertilisation of Orchids, *The* (Darwin), 229
Feuerbach, Ludwig, 133, 186
Field, *The*, 267
Figuier, Louis, 308
finches, 107
Fisgard, HMS, 158
FitzRoy, Robert, 90–93, 106, 107
fixity of species, 82–4, 205
flat-earth theory, 264–8
Flesher, James, 5
Flood, *see* Noachian flood myth
Florence, Italy, 170
Flower, William Henry, 219
football, 324
Foote, George William, 291, 295
Forbes, Edward, 157
Fordyce, John, 321
Forster, William, 264
Fossil Man of Denise, 223
fossils
  in Andes region, 88, 107
  Anning's discoveries, *see* Anning, Mary
  Buckland's study of, *see* Buckland, William
  Cambridgeshire, 280
  Conybeare's study of, *see* Conybeare, William
  Cuvier's study of, *see* Cuvier, Georges
  disinterring procedures, 17–18
  feathers, xvii, 227, 240–44, 325
  leaves, 261
  Lyme Regis, *see* Lyme Regis
  Mantell's study of, *see* Mantell, Gideon

Owen's study of, *see* Owen, Richard
Seeley's study of, 307–8
Solnhofen, 240–41, 248, 250
in United States, *see under* United States
Fossils of the South Downs, The (Mantell), 53–5, 56, 59
Foulke, William Parker, 246–9
Fourth Party, 319
Fowke, Francis, 312
Fox, Caroline, 38, 99
Fox, William, 329
France, xiv, xix, 11, 64–5, 79, 258
Bourbon Restoration (1814–30), 33
Catholicism in, 62
Crimean War (1853–6), 164–5
Fossil Man of Denise, 223
Napoleonic Wars (1803–15), 26, 165, 217
Paris Commune (1871), 273, 291
Prussian War (1870–71), 253, 271
Republic of Letters in, 3
Revolution (1789–99), 22–3, 30, 31, 33, 60
Revolutionary Wars (1792–1802), 63
Suez Canal construction (1859–69), 252
Frankenstein (Shelley), 61
Frankland, Edward, 231, 251
Franklin, Benjamin, 232
Franz Joseph I, Emperor of Austria, 332–3
Frederick III, King of Denmark, 8
Frederick Augustus II, King of Saxony, 146
Freethinker, 295
Freethought Publishing Company, 293
Freiburg, Germany, 26
French Academy of Sciences, 83
Friedrich III, German Emperor, 152
Friedrich, Caspar David, 146
Frolic, 180
Froude, Richard Hurrell, 102–3, 132–6, 144, 166, 196, 237, 239, 257
Fruits of Philosophy, The (Knowlton), 293–4
Fulham FC, 324
'Future of Secularism, The' (Nicholls), 159

Galapagos Islands, 92–3, 107, 183, 193
Galilei, Galileo, xix, 197, 206, 288
Gall, Franz Joseph, 124
Galton, Francis, 210
Gardeners' Chronicle, 198
Garibaldi, Giuseppe, 258
Garrison, William Lloyd, 159–60
Gaskell, Elizabeth, 164
Geikie, Archibald, 332
gemmules, 270–71
General Evening Post, 18
Generelle Morphologie der Organismen (Haeckel), 245–6
Genesis, Book of, *see* Book of Genesis
genetics, 271
Geoffroy Saint-Hilaire, Étienne, 83–4, 108, 117, 170
Geographical Society, 180, 251
Geological Essay (Kidd), 30
Geological Evidences of the Antiquity of Man (Lyell), 222–3
Geological Magazine, 279
'Geological Reform' (Huxley), 252
Geological Society
Anning's death (1847), 146
Bensted and, 97
Buckland and, 50
Cave Committee, 217
Conybeare and, 42, 45–6

Geological Society – *cont.*
  De la Beche and, 42, 146
  Grant and, 108–9, 117
  Huxley and, 263, 277
  Kidd and, 26
  Lyell and, 105
  Mantell and, 57
  Owen and, 116, 117
  Smith and, 55, 76, 80
  Somerset House, 8, 72, 105
  Stonesfield relics, 49
  Temple and, 44–5
  Wollaston Medal, 240
Geological Survey, 315
geology, xiv, xv, xvii, 10, 181, 206
  Buckland, xiv, xv, 11, 23, 24–8, 31–4, 103–4
  Conybeare, 45
  Hutton, xv, 10, 21, 32, 54, 68, 127, 153, 334
  Kidd, 25–6, 30
  Lyell, xvii–xviii, 63–70, 76, 100, 101, 104–5, 124, 162, 207, 325, 334
  Mantell, 98, 124, 160, 161–2
  Temple, 206
  Thomson, 262–3
  uniformitarianism, xvii, 63–70, 84, 263, 334
Geology of Oxford (Phillips), 279
George III, King of the United Kingdom, 18
George IV, King of the United Kingdom, 27, 28, 53
Geraldton, Western Australia, 334
Germany
  academies in, 76, 77
  Biblical criticism in, xvi, 61, 133–4, 139, 213, 215, 257
  Buckland's visit (1816), 26
  Carnegie *Diplodocus* in, 332
  cell theory in, 270
  Darwin's popularity in, 269, 321
  fossils in, xvii, xix, 36, 171, 240–41, 248, 250, 281
  Neander Valley, 217–18, 222, 230, 262, 325
  Republic of Letters in, 3
Gertie the Dinosaur (McCay), 329
Giant's Causeway, Northern Ireland, xii, 187
giants, 298
Gibbon, Edward, 8, 25
Gibeon, 7, 62
Gibraltar, xvii, 180, 230, 262, 325
Gigantosaurus, 280
Gilbert, Davies, 53, 58, 99
giraffes, 82
Gladstone, William, 255, 257, 311, 316, 317, 319, 320
Glasgow, Scotland, 263
Gloucester, Gloucestershire, 120, 158
Glyndebourne, East Sussex, 97
Goat's Hole Cave, Paviland, 37–9, 41, 50
Godalming, Surrey, 268
Godwin, William, 120
Goethe, Johann Wolfgang, 26, 84, 120, 146
Goodwin, Charles Wycliffe, 213
gorilla war, xx, 210–11, 212, 221, 241, 244, 325
gorillas, 210–12, 224, 244, 272
Gospel of John, 168, 256
Gosport, Hampshire, 154
Gould, John, 107
Gower Peninsula, Wales, 37–9, 41
Grahamstown, Cape Colony, 238
granite, 10, 26
Grant, Robert, 84, 88, 108, 117, 128

Grant, Ulysses, 303, 313
Gray, Robert, 237–8
Great Devonian Controversy (1834–40), 64
Great Exhibition (1851), xvi, 151–2, 164, 169, 310
'great man' theory, 124
Great Northern Railway, 138
Great Reform Act (1832), 101, 129
Great Stink (1858), 191
Greece, ancient, 273, 288, 296
Greek War of Independence (1821–9), 43
Green Sand, 72, 326
Greenhill, William, 1
Greenland, 182
Gregorian calendar, 8
Grenville, William Wyndham, 1st Baron, 27–8, 33
Grey, Charles, 2nd Earl, 101
Grey, Henry George, 3rd Earl, 115
Griffith, David Wark, 328
Grotius, Hugo, 216
gryphons, xiii
Guardian, 257
Guinea, 224
Gully, James, 137, 140
Gunpowder Plot (1605), 62
Gutenberg, Johannes, 67

Haberlein, Karl, 241, 242
Haddonfield, New Jersey, 246–9, 301
Hadrosaurus, 246–9, 248, 250, 325
Haeckel, Ernst, 245–6, 251, 269, 272
Hall of Science, London, 260
Hampden, John, 264–8
Handel, George Frideric, 321
Hanna, Hugh, 195
Harcourt, William Vernon, 51, 77
Hard Times (Dickens), 140
Hardy, Thomas, 22, 285, 296
Harley Street, London, 221
Harper's, 101, 299
Harrison, Thomas, 39
Harrow School, Middlesex, 234
Harryhausen, Ray, 330
Hawkins, Benjamin Waterhouse, 171–4, 223–4, 247–8
Hawkins, Thomas, 45, 112
Hayes, Rutherford, 313
Heathorn, Henrietta, 155, 156, 158, 188, 209, 210, 276
Hebrides, Scotland, 261
Helen, 178
helium, 231, 251
Hell, 139
Henslow, John Stevens, 89, 105, 204
heredity, 270
Hereford Times, 225
Herod, King of Judea, 7
Herschel, John, 271
Herschel, William, 57
Hesketh, Ian, 258
Hesperornis, 300, 304, 325
Hezekiah, 7
Himalayas, 139
hippocampus, 207–8, 210–12, 218–21
hippopotamuses, 39
Hirst, Thomas Archer, 231
History of the Conflict (Draper), 286
History of the Hebrew Monarchy, A (Newman), 139
History of the Jews (Milman), 61, 69–70, 135
Hitchcock, Edward, 111
Hodgkin, Thomas, 162
Holyoake, George, xvi, 118–21, 126, 158–60, 186, 197, 258
  secularism, coining of term, 159, 258

Holyoake, Madeline, 120
Home, Everard, 17, 18–19, 42, 44–5
Homer, 273
Hone, William, 28–9, 239
Honourable Artillery Company, 187
Hood, Thomas, 162
Hooke, Robert, 19, 104
Hooker, Joseph Dalton, 138–9
  Belfast meeting (1874), 288
  Darwin, relationship with, 138–9, 182–3, 192–4, 198, 208, 228, 229, 269, 274
  Darwin's death (1882), 321
  Huxley, relationship with, 188, 225
  Linnean Society readings (1858), 191
  Lyell, relationship with, 223
  Origin of Species, views on, 192, 193, 198
  Owen, relationship with, 199
  Oxford conference (1860), 204, 206
  plant species distribution, study of, 183
  Ross expedition (1839–43), 138
  Wallace, relationship with, 191
  Wilberforce, relationship with, 206
  X Club, 230–33
Horner, Leonard, 222
Hoste Henley, Henry, 18
'Hottentot Venus', 31–2
Hughes, Thomas, 196–7, 245
von Humboldt, Alexander, 38, 79, 89, 106, 177, 271
Hungary, 26, 258, 274
Hunt, William Holman, 166
Hunter, John, 42, 114
Hunterian Museum, 57
Hutcheson, Francis, 125
Hutchinson, Henry Neville, 323, 328
Hutton, James, xv, 10, 21, 32, 54, 68, 127, 153, 334
Huxley, George, 245
Huxley, Henrietta 'Nettie', 155, 156, 158, 188, 209, 210, 276
Huxley, Noel, 209
Huxley, Rachel, 245
Huxley, Thomas, xvi–xx, 152–8, 179, 182–8, 199, 203–12, 216, 244–51, 276–7
  Acanthopholis, 280
  Alpine sojourns, 287
  Archaeopteryx, 244–6, 277
  Argyll, relationship with, 261–2
  avian-hipped dinosaurs, study of, 244–6, 249–51, 277–8
  Bathybius haeckelii, 251
  Belfast meeting (1874), 288, 290
  Darwin, relationship with, 192, 194, 195, 198, 205, 272
  Darwin's death (1882), 320
  Egypt trip (1872), 276
  Evidence as to Man's Place in Nature (1863), 223–5
  flowers, views on, 229
  'Geological Reform' (1869), 252
  hippocampus debate, 208, 210–12, 218–21
  Hysilophodon, 276–7
  Jamaica Committee (1865–9), 245
  Lay Sermons (1870), 253
  London School Board, 276
  Mivart, relationship with, 274
  Natural History Museum and, 310, 311, 312
  Noel's death (1860), 209
  Normal School of Science, 254, 276, 312
  Origin of Species, views on, 192, 194, 195, 198, 205
  Owen, relationship with, 186–7, 208, 209–12, 218–21, 222, 241, 251, 325

Oxford conference (1860), 203–6, 285
pangenesis, views on, 270
Pusey, relationship with, 271
Rattlesnake voyage (1846–50), 154–8, 185, 205, 208
Reader, 238, 265
Royal Commission (1870–75), 254
Thomson, relationship with, 263
Tyndall, relationship with, 289, 290
United States tour (1876), 304–5
Vestiges, views on, 187, 205, 208
X Club, 230–33, 244–5, 251, 252
Hyde Park, London, 151, 152, 169
hyenas, 39–41
Hylaeosaurus Crystal Palace sculpture, 171
Hylaeosaurus, xvi, 95–6, *95*, 111, 112, 117, 161, 280
Hysilophodon, 276–7, *277*

Ichthyornis, 300, 304, 325
Ichthyosaurus, 17–20, *17*, 35, 44, 45, 56, 75, 111
Anning's discoveries, xiii, xv, 17–20, 33, 36, 42–3, 46, 73, 111, 141, 147, 323
Crystal Palace sculptures, 171
De la Beche and Conybeare's paper (1821), 49
Frederick Augustus II's purchase (1844), 146
Home's study of, 17, 18–19, 42
Iguanodon, xvi, 55–8, *58*, 63, 81, 94, 97–8, 111, 112, 116, 117, 161, 247
bipedalism, 250
Buckland and, 56, 57
Chambers and, 128
Conybeare and, 56, 57, 163
Crystal Palace sculptures, 171, 172
Cuvier and, 56, 57, 63, 162–3
Hysilophodon and, 276–7
Mantell and, xiv, 55–8, 63, 81, 94, 97–8, 111, 116, 161, 171, 247, 324–5
Owen and, 116, 117, 161, 171, 247
spike, 116, 171
Iguanodon dinner (1853), 172–3
Ilfracombe, Devon, 136
Ilkley, West Yorkshire, 194, 197
Illuminati, 20
Illustrated London News, 123, 173, 308
Illustrations of Political Economy (Martineau), 105
Impeachment of the House of Brunswick, The (Bradlaugh), 291
Imperial Academy, St Petersburg, 271
Imperial College London, 254, 311, 313
In Memoriam, A.H.H. (Tennyson), 132
India, 195, 234, 252
Indonesia, 179–81, 184, 189–90, 197, 230, 264–5
Industrial Revolution (*c.* 1760 – *c.* 1840), 55, 111, 118
Inferno (Dante), 155
Intellectual Observer, 244
Ipswich conference (1851), 185
Ireland, 3
Catholicism in, 62, 255
Church Act (1869), 255
Church Temporalities Act (1833), 101–2
Fenian Rising (1867), 255, 291
Great Famine (1845–52), 134, 323
Home Rule movement (1870–1918), 316
Rebellion (1641), 4

Ireland, Alexander, 130
Islam, 68
Isle of Wight, England, 276–7
Islip, Oxfordshire, 176
Israel in Egypt (Handel), 321
Italy, 26, 65–7, 79, 258, 333

J. P. Morgan, 330
Jabalpur, Central Provinces, 307
Jack Hills, Western Australia, 334
Jamaica, 11, 35, 44, 75
Jamaica Committee (1865–9), 245
James I and VI, King of England, Scotland and Ireland, 3
James II and VII, King of England, Scotland and Ireland, 13
Jameson, Robert, 32–3
Jane Eyre (Brontë), 123
Jardin des Plantes, Paris, 31
Jelf, Richard William, 168
jellyfish, xvi, 154, 185
Jericho, 290–91
Jerusalem, 6
Jessica's First Prayer (Smith), 263–4
Jesus, 6, 108, 133, 134, 168, 256–8
Job, Book of, xi, 267
John Bull, 257
John Dory's fingers, 14
John Murray
  Descent of Man, The (Darwin), 275
  Fertilisation of Orchids, The (Darwin), 229
  History of the Jews (Milman), 61, 69–70
  On the Origin of Species (Darwin), 192, 193, 197, 204, 215
  Quarterly Review, 41, 68, 104, 204
  Principles of Geology (Lyell), 192
  Reliquiae Diluvianae (Buckland), 41
  Self-Help (Smiles), 195
  Variation of Plants and Animals, The (Darwin), 270
John Wilkes's Catechism (Hone), 28
John, Gospel of, 168, 256
Johnson, James, 14, 18
Johnson, Samuel, 8
Jordeson, 179
Josephus, Flavius, 256
Joshua, Book of, 7, 62, 236
Journal of Horticulture, 228
Journey to the Centre of the Earth (Verne), 228
Jowett, Benjamin, 213–14
Judah, Kingdom of (*c.* 930 – *c.* 587 BC), 7
Judaism, 102, 118, 139
  anti-Semitism, 195
  census (1851), 143
  New Year, 8
  Passover, 237
  Sabbath, 8
  Torah, 8
Julian Period, 6, 8
Jurassic Park (1993 film), xi, xii, 327
Jurassic period, 145, 230, 241, 244, 305, 314

Kansas, United States, 300, 301
Kant, Immanuel, 20, 289
Kean, Edmund, 96
Keats, John, 29
Keble, John, 102
Kennedy, Edmund, 155
Ketchwayo, 235
Kidd, John, 25–6, 30, 100
Kimmeridge clay, 280
King James Bible, xiv

King, Dick, 234
King, Henry, 287
King, Philip Parker, 106
King's College London, 69, 167, 168–9
Kingdom of Judah (*c.* 930 – *c.* 587 BC), 7
Kingdon, John, 116
Kings, Books of, 6, 176
Kingsley, Charles, x, 136, 196, 197, 209, 216, 225–7, 245, 297, 325
Kirby, William, 104
Kirkdale, North Yorkshire, 39–41, 50, 56, 78, 232
Kirwan, Richard, 21
Knight, Charles, 329–30
Knowlton, Charles, 293–4
Knox, Robert, 211
Koh-i-Noor diamond, 151
Konig, Charles, 36, 43, 45

Labyrinthodon, 171
'Lady of Shalott, The' (Tennyson), 123
Lady's Newspaper and Pictorial Times, 173
Laelaps, 300
Lakes, Arthur, 305
Lakota people, 302–3
Lamarck, Jean-Baptiste, 82–4, 107, 153, 193, 198, 325
  Chambers on, 128
  Cope on, 303
  Grant on, 88, 108
  Haeckel on, 246
  Owen on, 117, 220, 244
  pangenesis and, 271
  Wallace on, 190
  Wilberforce on, 205
Lambeth Conference (1867), 238
Lancaster, Lancashire, 112
Lancet, 161
Land, The (Bradlaugh), 291
Lanham, Url, 302
Laosaurus, 314
Laplace, Pierre-Simon, 80
Lay Sermons (Huxley), 253
Le Puy, France, 223
Leader, 186
Leaping Laelaps (Knight), 329
Leeds Intelligencer, 256
Leeds Times, 225
Leeds, West Yorkshire, 141, 324
Leicester, Leicestershire, 125
Leicester Daily Post, 293
Leidy, Joseph, 247, 250, 298
Leisure Hour, The, 176
lemurs, 220, 262
Leopold I, King of the Belgians, 152
Levant Company, 6
Lewes, East Sussex, 11, 51, 94, 95, 96, 98, 160
Lhuyd, Edward, 49
lias, 24, 43, 81, 145, 146, 171
Liberal Party, 316–20
Life in the Primeval World (Adams), 280
Life of Jesus, The (Renan), 256
Life of Jesus, The (Strauss), 133, 134
Life on the Earth (Phillips), 279
Lifted Veil, The (Eliot), 189
Light of the World, The (Hunt), 166
limestone, 77, 128, 176
  Altmühl Valley, 240–43, 325
  Cuckfield, 53, 95
  Kirkdale, 39–40
  Lyme Regis, 18, 35, 36, 44
  Maidstone, 97–8
  Paviland, 37
  Scotland, 63
  Weald, 52–3, 326

Lincoln, Abraham, 221
Lincoln's Inn, London, 5
Linnean Society, 18, 155, 172, 191, 207, 264
Literary and Philosophical Institution, Bristol, 42
Literary Gazette, 162
literature, 123
lithophaga, 65
Liverpool, Merseyside, 252, 253
Liverpool, Robert Banks Jenkinson, 2nd Earl, 28, 260
Llyn Tegid, Snowdonia, 265
Loch Ness monster, 46
Lockyer, Norman, 251, 254
Lombok, 189
London, Charles Blomfield, Bishop of, 100
London, England, 11, 72–3, 141
  England East End, 153
  Great Stink (1858), 191
  School Board, 276
  Zoo, 109–10, 325
London Corresponding Society, 22
London Investigator, 206–7
Longfellow, Henry Wadsworth, 112
Longley, Charles, 215, 238
Lord's Prayer, 167
Lost World, The (1997 film), 327
Lost World, The (Doyle), 326–7
Lot, 290
Loughborough Monitor, 244
Louis XIII, King of France, 4
Louis XIV, King of France, 79
Louis XVI, King of France, 22
Lovelace, Ada, 130
Lubbock, John, 231, 254, 321, 332
Luck of Barry Lyndon, The (Thackeray), 123
Luddite movement (*c.* 1811–17), 28
Lushington, Stephen, 215
Luther, Martin, 286
Lydekker, Richard, 307
Lyell, Charles, xv–xviii, 11, 53, 56–7, 63–70, 76, 101, 177, 221–3, 325, 334
  Antiquity of Man (1863), 222–3
  Bridgewater Treatises and, 104–5
  Chambers' speech (1847), 131
  Copley Medal, 232
  Darwin, relationship with, 105, 191, 193, 207, 222, 223
  Elements of Geology (1838), 221
  European expedition (1828), 64–5
  flat-earth theory, views on, 265
  fossilised leaves, views on, 261
  Huxley, relationship with, 157
  Lamarck, views on, 83, 84
  Linnean Society readings (1858), 191, 192
  Lowell lectures, 222
  Origin of Species, views on, 193, 207, 222, 223
  Owen, relationship with, 115
  Oxford conference (1832), 81
  Principles of Geology (1830–33), 67–70, 84, 87–8, 100, 101, 162, 177, 192
  Tyndall, relationship with, 290
  uniformitarian geology, xvii, 63–70, 84, 325
Lym river, 14
Lyme Regis, Dorset, 11, 13–20, 26, 33, 34–6, 42–6, 71–5, 80, 111, 145–7
  coprolites, 80
  Ichthyosaurus xiii, xv, 17–20, 33, 36, 42–3, 46, 73, 111, 141, 147, 171
  Plesiosaurus, xiv, xv, 44–6, 46, 50, 56, 72, 75, 104, 111, 121, 171
  Squaloraja, 73
Lyons, Robert, 319–20

Maastricht, Netherlands, 32
Macaulay, Thomas Babington, 1st Baron, 271
Macmillan, Alexander, 256, 257–8
Macmillan's Magazine, 226
Madagascar, 220
Madeira, 91, 154
Madras Courier, 18
Madrid, Spain, 313
Maidstone, Kent, 97–8
Malay Archipelago, The (Wallace), 264–5
Malaya, 180, 184
Malta, 180
Malthus, Thomas, 109, 190, 259, 293, 294
Malvern, Worcestershire, 137, 140
Manaus, Brazil, 327
Manchester, England, 28, 78, 141, 151, 263, 324
Manchester City FC, 324
mandrills, 224
Mann, Horace, 142–4
Manning, Henry Edward, 253
Mantell, Gideon, xiv, xv, xvi, 11, 51–8, 75, 76, 81, 83, 94–8, 160–63
  Anning, visit to (1832), 145
  avian-hipped dinosaurs, 250
  death (1852), 162
  Fossils of the South Downs, The (1822), 53–5, 56, 59
  Gilbert's sponsorship, 99
  horse riding accident (1842), 161
  Hylaeosaurus, xviii, 95–6, 111, 161, 171
  Iguanodon, xiv, 55–8, 63, 81, 94, 97–8, 111, 116, 161–3, 171, 247
  Medals of Creation (1844), 161
  Owen, relationship with, 116, 161, 162–3, 171, 187, 208
  religious views, 54, 163
  Thoughts on a Pebble (1836), 161, 163
  Vestiges, views on, 130
  Wonders of Geology, The (1838), 98, 124
Mantell, Hannah, 160
Mantell, Mary Ann, 55–6, 160
Mantell, Walter, 160
March of Progress, The (Zallinger), 224
Marryat, Frederick, 180
Marsh, Catherine, 166
Marsh, Othniel Charles, 296–306, 313–16, 325, 331
Martin, John, 62, 98, 308
Martineau, Harriet, 105, 106, 130, 159–60, 173, 197
Marx, Karl, 225
Mary, Mother of Jesus, 256
mastodons, 223
Matterhorn, 287
Matthew, Gospel of, 7
Maurice, Frederick Denison, 166–9, 226, 235, 236, 239, 253
Mauritius, 154, 155
Maxwell, James Clerk, 289
Mayhew, Henry, 123, 152
Mazarin, Jules Cardinal, 8
Mazzini, Giuseppe, 258
McCay, Winsor, 329
Mechanics' Institute, 118, 125
Medals of Creation (Mantell), 161
Medical Gazette, 154
Medicine Bow, Wyoming, 306, 313–15, 316
Megalosaurus, xiv, 46, 49–51, *50*, 81, 82, 94, 111, 112, 117, 128, 151, 280, 325
  bipedalism, 245, 250, 277, 279
  Crystal Palace sculptures, 171
  Lost World depiction, 327

Megatherium, 32, 81, 92, 111, 123, 161, 171
Mendel, Gregor, 271
Mesozoic period, 279, 304, 305
metamorphic rock, 10
Metaphysical Society, 253
Methodists, 21, 143, 263
Mexico, 90
von Meyer, Hermann, 240–41, 242
Midlands, England, 26
van Mildert, William, 20
Mill on the Floss, The (Eliot), 203
Mill, John Stuart, 186, 261
Millais, John Everett, 166
Milman, Henry Hart, 61, 69–70, 135
Milton, John, 62, 104
miracles, 213
Mitsubishi, 334
Mivart, St George, 262, 273–4
mockingbirds, 93, 107
Moldavia, 165
mole rats, 82
Moltke, Helmuth, 287
monitor lizards, 49, 94
'Monkeyana', 212
monkeys, 179
Monmouth Rebellion (1685), 13
monogenesis, xviii
Montana, United States, 296
Montano, Benito Arias, 6
Montevideo, Uruguay, 88, 91, 157
Monthly Magazine, 30–31
Moon, 232
Moore, James, 52, 140, 273
Morant Bay Rebellion (1865), 245
Moravians, 143
More, Hannah, 21–2
Morland, Mary, 78–9
Mormons, 143
Morning Herald, 135
Morning Post, 187, 307
Morrison Formation, Colorado, 331
Mosasaurus, 32, 297
Moses, 31, 33, 34, 45, 59, 63, 67, 101, 236–7
Moulton, Richard, 4
Mount Etna, Sicily, 9, 261
Movement, 159
Müller, Johannes Peter, 153
Murchison, Charlotte, 63–7, 72, 145
Murchison, Roderick, xvi, 11, 63–7, 71, 100, 216
  Anning, relationship with, 72, 80
  British Association establishment (1831), 76, 77
  British Museum and, 310
  Chambers' speech (1847), 131
  European expedition (1828), 64–7
  Huxley, relationship with, 157
  Permian system, 125
  Wallace, relationship with, 180
Murray, John, 41, 61, 68, 69–70, 104
Murray III, John 192, 193, 195, 197, 215, 229, 270, 275, 321
muscular Christianity, 195–7, 297, 324
Museum of Natural History, Oxford, 204, 245, 308
'My Last Duchess' (Browning), 123

Nabonassar, King of Babylon, 7
Naples, Italy, 65
Napoleon I, emperor of the French, 23, 26, 83, 90, 165, 217, 287
Napoleon III, Emperor of the French, 165, 257
Nashville, Tennessee, 304
Natal Colony (1843–1910), 234–9

National Museum of Natural History, Paris, 31, 332
National Reformer, 259–61, 292, 317
National Republican League, 291
National Secular Society, 259–60, 292, 317
National Trust, xii
Nativity, 6
Natural History Museum, London, xvii, 308, 309–13, 324, 332
Natural History of Oxfordshire, The (Plot), 47–8
Natural History Review, 217
natural selection, 121–2, 138–9, 144, 183–5, 192–9, 204–6, 230, 255, 325
  Cope on, 304
  Kingsley and, 226
  Lyell and, 223
  Owen on, 220, 243
  sexual selection, 273
  Thomson on, 263
  Wallace on, 265
natural theology, xix, 33, 100
Nature, 252, 254, 320
Naturphilosophie, 246
nautiluses, 114
Neander Valley, Rhineland, 217–18, 222, 230, 262, 325
Neander, Joachim, 216–17
Neanderthals, xvii, xviii, 217–18, *218*, 222, 230, 262, 325
Nebraska, United States, 298
Nebuchadnezzar II, King of Babylon, 7
Nehemiah, Book of, 7
Neill, Sam, xii
Nemesis of Faith, The (Froude), 134–6, 166
Neo-Babylonian Empire (626–539 BC), 7, 62
Neolithic period, 254
Nephilim, 298
Neptunists, 32, 68
Netherlands, 32, 111
New England, 297
New Guinea, 154, 156, 180, 189
New Jersey, United States, xvii, 246–9, 298, 301
New Model Army, 2
New Monthly Magazine, 69
New System of Geology, A (Ure), 61
New Testament, 7, 21, 126–7, 139, 167, 214
  Corinthians, 126–7, 139
  Epistle to the Hebrews, 167
  Gospel of John, 168, 256
  Gospel of Matthew, 7
New York, United States, 297–8, 300, 301–2, 304, 313
New York Journal, 331
New York Lyceum of Natural History, 71
New York Post, 331
New York Times, 274
New York University, 204, 285
New Zealand, 160
Newcomes, The (Thackeray), 177
Newman, Francis William, 139
Newman, John Henry, 102–3
Newspaper and Stamp Duties Act (1819), 28, 258, 260
Newton, Isaac, 20, 58, 100, 237, 272, 288, 289, 321
Niagara Falls, 297
Nicholas II, Emperor of Russia, 333
Nicholls, Charles Frederick, 159
Nightingale, Florence, 165
Nineveh, 62

Noachian flood myth, xv, 19–20, 39–41, 63, 101, 173–4
  Buckland and, 39, 40–41, 99
  Bugg and, 60
  Cuvier and, 32
  FitzRoy and, 107
  Home and, 19–20
  Hutton and, 10
  Lyell and, 67, 69
  Neptunists and, 32, 68
Nolan, Frederick, 101
Nonconformist Protestants, *see* Dissenting Protestants
Norfolk Fens, 265
Normal School of Science, London, xviii, 313
North and South (Gaskell), 164
North British Review, 257
Northampton, Northamptonshire, 317–18
Northamptonshire, England, 116
Northern Ireland, xii, xiii, 186–7, 195
Norton, Charles Adderley, 1st Baron, 320
Norway, 333
Norwich, Norfolk, 251
Nova Scotia, 297

O'Connell, Daniel, 62, 318
O'Connor, Thomas Power, 292
Obscene Publications Act (1857), 195
Observer, 174
Ohio, United States, 222
Old Curiosity Shop, The (Dickens), 123
Old Testament, xii–xiii, 6, 21, 62–3, 135, 158–9, 213, 214
  Book of Esther, 7
  Book of Ezra, 7
  Book of Genesis, *see* Book of Genesis
  Book of Job, xi, 267
  Book of Joshua, 7, 62, 236
  Book of Nehemiah, 7
  Book of Zechariah, 236
  Books of Kings, 6, 176
  Colenso's study of, 235–9
  Pentateuch, 236–8
Olympic Games, 92, 96
Omosaurus, 307
On the Genesis of Species (Mivart), 273
On the Origin of Species (Darwin), xvi, 122, 192–9, 204–6, 222, 228, 261, 278–9, 306
'On the Tendency of Varieties . . .' (Wallace), 190–91
'On the Woman in Paviland Cavern' (Duncan), 37
Opium War, First (1839–42), 154
opossums, 108–9
Oracle of Reason, The, 111, 118, 158
Orange Free State (1854–1902), 238
orangutans, 83, 109–10, 114, 180–81, 207, 210, 222, 325
orchids, 228–9, 270
Organic Remains (Parkinson), 52
Oriel College, Oxford, 102
Orinoco river, 89
Ornithischia, 280
Osborn, Henry Fairfield, 316, 328
Ottoman Empire (1299–1922), 165, 316, 317
Our American Cousin (Taylor), 221
Outlines of the Geology of England and Wales (Conybeare), 45
Overstone, Samuel Jones-Loyd, 1st Baron, 238
Owen, Caroline, 114–15

Owen, Richard, xvi–xx, 112–18, 126, 157, 162–3, 216, 241–2, 252, 279, 281
  Anning, relationship with, 145
  Archaeopteryx, 241–4
  aye-aye paper (1862), 220
  brain, study of, 207–8, 210–12, 218–21
  Copley Medal, 232
  creation, views on, 116–17, 187, 194
  Crystal Palace sculptures, 172, 247, 250
  Darwin, relationship with, 137, 198–9, 207, 244
  dinosaur, coining of term, 117, 141
  'Gorilla and the Negro, The' (1861), 211
  Hadrosaurus, 248
  Huxley, relationship with, 186–7, 208, 209–12, 218–21, 222, 241, 251, 325
  Iguanodon and, 116, 117, 161, 171, 247, 276
  Iguanodon dinner (1853), 172–3
  Kingsley, relationship with, 227
  Mantell, relationship with, 116, 161, 162–3, 171, 187, 208
  Natural History Museum, 308, 309–13
  Omosaurus, 307
  Pour le Mérite, 271
Owen, Robert, 118, 126
Oxford, Oxfordshire, 4, 47–51, 78, 80–81
  British Association meeting (1832), 80–81
  British Association meeting (1847), 131
  British Association meeting (1860), 203–6, 285
  University, *see* University of Oxford
Oxford, Samuel Wilberforce, Bishop, 143, 174, 204–6, 214–15, 245, 255, 285
Oxford Movement, 102, 132, 133, 139, 168, 271

Paine, Thomas, xiv, 29–30, 107–8, 135
Pair of Blue Eyes, A (Hardy), 285
Palace of Westminster, London, 324
Palaeolithic period, 254
Palaeologica (Meyer), 240
Palaeontographical Society, 279
palaeontology, xiv, xix, xvii, 81, 104, 111
Palaeotherium, 171
Palermo, Sicily, 79
Paley, William, xix, 89, 233
palm trees, 179
pangenesis, 270–71
Panizzi, Anthony, 310
Paradise Lost (Milton), 62
Paris, France, xiv, 11, 22–3, 26, 79, 99, 313, 332
Paris Commune (1871), 273, 291
Parker, John William, 213
Parkinson, James, 21, 49, 52, 43
Parsons, John, 25
Passover, 237
patriarchs, 7
Pattison, Mark, 213
Paul, Saint, 126–7, 139, 167, 213–14
Paviland, Gower, 37–9, 41, 50
Paxton, Joseph, 151, 169, 170
Peabody, George, 297, 315
Peabody Museum of Natural History, 297, 302, 304, 313, 315, 328
Peaceable Army, 2
peacocks, 273
pearly nautiluses, 114

Pedro II, Emperor of Brazil, 312
Peel, Robert, 78, 94, 118, 174–5, 234
Penang, Malaya, 180
Penn, Granville, 60
Penrice Castle, Gower, 37
Pentateuch, 236–8
People's Charter (1838), 118
Permian period, 125, 252
Persian cats, 270
Persuasion (Austen), 13
Perthshire Courier, 33
Peter I, Emperor of Russia, 271
Peterborough, Elizabeth Mordaunt, Countess, 5
Peterloo Massacre (1820), 28
Phases of Faith (Newman), 139
Philadelphia, Pennsylvania, 246, 247, 248
Philippines, 182
Phillips, John, 279
Phillips Academy, 297
Phillpotts, Henry, 118, 136
Philosophical Institution, Bristol, 44
Philosophie Anatomique (Saint-Hilaire), 84
Philosophie Zoologique (Lamarck), 82
phosphorescence, 333
phrenology, 113, 124–7, 265
Physiologie des Menschen (Müller), 153
'Pied Piper of Hamelin, The' (Browning), 123
Pietermaritzburg, Natal, 235, 236
pigeons, 184, 193, 228
Pilot Boat Inn, Charmouth, 16
Pinney, Anna Maria, 73, 145
Pinney, William, 73
Pitt, William, 22
Pittsburgh, Pennsylvania, 331
Pius IX, Pope, 256, 286
plankton, 205
plants, 183, 228–30, 269, 270, 321
Plateosaurus, 240
Plato, 215
platypuses, 114
Playfair, Lyon, 79
Plesiosaurus, xiv, xv, 44–6, 46, 50, 56, 72, 75, 104, 111, 121, 280
  Crystal Palace sculptures, 171, 172
Plot, Robert, 47–8
Plutonists, 32
Plymouth, Devon, 112, 157
Plymouth Brethren, 143
Poems (Tennyson), 123
Poland, 258
Political Litany, The (Hone), 28–9
polygenism, 272–3
Polynesia, 182
Pompeii, Italy, 170
Popular Science Monthly, 286
pornography, 195
Portuguese men-of-war, 154, 185
Pour le Mérite, 271
Poyais, 90
Pre-Historic Times (Lubbock), 254
Pre-Raphaelites, 166
Prehistoric Peeps (1905 film), 328
Presbyterianism, 289
press-stamps, 28, 258, 260
Priest, Robert, 256
Priestley, Joseph, 22
Primeval Man (Argyll), 261
Prince Edward Island, 252
Princess, The (Tennyson), 123
Princeton University, 315–16
Principles of Biology, The (Spencer), 231
Principles of Geology (Lyell), 67–70, 84, 87–8, 100, 101, 162, 177, 192

Privy Council, 238
Proteosaurus, 17–20, *17*, 33, 35, 36, 42–3, 53
see also *Ichthyosaurus*
Protestantism, xiv, 9, 21, 51, 61, 62, 101, 142–4, 158, 264, 323
protoplasm, 251
Prussia, Kingdom of (1701–1918), 76, 217, 253, 271
Pterodactylus, 111, 172, 307, *308*, 327
pterosaurs, 59, 104, 128, 328
Dimorphodon, xvi, 74–5, *74*, 111, 145, 307, *308*, 325
Pterodactylus, 111, 172, 307, *308*, 327
Rhamphorynchus, 240
Puerto Deseado, Argentina, 92, 96
Punch, 123, 151, 169, 328
Punjab, 151
Pusey, Edward Bouverie, 102, 103, 133, 237, 271
pyrosomes, 205

Quakers, 142, 143, 231, 248, 303
Quarterly Review, 41, 59, 68, 69, 104, 204, 214–15, 265, 273
Queen's College London, 169
Queen's Medal, 158

R.E.M., xii
racism, 211, 230, 220, 245, 272–3
radicalism, xviii, 22, 25, 30, 107–8, 119, 129, 131, 141, 158–9
Raglan, FitzRoy Somerset, 1st Baron, 165
railways, 138
Rattlesnake, HMS, xvi, 154–8, 185, 205, 208
Rawlins, Frank, 20
Reader, 238, 265
Reasoner, The, 159, 258
Recherches sur les Ossemens Fossiles (Cuvier), 31, 41, 57
Red Cloud, 302–3
Red Deer River, 307
Red Lady of Paviland, 37–9, 41, 50
Reed, William Harlow, 313–14
Reeves Associations, 22
Reform Act (1832), 101, 129, 323
Reform Act (1867), 317, 323
Reform Act (1884), 320, 323
Regent's Park, London, 72, 105, 110, 184
Règne Animal, Le (Cuvier), 31
Reign of Law, The (Argyll), 261
Religious Tract Society, 166
Reliquiae Diluvianae (Buckland), 41
Renan, Ernest, 256
Republic of Letters, 3
republicanism, 259, 291
Return of the Native, The (Hardy), 296
Retzius, Anders, 211
Reuters, 165
Rhamphorynchus, 240
rheas, 92, 93
Rhineland, Germany, xvii, 217
rhinoceros, 39, 56
Rhopalodon, 252
Richelieu, Armand du Plessis, 1st Duke, 4
Richmond, Surrey, 158
Rio de Janeiro, Brazil, 91
Rio Negro, 178
Riou, Édouard, 308
Ripon, North Yorkshire, 141
Robespierre, Maximilien, 23, 31
Rochdale Pioneers, 258
Roman Catholic Relief Act (1829), 62, 101, 323

Roman Catholicism, *see* Catholicism
Roman Empire (27 BC–AD 395), 53, 252
Ross expedition (1839–43), 138
Ross, James, 182
Rothschild, Lionel, 318
Rothschild, Nathan, 316
Rowbotham, Samuel, 264, 265
Royal College of Science, London, 313
Royal College of Surgeons, 18, 42, 56, 57, 115, 118, 154
Royal Commission for Scientific Instruction (1870–75), 254
Royal Geographical Society, 210, 268
Royal Horticultural Gardens, Chiswick, 161
Royal Institution, xvii, 78, 115, 116, 193, 199, 212, 231, 287
Royal Medal, 185
Royal Military College, 34
Royal Navy, 164, 230
Royal Society, 18, 40–43, 57, 76, 99, 116, 130, 158, 182, 231, 249
  Copley Medal, 41, 232–3, 269
Royal Society of Edinburgh, 10
Royle, Edward, 259
Rugby School, Warwickshire, 196, 206
Ruskin, John, 79, 166, 242, 312
Russell, Hannah, 94
Russell, Richard, 13
Russian Empire (1721–1917)
  armour, 152
  Crimean War (1853–6), 164–6, 180, 196, 291
  Imperial Academy of Sciences, 116, 333
  Jessica's First Prayer in, 264
  museums in, 313
  Napoleonic Wars (1803–15), 217
  Ottomans, relations with, 317
  Permian system, 125–6
Rutherford, Ernest, 333–4

Sacred History of the World, The (Turner), 100–101
Salisbury and Winchester Journal, 35, 73
Salisbury, Wiltshire, 324
Salomons, David, 318
Sandringham House, Norfolk, 18
Sarawak, 181
Satan, 62, 112, 153, 236
Saturday Review, 196
Saurischia, 280
sauropods, 315
de Saussure, Horace Bénédict, 9
Savile, Frank Mackenzie, 327
Savill, Thomas Choate, 129
Scaliger, Joseph Justus, 6, 7, 9
Scarborough, North Yorkshire, 39
scarlet fever, 191
Schaaffhausen, Hermann, 217
schist, 10
School of Mines, Piccadilly, 187, 210, 253, 280
Science Schools Building, Kensington, 254, 276, 312
Scientific Opinion, 264
Scotland, 10, 63–4, 105, 261, 331
Scott, George Gilbert, 324
scriptural geologists, 60–61, 63
Scriptural Geology (Bugg), 60
Scrope, George Poulett, 68
scrotum humanum, 48
séances, 265
Second Temple, 6
Secord, James, 128
secularism, xvi, xvii, 20, 159–60, 207, 258

National Secular Society, 259–60, 292, 317
Sedgwick, Adam
British Association establishment (1831), 76
Copley Medal (1863), 232
Darwin, relationship with, 89, 105, 107, 198
Kingsley on, 216
Lyell, relationship with, 69, 105
Seeley, relationship with, 280, 307
Vestiges, views on, 129, 130
Seditious Meetings Act (1819), 28
seeds, 183
Seeley, Harry, 280, 307–8, 328
Seeley, John Ronald, 258, 261, 280, 324
Self-Help (Smiles), 195
Self-Help by the People (Holyoake), 258
self-improvement, 195
Sevastopol, siege of (1854–5), 165, 166
Seven Years' War (1756–63), 82
Sèvres porcelain, 152
Sewell, William, 135–6
sewers, 175
sexual selection, 273
Shadow of the Clouds (Froude), 134
Shaftesbury, Anthony Cooper, 7th Earl, 257
Shakespeare, William, 289
Sheerness, Kent, 157
Sheffield, South Yorkshire, 118, 151, 324
Sheldonian Theatre, Oxford, 80
Shelley, Mary, 61
Shelley, Percy Bysshe, 259
Sheridan, Philip, 299
Sherlock Holmes stories (Doyle), 326
Shrewsbury, Shropshire, 88, 105, 121, 137, 228
Siccar Point, Berwickshire, 10
Sicily, 79, 258
Siege of Sevastopol (1854–5), 165, 166
Silas Marner (Eliot), 144
Silk Roads, xiii
Silliman, Benjamin, 163
Silvester, Harriet, 71
Sinecurist's Creed, The (Hone), 28
Singapore, 180
Sioux people, 299, 303
Six Acts (1819), 28, 260–61
Skibo Castle, Scottish Highlands, 331
slavery, xviii, 75, 90, 102, 113, 160, 212, 215, 245, 323
smallpox, 164
Smiles, Samuel, 195
Smith, Adam, 10
Smith, Horace, 98
Smith, Sarah, 263–4
Smith, William, 54–5, 76, 80
Smoky Hill River, Kansas, 300
Snowdonia, Wales, 265
Social Statics (Spencer), 186
socialism, 118, 126, 167, 169, 253
Society for the Diffusion of Useful Knowledge, 126
Society for the Propagation of the Gospel, 235
Sodom, 290
Solnhofen, Bavaria, 240–41, 248, 250, 281, 325
Somerset House, London, 8, 72, 105
Somerville, Mary, 64, 80
Sotheby, Samuel Leigh, 173
Soul, The (Newman), 139
South America, 89–90, 126, 177–8
South Downs, England, 51–8, 59

Southwark, London, 324
Southwell, Charles, 118, 119
Sowerby, George, 106
Spain, 313
Spanish Armada (1588), 13, 62
Spectator, 198, 253, 274
Spencer, Herbert, 185–6, 231, 232, 245, 286, 295
Spielberg, Steven, xi
Spinoza, Baruch, 186, 259
spiritualism, 265
Spottiswoode, William, 231, 288, 321
Springtime of Nations (1848), 167
Spurzheim, Johann, 124
Squaloraja, 73
Sri Lanka, 180, 333
St Bartholomew's Hospital, London, 113
St Donat's Castle, Vale of Glamorgan, 2, 4
St George's Hotel, Mayfair, 231
St James's Gazette, 328
St John's College, Cambridge, 234
St John's Wood, London, 276
St Pancras station, London, 324
St Petersburg, Russia, 271, 313, 333
stags, 273
Stanley, Dean Arthur, 196, 257
Stanley, Owen, 154–5, 156
Statesman, 18
Stegosaurus, 314, 328
Sternberg, Charles, 303
Stewart, Dugald, 125
Stonesfield, Oxfordshire, 49–51
Stonyhurst College, 326
Stowe House, Buckinghamshire, 44
Stradivari, Antonio, 216
Strand Magazine, 326
stratigraphy, 55
Strauss, David, 133, 134, 257
Student's Darwin, The (Aveling), 321
Study of History, A (Toynbee), 1
Stutchbury, Samuel, 57
Suez Canal, 252
Sunday at Home, 166, 264
Sundays, 118
supernatural, 265
Sussex Advertiser, 162
Swan, John, 6
Swindon Brick and Tile Company, 307
Sydenham College, 153
Sydenham Hill, London, 170
Sydney, New South Wales, 93, 155, 156, 188
synapsids, 252
Syriac Old Testament, 6
System of Natural History (Brookes), 48
System of Philosophy (Spencer), 249

Table Mountain, Cape Colony, 93
Tahiti, 93
Tasmania, 136, 155, 183
Taunton Courier, 43, 73
taxes on knowledge, 258, 260
Taylor, Robert, 107–8
Taylor, Tom, 221
telegraph, 164
Telemark, Norway, 333
Teleosaurus, 171, 280
telescopes, 152
Temple, Frederick, 206, 213, 214
Temple, Richard, Marquess of Chandos, 44
Temple of Serapis, Pozzuoli, 65–7
Ten Weeks in Natal (Colenso), 235
tendrils, 229–30
Tenerife, Canary Islands, 89, 91, 115, 261

Tennyson, Alfred, 1st Baron, 123, 129, 132, 253
Ternate, 189, 191
Terre avant le Déluge, La (Figuier), 308
Thackeray, William Makepeace, 123, 177
Thelwall, John, 22
Theological Essays (Maurice), 167
Theological Magazine, The, 21
Theophilus of Antioch, 6
Theory of the Earth (Cuvier), 32
Theory of the Earth (Hutton), 153
therapsids, 252
thermodynamics, 263
theropods, 315
Thirty-Nine Articles (1571), 136, 167, 213, 238, 259
Thomas Cook, 152
Thomson, William, 262–3, 281, 334
thorianite, 333
Thoughts on a Pebble (Mantell), 161, 163
Throne and Altar riots (1791), 22
thunnosaurs, xiii
Tierra del Fuego, 92, 183, 193
tigers, 39
Tilgate Forest, West Sussex, 55–8
Times, The, 121, 143, 165, 197, 214, 242, 252, 309, 318
Titanosaurus, 307
Tom Brown's School Days (Hughes), 196–7
Tooke, John Horne, 22
Torah, 8
Torquay, Devon, 217, 228
tortoises, 93, 94
Toynbee, Arnold, 1
Tractarians, 102–3, 132, 133, 139, 168, 271
transmutation, 82–4, 107, 127, 144, 193, 198, 325
  Chambers on, 128
  Cope on, 303
  Grant on, 88, 108, 139
  Haeckel on, 246
  Owen on, 116, 117, 220, 244
  pangenesis and, 271
  Wallace on, 190
  Wilberforce on, 205
trephining, 94
Trevelyan, George Otto, 317, 332
Triassic period, 125, 306
Triceratops, 328
trilobites, 26
Trinity College, Cambridge, 256
Trinity College, Dublin, 3
Truro, Cornwall, 324
Turgenev, Ivan, 276
Turner, Sharon, 100–101
turtles, 93
Two Years Ago (Kingsley), 196
Tyndall, John, 185, 188, 231, 245, 281, 287–90, 301
Tyrannosaurus rex, xii, 316

Ulster Bible Belt, xii, xiii
Ulster Examiner, 289
Ulster Hall, Belfast, 288
'Ulysses' (Tennyson), 123
Unconscious Memory (Butler), 309
undergroundology, 28
unicorns, 41
uniformitarian geology, xvii, 63–70, 84, 263, 325, 334
Unitarianism, 167
United Kingdom
  Catholicism in, 62, 101, 141, 195, 323
  census (1851), 140–45

United Kingdom – *cont.*
Christianity in, xiv, xix, 62, 101, 141, 166, 323
Crimean War (1853–6), 164–6, 180, 196, 291
Empire, xvi, 78, 126, 140, 164, 180, 234, 323
General Election (1880), 316–20
geology of, 10, 52
Great Exhibition (1851), xvi, 151–2, 164, 169, 310
Lyme Regis fossils, *see* Lyme Regis
Reform Acts, 101, 129, 317, 320, 323
Six Acts (1819), 28, 260–61
South Downs fossils, 51–8, 59
United States, xi, xvii, xix, 222, 246–9, 281, 285, 295, 296–306, 313–16, 328
American Museum of Natural History, 313, 328, 329
Bone Wars (*c.* 1872–92), xvii, 300–306, 313–16, 333
Cardiff Giant, 298
Carnegie *Diplodocus*, 330–33
Civil War (1861–5), 247, 285, 297, 299, 330
Cope's fossil hunting, 248, 277, 297, 300–306, 313–16
Geological Survey, 315
Hadrosaurus discovery, 246–9, 298, 325
Huxley's tour (1876), 304–5
Marsh's fossil hunting, 296–306, 313–16
Universities Tests Act (1871), 256
University College London, 69, 136, 139, 167, 258
University of Cambridge, 69, 76, 88–9, 196, 232–3, 234, 271, 308
Carlile and Taylor's campaign (1829), 107–8
Universities Tests Act (1871), 256
University of Edinburgh, 88, 108, 113
University of Erlangen, 133
University of Göttingen, 133
University of Jena, 274
University of Oxford, 23, 25–8, 34, 38, 46, 47–9, 78, 271, 308
Ashmolean Museum, 25, 47, 49, 308, 309
Museum of Natural History, 204, 245, 308
Nemesis of Faith publication (1849), 135–6
Tractarian movement, 102–3, 132, 133, 139, 168, 271
Universities Tests Act (1871), 256
University of Paris, 84
University of Pennsylvania, 247
University of Tübingen, 133
Ural Mountains, 252
uranium, 333–4
urbanisation, 141, 323–4
Ure, Andrew, 60–61
Uruguay, 88, 91, 157
Ussher, James xii, xiii, xv, xvii, 1–9, 10, 101, 254, 263, 321, 334
Utah, United States, 296

vaccinations, 164
Vale of Glamorgan, Wales, 1–2, 4
Vanini, Lucilio, 291
Vanity Fair, 279
Variation of Plants and Animals, The (Darwin), 270
Venezuela, 178
Verne, Jules, 46, 228, 308
Vestiges (Chambers), 122, 127–31, 144, 186, 190, 193–4, 198

Darwin and, 139, 177
Huxley and, 187, 205, 208
Vicars, Hedley, 166
Victor Emmanuel II, King of Italy, 242
Victor Emmanuel III, King of Italy, 333
Victoria, Princess Royal, 152
Victoria, Queen of the United Kingdom, 129, 152, 164, 312
Vienna, Austria, 26, 313, 332
Vindiciae Geologicae (Buckland), 33, 42
Virgin Islands, 179
Virginian, The (Wister), 313
Vladimir, Archduke, 333
volcanoes, 26, 32, 64–5, 84, 91, 92–3, 138
Voltaire, 20
Vulcanists, 32
Vyvyan, Richard, 130

W. H. Smith, 257
Wales, 1–2, 4, 37–9, 41, 265
Wallace, Alfred Russel, xvi, 177–81, 184–5, 189–92, 197, 222, 230, 255, 264–8
Darwin, relationship with, 191–2, 197, 228, 274
Darwin's death (1882), 321
Descent of Man, views on, 274
East Indies exploration (1854–62), 180–81, 189–90, 197, 230, 264–5
flat-earth challenge (1870), 264–8
Linnean Society readings (1858), 191–2
Malay Archipelago, The (1869), 264–5
natural selection, doubts over, 265
'On the Tendency of Varieties . . .' (1858), 190–91, 228
Origin of Species, views on, 197
phrenology, belief in, 265
supernatural, belief in, 265
survival of the fittest, 190
tree analogy, 181, 184–5
Wallace Line, 189
Wallachia, 165
war, use of term, xix
Warfare of Science, The (White), 287
Waring, Henry, 43
Water-Babies, The (Kingsley), xl, 216, 225–7
Waterhouse, Alfred, 310, 312
Waterloo station, London, 197
Watson, James, 159
Watt, James, 10
Weald, England, 52–8, 64, 326
Webb, Beatrice, 292
Wedgwood family, 90, 106
Wedgwood, Josiah, 212
Weekly Herald, 299
Weimar, Germany, 26
Weisshorn, 287
Wellington, Arthur Wellesley, 1st Duke, 62
Wellnhofer, Peter, 240
Wells House, Ilkley, 194, 197
Wells, Herbert George, 171
Western Advertiser, 43
Westminster Abbey, London, xix, 8, 175, 320–22
Westminster Review, 169, 186, 187
Westward Ho! (Kingsley), 196
Whately, Richard, 198
Whewell, William, 76, 100, 112, 115, 129
Whigs, 68, 73, 76, 90, 101
White, Andrew Dickson, 286–8

White, Henry Kirke, 146
Whitworth, Joseph, 254
Wilberforce, Samuel, 143, 174, 204–6, 214–15, 245, 255, 285
Wilhelm II, German Emperor, 332
Wilkins, John, 232
William IV, King of the United Kingdom, 94, 96, 160
Williams, Rowland, 213, 215
Williams, Thomas, 30
Williston, Samuel Wendell, 305, 314–15, 331
Wilson, Henry Bristow, 213, 215
Wiltshire, England, 26
Wister, Owen, 313
Wolfe, Charles, 63
Wollaston Medal, 240
Wonders of Geology, The (Mantell), 98
Woodville, Elizabeth, 175
Woodwardian Museum, Cambridge, 280
Wordsworth, William, 24, 251–2
Working Men's College, 169, 253, 280
Wuthering Heights (Brontë), 123
Wyoming, United States, 296, 301, 306, 313–15, 316, 328, 331

X Club, xx, 230–33, 237, 249, 252, 287, 288, 332
  Archaeopteryx debate, 243
  Darwin's funeral (1882), 321
  Darwin's Copley Medal (1864), 232–3, 269
  manifesto, 244–5
  Nature and, 251

Yale University, 297–9, 302, 313, 314, 315, 328, 333
York, Edward Venables-Vernon-Harcourt, Archbishop, 41
York, North Yorkshire, 77
York, Thomas Musgrave, Archbishop, 214
Young Men's Christian Association (YMCA), 324
Young, Thomas, 247

Z (lost city), 326
Zallinger, Rudolph, 224
Zechariah, Book of, 236
zetetic astronomy, 264
zircon dating, 333–4
Zoological Society, 105, 179, 219, 264, 327
zoology, xiv, 31, 82, 88, 93, 108, 112, 115, 154
Zulu people, 234, 235–6

penguin.co.uk/vintage